Energy Cogeneration Handbook

Energy Cogeneration Handbook

Criteria for Central Plant Design

Design criteria for central plants that facilitate energy conversion, utilization, and conservation; an evaluation of project alternatives and an examination of systems and their functions to achieve optimum overall design in the generation of heating, cooling, and electricity.

by

GEORGE POLIMEROS

INDUSTRIAL PRESS INC.
200 Madison Avenue, New York, N.Y. 10157

Library of Congress Cataloging in Publication Data

Polimeros, George.
    Energy cogeneration handbook.

    Includes bibliographies and index.
    1. Cogeneration of electric power and heat.
I. Title.
TK1041.P64   1981   621.1′9   80-27186
ISBN 0-8311-1130-5

FIRST PRINTING

ENERGY COGENERATION HANDBOOK

Foreword

George Polimeros, a native of Greece, earned his Bachelor's and Master's degrees in Mechanical Engineering in the United States. I first met him in the late 1950s when we both worked for the same HVAC consulting firm. Since then, we have collaborated on several major engineering projects, principally dealing with industrial refrigeration. Knowing him for as long as I have and working with him on those projects, I can appreciate his intelligence and technical competance.

For the past 10 years George devoted most of his free time to preparing this handbook, drawing on his many years of engineering experience to offer the reader an array of new and useful information on energy cogeneration—a most important field in this age of energy shortages and increased energy cost.

When the manuscript was first sent to Industrial Press for publication, they recognized it as an important contribution to the energy engineering field, and, to confirm this impression, they asked me to read the preliminary manuscript; my view completely concurred with theirs.

Unfortunately, in the fall of 1979, before he could see his work in print, George Polimeros died. I was asked by his widow, Demetra, to assist Industrial Press in bringing this book to fruition. I accepted since I was delighted to be involved in a book that would surely be a valuable contribution to the cogeneration literature.

In life, George thoroughly enjoyed teaching young engineers the fine points of the "trade." His book will perpetuate this effort to benefit the whole profession.

Harrison D. Goodman
New York, 1980

Preface

As the energy crisis emerged in recent years, many conflicting explanations and proposed solutions were attempted, but from the outset it was clear that whatever mix of ideas was adopted, Conservation, the more efficient use of energy, would be central. As potential savings associated with a variety of conservation measures were better understood, it became apparent that cogeneration was a very attractive option and opened a rich and complex field for exploration. (A cogeneration plant is designed to capture waste heat that usually escapes to the atmosphere. This waste heat is then used to generate additional energy, process steam, etc.) This book deals with energy conversion, utilization and conservation. It attempts to gauge the experience of the last thirty years in central plant design and suggests how we may improve on past experience. It emphasizes the proper application of criteria as a means of controlling initial and operating costs and securing satisfactory plant operation that fulfills its main design objectives. Its general philosophy is to promote central plants, thus discouraging disorganized and inefficient utilization of energy at various dispersed, smaller plants.

Conservation of energy presupposes full knowledge of the critical plant processes as a basis for intelligent use of energy. The objective of the subject matter and its arrangement is to acquaint the reader with the critical aspects of the plant and its various systems and to study prime equipment arrangement and modes of operation and service for maximum performance at minimum cost. The purpose of this text is not merely to describe equipment but rather to explain the dynamic relationship between combinations of various pieces of equipment and how these dynamics actually affect plant performance.

How attractive and viable is the central plant approach in our future? Unfortunately, the current energy production and distribution system is not geared to the exigencies of energy shortages. A good percentage of our national energy is rejected through cooling towers. The central plant is coming to the forefront because it is capable of plant efficiencies as high as 65-75%, compared to the usual 32-35% of the public utilities, provided there is constant demand for its services. The central plant can offer its energy at substantially lower rates and first costs. Therefore, this is an area of huge potential for profits for the investor and savings for the consumer and is a meeting point for both groups. The central plant concept, which incorporates the concept of cogeneration as it is currently known, is being promoted by the government through tax credits because it is conducive to energy conservation and can reach into areas that the public utilities do not. *Power* magazine, among others, says that central plants will fill the blank spaces in the electrical distribution networks. Central plants have the advantage of flexibility, the ability to serve their surrounding communities in complete harmony with their needs. In addition to electricity, which is the only product of the utility plant, central plants can provide chilled and hot water, steam and, in some cases, domestic hot water and can interface with local industries by absorbing or transmitting energy to industrial processes. Local economies will benefit by creation of jobs in the central plants and by additional employment in supporting industries and in in manufacturing plants producing equipment for central plants.

What kind of legal framework can aid construction of central plants? The recent passage of the National Energy

Act (NEA) constitutes the basic legal foundation. It is difficult to predict the exact form that the implementation and approval procedures will take. However, State or, preferably, local commissions will be necessary to license the construction of plants after ascertaining their economic viability. As part of their work, the commissions will supervise plant activities and rate structure and will create the legal basis for electrical interconnections between utility networks and central plants. Central plants will then be able not only to generate their own commodities but to exchange them with anyone, including the utility, to improve load factor and dispose of excess capacity. Chapter 2 explains the rationale for the central plant and points out that engineering and management techniques can be employed to optimize energy use, promote safety, effect economies, recapture waste heat and protect the environment. The provisions of NEA pertain not only to space heating but to industrial applications as well.

The applications of this book lie in several areas:

1. Large, single buildings where some people live and work.
2. Institutional or civic complexes such as campuses, hospitals, airports, government offices, shopping centers, apartment house complexes and military bases.
3. New cities, where the plant itself and all piping and underground cables can be incorporated into the master plan. A prototype of this application is the new town of Reston, Virginia.
4. Old cities, involving a combination of community facilities. Digging up streets in an old city is an expensive affair, since interference with other utilities may be expected and a thorough analysis of underground obstacles and cost factors is required.
5. Downtown commercial districts, composed mainly of large office buildings and department stores. A perfect example of that can be seen in Hartford, Connecticut.
6. A combination of residential, commercial and industrial uses, where the probability of obtaining a more constant load factor and balanced load is improved.
7. Industrial establishments where heavy refrigeration, large boiler installations and electric generation facilities prevail such as breweries, pharmaceutical plants, paper mills, etc.
8. Industrial complexes where interchange of energy between the various units of the complex is feasible and profitable. Undoubtedly this area constitutes the main arena for industrial cogeneration.
9. Chemical and petrochemical industries where, in some plants, central cooling is already being practiced.
10. Some heat-producing industries where energy conservation has not been prevalent in the past. Among industries having considerable potential are steel, copper, aluminum, glass, synthetic rubber, textiles, the food industries, selected plastics and paper. A fine example is the hot water district heating system of Bellingham, Wash., which derives its energy from the exhaust gases of the nearby Intalco Co. aluminum plant.

What kind of subject matter does this book examine and how does it treat this material?

A great deal of information on central plants has been published in many magazines, much of it contradictory and dated. Many of the systems discussed in this book have been in existence for many years and some limited historical data are available. There are some books on district heating and, at least, one on aspects of total energy. However, for the first time, an attempt is being made here to deal comprehensively with central plant activities, be they heating, cooling generation or process, from an engineering and design point of view. The aim is to intergrate certain aspects of what we know about the central plant and equipment with practical experience into useful design approaches and to examine various design and engineering philosophies to stimulate original solutions. No step-by-step plant design procedure is attempted. There are literally hundreds of plant designs; any effort to cover all possible combinations would be inadequate and, in the final analysis, counterproductive.

The economic aspects of cogeneration plants are not being examined for five plausible reasons. First, up to November, 1978 (National Energy Act), it was illegal to tie into the local electrical network for the purpose of selling electricity. The financial aspects of cogeneration are not on record yet. Second, the high rate of inflation and the rapidly-changing relationship between material and labor costs invalidates financial conclusions and data very quickly. The third reason is that our economy is designed around fossil fuels of uncertain availability and whose price structures are unpredictable. A fourth reason, although not of primary importance, is that most plants belong to the one-of-a-kind category and are designed on a tailored basis to fit particular conditions and to accommodate predetermined performance and efficiency limits. Data derived from one plant may not be entirely applicable to another. Lastly, cogeneration has diminished from 20% of our total electric output in the 1930s to the current 5%. This experience has not been systematically gathered, is of limited nature, and is historically dated. However, a number of short economic analyses are carried out for specific applications.

Here criteria are stressed. Subject matter, project alternatives and examination of systems and functions are consistently evaluated in the text. The details of boilers, cooling towers, pumps, absorption machines, condensers and other equipment are thus not discussed since adequate literature abounds.

Setting up criteria and parameters of a project is very difficult for most people. Most textbooks avoid this most critical area, which is precisely the area we stress here. The main concern of the central plant engineer is to design a workable plant based on sound criteria and concepts and to make the plant operationally attractive. Therefore, we concentrate on criteria for basic system design and primary equipment selection and examine those factors which affect plant efficiency. We embrace this point of view as a practical means to an effective plant. Because properly specified controls, pumping systems, equipment arrangement and drives contribute so importantly to a well-conceived plant, we deemed these subjects each worthy of a chapter.

These are some of the key subjects that we will examine:

1. *Quick energy estimating* data for preliminary load and equipment sizing have been assembled from many sources.
2. The pros and cons of *steam versus hot water,* with all related criteria, so that a balanced decision between them can be arrived at.
3. *Economical use of coolants or heating media.* Temperature and velocity parameters, control philosophy, thermodynamic advantages of hot and chilled water or steam are examined, so that one may find quickly what is good practice.
4. *Energy conservation and heat recovery* opportunities in a central plant are examined in some detail.
5. *Chiller arrangement*, which has a very important effect on energy conservation, is amply discussed.
6. *Power cycles* of all descriptions. Special emphasis is placed on combined cycles and their importance to energy conservation. Combination centrifugal-absorption machine cycles are studied from many points of view.
7. *Heat balance* techniques are elaborated in detail and a step-by-step method is worked out for an industrial plant-district heating combination.
8. *Prime movers* receive attention because they can affect plant efficiency adversely when improperly selected. Quick-estimating turbine performance curves are given that serve most preliminary design purposes. Variable-speed drives are examined from the standpoint of improving system hydraulic performance. Industrial gas turbines and their critically important role in heat recovery ~~are~~ are analyzed.
9. *Controls* for major equipment and plant processes are examined extensively as the key to good plant performance. Items such as compressor surge and efficiency are dealt with as are the complexities of hot gas bypass. The role of plant automation and computer programs in plant operation and their benefits are surveyed.
10. *Pumping systems* of all descriptions are examined as part of an effective equipment arrangement and the main means of successful energy delivery. The interaction of the various components of the pumping system is looked at carefully from the standpoint of system stability.
11. *Distribution systems* are the delivery arms of the pumping systems. Hydraulic circuit performance of various types are analyzed.
12. *Waste disposal* as a source of energy and its significance in the power picture are examined.
13. *Organic Rankine cycles* are surveyed to determine their current and future potential in central plants.
14. *Pressurizers* and water expansion of chilled water and particularly high temperature water systems are analyzed.
15. *A bibliography* at the end of each chapter enables the reader to look further into various aspects of each subject.

The West Germans are creating a central national heating network tied to their nuclear generating plants. On the other hand, some engineering firms, mostly Canadian, propose to utilize purchased electricity and local heat pumps in connection with chillers, and solar energy where applicable, to achieve energy efficiency. Undoubtedly this approach has its applications, with some limitations to large projects. The main objections are that self-generated electricity is ordinarily cheaper than purchased and that substantial uses must exist in the summer for the heat generated by the heat pump. In addition, heat pumps have climatic limitations and because they cannot produce process steam, do not easily fit industrial heat recovery into their arrangement. However, the U.S. Department of Energy has recently let out contracts for the incorporation of the heat pump concept into central plants, using various working fluids and steam as a source. The flexibility of the central plant in accommodating the needs for electric generation, industrial process steam, building facilities and industrial heat recovery, give the central plant concept a wider range of application and acceptability. Of course, all proposals, including the central plant, must be proven economically justifiable.

This book will be of interest to a number of people of diverse technical interests and levels. However, it is written principally for professionals rather than as a textbook or primer, but is certainly full of information for the serious student of plant cogeneration.

The author would like to express his appreciation to the many publishing, manufacturing and other organizations who gave permission to use graphs and other items, and to the librarians of the United Engineering Center, New York for their helpfulness. The following companies have provided assistance in various forms: American Society of Heating, Refrigerating and Air Conditioning Engineers (ASHRAE); *ASHRAE Journal; Building Systems Design* magazine; Carrier Corp.; Cleaver-Brooks Div. of Aqud-Chem. Inc., Cosentini Associates; Electric Machinery Mfg. Company; Elliott Company, Division Carrier Corp.; Gas Turbine Publications, Inc.; General Electric Company; *Heating/Piping/Air Conditioning* magazine; International District Heating Association; Johns-Mansville Corp.; John Wiley & Sons, Inc.; Kinetics Corp.; McGraw-Hill Book Company; National Climatic Service; *Plant Engineering* magazine; *Power* magazine; *Power Engineering* magazine; Solar Turbines International, an International Harvester Group; The Trane Company; Thermo Electron Corp; U.S. Electric Motors, Div. of Emerson Electric Company; Woodward Governor Company; York, Division of Borg-Warner Corp.

The author is indebted to Harland E. Rex and Harrison D. Goodman for reviewing the entire manuscript and offering helpful suggestions, to Kadir Karakus for checking heat balance and other calculations and wishes to thank, for various reasons, Patricia Flanagan, Charles Egbert, Thomas Blatner, Edward F. Doyle, Rohit Desai and Edward L. Von Stein. I wish to acknowledge the contributions of my editors, Richard Koral, William Semioli, and Maryanne Colas. The editing of Richard L. Koral improved the cohesion, clarity and arrangement of this book. I am especially indebted to my wife, Demetra, for the encouragement she offered and to my sons, Mark and Nicholas, for their patience over the years when I was unavailable to them.

George Polimeros
Brooklyn, N.Y.

Introduction

The recent emergence of the energy crisis gave rise to numerous and often conflicting explanations and proposed solutions. But from the very outset one thing was clear—whatever mix of solutions we adopted, conservation, which above all means the more efficient use of energy, would be central. And as the potential savings associated with a variety of conservation measures became better understood, it also became apparent that cogeneration was an especially attractive option.

The study of cogeneration opens a rich and complex world to exploration. Here, I would like to focus on how such a study offers a number of rewarding insights into three salient dimensions of the energy crisis. First, in cogeneration we have an example of a readily available conservation measure that can yield immediate and dramatic savings; it reminds us that conservation is a vast and untapped source of energy. Second, the recent experience of cogeneration in the United States confirms what some shrewd observers have known for some time—namely that in forging a new energy base the nontechnological problems would turn out to be the most difficult of all. And third, cogeneration points to the importance of placing each technology in the context of an overall energy system and incorporating "external" costs into decisions about the desirability of each technology.

If we have learned anything about energy in the past six years, it is surely that the past holds valuable lessons and that we should not count on quick technological fixes to resolve our problems. Consider the case of cogeneration, a proven technology dating back to the turn of the century.

In the early 1900s, some 58 percent of the total power that was produced by on-site industrial power plants was cogenerated.[1] By 1950, however, on-site industrial generation accounted for only 15 percent of total U.S. electrical generation; and by 1974 this figure had dropped to about 5 percent.

This dramatic decline may be attributed to a number of factors. Increasing regulation at both the state and federal levels over all forms of electrical generation, the extent to which most utilities discouraged the on-site generation of electricity, the tendency of industry to favor market-oriented over cost-cutting investments—all these contributed to cogeneration's decreasing role. But the most powerful force was that, increasingly, energy costs came to represent a declining percentage of industry's expenses. Indeed, between 1940 and 1950, the real price of electricity for industrial consumers was cut in half.[2]

The European experience has been very different. "Historically, industrial cogeneration has been five to six times more common in some parts of Europe than in the United States. . . ." In 1972, for example, "16 percent of West Germany's total power production was cogenerated by industries; in Italy, 18 percent; in France, 16 percent; and in the Netherlands, 10 percent."[3]

One factor accounting for the greater success of cogeneration in Europe was that prior to 1945 Europe did not have a utility grid system capable of supplying easily available and reliable power; another was that there were fewer regulations restricting the sale of electricity than in

---

[1] U.S. Department of Energy, *Cogeneration: Technical Concepts Trends Prospects,* September, 1978, p. 22.

[2] *Energy Future,* Report of the Energy Project at the Harvard Business School, edited by Robert Stobaugh and Daniel Yergin, New York, 1979, p. 159.

[3] U.S. Department of Energy, *op. cit.,* p. 23.

the United States. But, again, the principal reason had to do with the price and supply of energy resources. Europe has never been endowed with plentiful and therefore cheap energy resources. Necessity, if not the mother of invention, certainly became the mother of efficiency. It is no mere coincidence that West Germany's industries, the leading users of cogeneration, use 38 percent less energy per unit of output than do industries in the United States.[4]

What the above comparison demonstrates is that cogeneration's rate of adoption and success has not been shaped by technological considerations. Rather, this technologically successful option has succeeded or failed in response to market forces: in the case of the United States, cheap and abundant energy resources; in the case of Europe, scarce and expensive energy resources. To the extent that market forces in the United States now resemble those in Europe, the future for cogeneration in the United States looks promising.

However favorable, market forces alone will not suffice to assure the widespread adoption of cogeneration systems, for the road to success is strewn with some formidable institutional, regulatory, and legal barriers. In the words of the Report of the Energy Project at the Harvard Business School, *Energy Future,* where cogeneration is called "Industry's North Slope," in cogeneration we have "a near-perfect example of obstacles being not technical, but almost entirely institutional and organizational."[5]

Among the most serious regulatory obstacles are "the uncertainty regarding the jurisdiction of the Federal Energy Regulatory Commission and state public utility commissions over different cogeneration arrangements (such as joint ventures) and the sale of excess power to utilities."[6] Additionally, private companies are reluctant to enter this highly regulated arena and risk the possibility of being classified as a public utility. By imposing restrictions on joint ventures, antitrust and tax laws also impede the adoption of cogeneration systems. And federal and state environmental protection laws pose additional costs, delays, and risks.

Even in an environment characterized by diminishing energy supplies and ever-higher prices, surmounting these obstacles will not be easy. To be sure, technical problems remain: for example, to develop cogeneration systems capable of using alternative fuels, to improve fuel efficiencies, to make up for the current lack of hardware to retrofit existing equipment. But the real challenge remains the resolution of nontechnological problems.

Finally, in advancing the cause of cogeneration it is important to consider all of its implications and to accord due attention to "external" costs. These are costs induced by the adoption of a technology or fuel and not borne by the individual consumer. For example, although "the effects of environmental pollution for a region will be reduced as a result of the fuel-saving benefits of a cogeneration installation, increases in localized pollution may occur."[7] Who will pay for the control of this local pollution? What effects will this pollution have on the local community? Do the savings to be derived from the cogeneration systems offset the environmental risks? These are just a few of the many questions which must be answered. And what they suggest is that fuel efficiency alone cannot be the final arbiter.

Notwithstanding all of the complex issues noted above, the case for cogeneration remains very strong. Here is a proven, adaptable, energy-efficient technology capable of tapping our principal energy resource—conservation.

Gerald S. Leighton

Director, Community Systems Division
Office of Conservation and Solar Energy,
Assistant Secretary
U.S. Department of Energy

---

[4] R. Goen and R. White, *Comparisons of Energy Consumption Between West Germany and the Untied States,* Government Printing Office, Washington, D.C., 1975.

[5] *Energy Future, op. cit.,* p. 160.

[6] U.S. Department of Energy, *op. cit.,* p. 31.

[7] U.S. Department of Energy, *op. cit.,* p. 33.

Contents

Contents

Energy Cogeneration Handbook

CHAPTER 1

Estimating the Need for Energy Cogeneration

Load and energy estimating in recent years have been largely taken over by computers. Many government agencies and the ASHRAE Standard 90–75 now require computer calculations. Some of the problems associated with computers emanate from the fact that most computer programs belong to outside services. The user most often does not have the time to evaluate or is not in a position to gauge the validity of such programs. Thus, many engineers feel resigned to accept computer printouts which they do not fully understand.

However, there are times when rapid estimates are needed before the costly and time-consuming computer programs can be run, as when, quite often, an energy estimate is required to size central plant equipment long before the surrounding structures the plant will serve have been designed. Most often, the central plant contract is let before contracts for the other structures. All these requirements demand quick-estimating methods, some of which are described here. Unfortunately, all manual methods were designed for either small and residential applications or for single, large structures.

Incomplete energy usage data are shown in Tables 1–19 and 1–20 as representative values of past energy utilization patterns. To present a complete energy use analysis is a monumental task that would require substantial funding and manpower.

1.1 Heating Degree-Day Method

The available degree-day data are, for the most part, connected with heating energy requirements of a single family residence. This method of heat estimating, by judicious extension, may be used for a complex of residential buildings. A degree-day method for cooling applications has been developed which will be examined in Section 1-5.

The American Gas Association,[1] the National (now International) District Heating Association[2] and the American Society of Heating and Ventilating Engineers (now ASHRAE) determined years ago that the heating re-

quirements of a residence are proportional to the difference between indoor and outdoor temperatures, if the outdoor base temperature is 65 F (18.3 C). For industrial buildings degree-day figures are included in this chapter with the base temperatures of 45 F (7.2 C) and 55 F (12.8 C) in Table 1-4, which should not be used if the building control systems are radically different from those of residential units.

Concerning the 65 F (18.3 C) base, if the day's mean outdoor temperature is 45 F (7.2 C), it predicts that twice as much fuel will be consumed than when the outdoor mean temperature is 55 F (12.8 C). At 65 F (18.3 C) outdoor temperature, no heating is necessary.

Fuel consumption is estimated by employing the equation

$$F = U \times N_b \times D \times C_f \qquad (1.1)$$

Where

F = Fuel or energy units (see Table 1-3)
U = Unit fuel consumption constant per degree-day (see Table 1-3)
N_b = Hourly heat loss, Btu/hr (W)
D = Number of degree-days in heating season
C_f = Temperature correction factor (see Table 1-1).

This method, like the calculated heat loss method described next, requires a calculation and implies that the number of degree-days per season and the heat load rate of the heated structure remain fixed. A previous knowledge of fuel consumption for similar buildings and a correction factor for other than 0 F (−17.8 C) outside design temperature are required.

---

[1] *House Heating*, Industrial Series. New York: American Gas Association, 1925?, pp. 10-16.

[2] Report of the Commercial Relations Committee "Steam Consumption and the Degree Day," *Proceedings of the National District Heating Association* (now International), 1932, pp. 177-204.

Example No. 1:

Estimate a) the natural gas, b) the No. 2 fuel oil, and c) the coal required to heat a large residence in New York City. The design heating loss is 500,000 Btu/hr (146 550 W). Design outdoor temperature is 11 F (−11.7C). Seasonal heating in degree-days is 4871. Assume system efficiency of utilization of 70%.

Solution (from Table 1-3):

U (gas) = 0.00490 therm/(deg-day)(1000 Btu heat loss)
U (oil) = 0.00347 gal/(deg-day)(1000 Btu heat loss)
U (coal) = 0.04080 lb/(deg-day)(1000 Btu heat loss)

From Table 1-1:

$$C_f = 1.167$$

Applying Equation 1.1:

$$F \text{ (gas)} = 0.00490 \, \frac{\text{therm}}{\text{deg-day} \times 1000 \text{ Btu/hr}}$$

$$\times \, 500{,}000 \, \frac{\text{Btu}}{\text{hr}} \times 4871 \, \frac{\text{deg-day}}{\text{year}} \times 1.167$$

$$= 13{,}840 \text{ therm/year}$$

$$F \text{ (oil)} = 0.00347 \, \frac{\text{gal}}{\text{deg-day} \times 1000 \text{ Btu/hr}}$$

$$\times \, 500{,}000 \, \frac{\text{Btu}}{\text{hr}} \times 4871 \, \frac{\text{deg-day}}{\text{year}} \times 1.167$$

$$= 9801 \text{ gal/year}$$

$$F \text{ (coal)} = 0.04080 \, \frac{\text{lb coal}}{\text{deg-day} \times 1000 \text{ Btu/hr}}$$

$$\times \, 500{,}000 \, \frac{\text{Btu}}{\text{hr}} \times 4871 \, \frac{\text{deg-day}}{\text{year}} \times 1.167$$

$$= 115{,}243 \text{ lb coal/year}$$

Efficiency of utilization reflects the amount of heat the system actually delivers to the house, not only by radiators but by pipes, chimney walls, boiler jacket,

Table 1-1
Correction Factors for Outdoor Design Temperature[3]

Outdoor design temperature, F temperature, C	−20 (−28.9)	−10 (−23.3)	0 (−17.8)	10 (−12.2)	20 (−6.7)
Correction factor, C_f	0.778	0.875	1.000	1.167	1.400

Reprinted with permission from *ASHRAE Guide and Data Book 1973*.

Table 1-2
Efficiency of Utilization (Residential Systems)[4]

Type of Unit	Efficiency %
Gas designed	70−80
Gas conversion	60−80
Oil design	65−80
Oil conversion	60−80
Bituminous coal, hand-fired, with controls	50−65
Bituminous coal, stoker-fired	50−70
Anthracite, hand-fired, with controls	60−80
Anthracite, stoker-fired	60−80
Coke, hand-fired with controls	60−80
Electric resistance	90−100

breeching, etc., over an entire heating system as compared to the heating value of the fuel consumed during the same period. (See Table 1-2.) Boiler efficiency, the efficiency of an apparatus, must not be confused with the efficiency of utilization.

Some investigators claim that actual efficiencies of utilization are more in the range of 35-50% rather than 75–80% and that this error is usually compensated for by overestimation of land and by neglecting to consider heat gains.[5,6]

Table 1-3 is valid between 3500 and 6500 degree-days. For cases where the degree-days are less than 3500 or more than 6000, subtract or add 10% accordingly.

In warehouses, industrial plants, and other structures where a temperature of 50F (10C) or 60F (15.6C) is to be maintained and where no heat is required until the mean daily temperature drops below 45F (7.2C) and 55F (12.8C), respectively, one cannot use the 65F (17.8C) base. The 45F (7.2C) and 55F (12.8C) base columns of Table 1-4 are for industrial installations. The 65F (12.8 C) base column is given only for comparison. The Deep South is not included, as these units do not apply.

ASHRAE 1976 Systems Handbook proposes a modified degree-day method as an interim approach until field-validated heating statistics become available from on-going projects. The degree-day method appeared in the

[3] *ASHRAE Guide and Data Book, Systems 1973*, Chapter 43, Energy Estimating Methods, p. 43.9. For later data, see note 8.

[4] Clayton B. Hershey, "Old and New Values in Space Heating," *ASHRAE Journal*, October, 1963, p. 54.

[5] *Ibid.*

[6] Gordon, William J. "Annual Boiler Efficiencies," *Proceedings, International District Heating Association*, Vol. LVI, 1965, pp. 100–101.

ASHRAE 1973 Systems Handbook[7] for the last time. ASHRAE hopes to prepare a simple method for future editions.

Table 1-3

Unit Fuel Consumption Constants—U[8]

Based on 0 F Design Outdoor Temperature, 70 F (21.1 C) Indoor Temperature

Fuel and Units, F	Efficiency of Utilization		
	60%	70%	80%
	Unit Fuel Consumption per Degree Day per 1000 Btu/hr Heat Loss		
Gas (therms)	0.00572	0.00490	0.00429
Gas (cubic feet)	0.572	0.490	0.429
Oil (gallons)	0.00405	0.00347	0.00304
Coal (pounds)	0.0476	0.0408	0.0357

Based on:
1 therm = 100,000 Btu (105.506 MJ)
1 gallon of fuel oil = 141,000 Btu (148.755 MJ)
1 lb of coal = 12,000 Btu (12.60 MJ)
1 cu ft of natural gas = 1000 Btu (1.055 MJ)

Reprinted with permission from *ASHRAE Guide and Data Book* 1973.

The modified degree-day method formula is as follows:

$$E = \frac{H_L \times D \times 24}{\Delta t \times n \times V} (C_p)(C_F) \tag{1.1a}$$

Where

E = Fuel or energy consumption for the estimated period
H_L = Design heat loss, including infiltration, Btu/hr
D = Number of degree-days in heating season
Δt = Design temperature difference, F
n = Rated full load efficiency, a dimensionless decimal
V = Heating value of fuel, consistent with H_L and E
C_p = Interim correction factor for heating effect vs. degree-days (dimensionless)
C_F = Interim part-load correction factor for fossil-fueled systems only (dimensionless); equals 1.0 only for electric resistance heating.

[7]ASHRAE Systems Handbook 1973, *op. cit.*, p. 43.10.
[8]Reprinted, by permission, from ASHRAE Systems Handbook 1976.

Following are ASHRAE C_p and C_F factors:

Outdoor Design Temp. F(C)	−20(−28.9)	−10(−23.3)	0(−17.8)	+10(−12.8)	+20(−6.7)
C_p	0.57	0.64	0.71	0.79	0.89

Percent Oversizing	0	20	40	60	80
C_F	1.36	1.56	1.79	2.04	2.32

Reprinted, by permission, from *ASHRAE Systems Handbook* 1976.

1.2 Calculated Heat Loss Method

The amount of energy necessary to heat a structure can be found by employing the formula

$$F = \frac{XN}{EC} \tag{1.2}$$

Where

F = Fuel or energy consumption, in units of C, for heating season
X = Average heating load through the entire heating season, Btu/hr
N = Number of heating hours in heating season
E = Efficiency of fuel utilization over heating season, expressed as a decimal
C = Heating value, Btu, of energy or fuel unit (*e.g.*, Btu/gal, Btu/bbl, Btu/cu ft, Btu/Mcf, etc.)
X = May be expressed by the equation

$$X = \frac{H(t-t_a)}{(t_d-t_o)} \tag{1.3}$$

Where

H = Calculated peak heat loss, including infiltration, based on t_o and t_d
t = Average indoor temperature, usually 70–71 F (21.1 C–22.2 C) but, at times, 65 F (18.3 C) (see Equation 1.5).
t_a = Average outdoor temperature, during heating season, F.
t_d = Indoor design temperature, F.
t_o = Outdoor design temperature, F.

Table 1-4
Industrial Degree-Days
45 and 55 F (7.22 and 12.77 C) Bases; 65 F (17.77 C) Base for Comparison

State and City	45 F Base	55 F Base	65 F Base	State and City	45 F Base	55 F Base	65 F Base	State and City	45 F Base	55 F Base	65 F Base
COLORADO				Houghton	4029	6112	–	OKLAHOMA			
Denver	1548	3440	6283	Lansing	2537	4444	6009	Oklahoma City	600	1835	3725
Grand Junction	1757	3433	5641	Sault Ste. Marie	4049	6575	9048	OREGON			
Pueblo	1499	3261	5462	MINNESOTA				Baker	2321	4307	–
CONNECTICUT				Duluth	4419	6774	10080	Portland	373	1911	4635
Meriden	–	734	–	Minneapolis	3309	5417	8382	Roseburg	272	1868	4491
New Haven	1769	3237	5897	Moorhead	4796	6572	–	PENNSYLVANIA			
DISTRICT OF COLUMBIA				St. Paul	2497	5497	–	Erie	2337	3837	6451
Washington	1041	2487	4224	MISSOURI				Harrisburg	1565	3236	5251
IDAHO				Kansas City	1463	2980	4711	Philadelphia	1122	2695	5144
Boise	1045	2814	5809	Saint Louis	1186	2745	4900	Pittsburg	1377	3028	5987
Lewiston	1034	2688	5542	Springfield	982	2423	4900	Scranton	1938	3755	6254
Pocatello	2161	4140	7033	MONTANA				RHODE ISLAND			
ILLINOIS				Havre	3736	5874	8700	Block Island	871	3388	5804
Cairo	749	2119	3821	Helena	2843	5071	8129	SOUTH DAKOTA			
Chicago	1969	3743	6639	Kalispell	2874	5131	8191	Yankton	2898	6045	–
Springfield	1677	3289	5429	NEBRASKA				TENNESSEE			
INDIANA				Lincoln	3023	3850	5864	Chattanooga	242	1398	3254
Evansville	799	2335	4435	North Platte	2291	4152	6684	Knoxville	431	1741	3494
Indianapolis	1397	2829	5699	Omaha	2284	3982	6612	Memphis	166	1284	3232
IOWA				Valentine	2833	4801	7425	Nashville	419	1678	3578
Davenport	2296	4142	–	NEVADA				UTAH			
Des Moines	2440	4180	6588	Winnemucca	1670	3468	6761	Modena	1978	3981	–
KANSAS				NEW HAMPSHIRE				Salt Lake City	1475	3202	6052
Dodge City	1385	2962	4986	Concord	2646	4640	7383	VERMONT			
Topeka	1518	1811	5182	NEW JERSEY				Burlington	3014	4984	8269
Wichita	1152	2587	4620	Atlantic City	1123	2904	4812	Northfield	3652	7121	–
KENTUCKY				NEW YORK				VIRGINIA			
Lexington	–	2557	4683	Albany	2018	4302	6875	Lynchburg	554	1928	4166
Louisville	1073	2294	4660	Binghamton	2073	4296	7286	Norfolk	260	1496	3421
MAINE				Buffalo	2359	4316	7062	Richmond	549	1895	3865
Eastport	2956	5236	–	Ithaca	2412	4023	–	WASHINGTON			
Portland	2530	4572	7511	New York	1412	3089	4871	North Head	184	2062	–
MARYLAND				Oswego	2274	4363	–	Seattle	408	2185	4424
Baltimore	986	2491	4654	Rochester	2341	4231	6748	Spokane	1741	3672	6655
MASSACHUSETTS				NORTH DAKOTA				WEST VIRGINIA			
Boston	1787	3603	5634	Bismarck	3831	6468	8851	Elkins	1506	3327	5675
Nantucket	1514	3419	5871	Williston	4616	6399	9243	Parkersburg	1147	2784	4754
MICHIGAN				OHIO				WISCONSIN			
Alpena	3131	5499	8506	Cincinnati	1376	3003	4410	Green Bay	3318	5331	8029
Detroit	2240	4089	6232	Cleveland	1525	3795	6357	La Crosse	3034	3992	7589
Escanaba	3699	5918	8481	Columbus	1600	3255	5660	Madison	3067	4850	7863
Grand Haven	2405	3435	–	Dayton	1487	3147	5622	Milwaukee	2657	4617	7635
Grand Rapids	2332	4177	6894	Sandusky	1949	3425	5796	WYOMING			
				Toledo	1990	3757	6494	Cheyenne	2500	4700	7381
								Lander	3208	5450	7870

Source: *Handbook of Air Conditioning, Heating and Ventilating*, 3rd Ed., Stamper & Koral, Eds. (New York: Industrial Press) 1979, p. 1–126.

Substituting Equation 1.3 in 1.2, we obtain

$$F = \frac{H(t-t_a)N}{E(t_d-t_o)C} \qquad (1.4)$$

There are severe limitations to Equation 1.4 because it is used primarily for residences and is based on the calculated peak heat loss. It implies that final design details, such as insulation and construction, are settled and that the engineer is thereby free to make final heat transfer calculations. Peak heating load calculations usually take into consideration abnormally high wind velocities and infiltration rates which, under average conditions, are less than assumed. Infiltration is a function of both wind velocity and outdoor temperature. In addition, solar radiation, as well as people, artificial lighting and other internal heat sources, tend to diminish the heating requirements without affecting the magnitude of the peak demand.

The classical degree-day total for any period is the sum of degree-days per day during that period. If average outdoor temperature for any day in that period equals or exceeds 65 F, there are no degree-days for that day. One cannot, by this method, calculate the number of degree-days for more than one day by using average temperatures for that period. Degree-days are expressed by a discontinuous function.

Equation 1.4, on the other hand, ignores the classical degree-day. It must assume that the effect on energy consumption of the occasional average 65 F or over outdoor temperature during the heating season is offset when outdoor temperature drops below 65 F. Thus, a heat storage effect is taken into account, perhaps not sufficiently.

In the sense of a continuous degree-day function as defined above, we note that $(t-t_a)N$ defines annual degree-hours. If we conveniently allow t to be 65 F (18.3C), and divide the entire expression by 24, a simplified shortcut to Equation 1.4 can be found. Lowering t to 65 F is permissible on the supposition that seasonal fuel consumption (F) is usually overestimated and H is a peak load. Thus, degree-days $(D) = (t-t_a)N/24$ and Equation 1.4 may be rewritten:

$$F = \frac{H(t-t_a)N}{E(t_d-t_o)C}$$

$$F = \frac{HD(24)}{E(t_d-t_o)C} \qquad (1.5)$$

The heating requirement drops by 3.6% per 1000 feet increase in elevation above sea level.

Efficiency of utilization varies in each structure and the reader is referred to Table 1-2 for a proper evaluation of this factor.

Critical judgment must be exercised in obtaining a more representative value of H. Many attempts have been made to incorporate all the factors mentioned above into a rational system of more accurate monthly or annual energy estimates.

The calculated heat loss method, used judiciously, will provide reasonable preliminary energy estimates of heating plant capacity for a residential community. A load factor of 95% may be used to estimate the aggregate load. Community stores, libraries, and schools should be added to the aggregate load to obtain the total projected annual energy requirements.

Some investigators have determined that t_a can be found within 2 F for the eastern half of the United States (east of 102 degrees west longitude and below 2000 ft elevation) if a cyclical pattern is used.[9,10] The harmonic

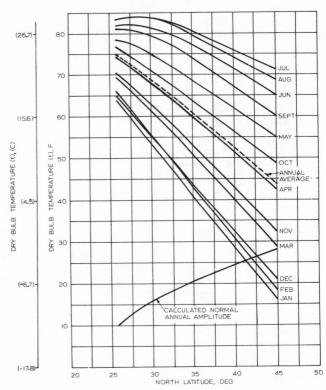

Courtesy Heating, Piping, Air Conditioning magazine.

Figure 1-1. Normal mean monthly temperatures as a function of latitude.

[9]Will K. Brown, Jr., "Typical Year Temperature Data," *Heating, Piping, Air Conditioning*, October, 1969, pp. 61–66.
[10]T. Kusuda and P. R. Eichenback, "Coincident Summer Weather Characteristics of Six Selected Cities in the United States," *ASHRAE Journal*, November, 1966, pp. 34–42.

time function, a least squares curve-fitting technique, can simulate mathematically cyclical climatological input data.

Figure 1-1 correlates normal mean (long term average) temperature with latitude. Coastal cities and cities at higher elevations tend to experience a greater departure from long term average temperatures.

Normal mean temperatures for various cities at 26, 32, 40 and 45 degrees north latitude are given in Tables 1-5 through 1-8.

Table 1-5
26 Degrees North

Normal Mean Temperatures, F

Month	Miami, Fla.	Brownsville, Tex.	Average, °F
Jan	66.9	61.4	64.2
Feb	67.9	64.0	66.0
Mar	70.5	67.9	69.2
Apr	74.2	73.9	74.1
May	77.6	79.0	78.3
Jun	80.8	82.7	81.8
Jul	81.8	84.0	82.9
Aug	82.3	84.1	83.2
Sep	81.3	81.2	81.3
Oct	77.8	75.9	76.9
Nov	72.4	67.6	70.0
Dec	68.1	62.9	65.5
Average	75.1	73.7	75.4

Courtesy of Heating, Piping, and Air Conditioning magazine.

1.3 Combination Method—Degree-Day and Calculated Heat Loss

Many attempts have been made to refine the degree-day method. H. C. S. Thom plotted degree-days against a summation of the differences between inside and actual outside temperatures for fifty-two weather stations.[11] The correlation is given by the expression:

$$F = \frac{24\,H\,z}{EC\,(t_d - t_o)} \qquad (1.6)$$

Where

H = Calculated peak heat loss based on t_o and t_d, Btu/hr
F = Fuel or energy, in units of C, for heating season
C = Heating vlaue of energy or fuel unit, e.g., But/gal, Btu/bbl, etc.
E = Efficiency of fuel utilization over heating season, expressed as a decimal
t_d = Indoor design temperature, F
t_o = Outdoor design temperature, F
z = A function of degree-days.

If z can be defined, then one can avoid the difficulty of using t and t_a in the calculated heat loss method described in Section 1.2.

[11] H.C.S. Thom, "A New Short Cut to Estimating Fuel Consumption," *Air Conditioning, Heating and Ventilating*, September, 1963, pp. 47–49.

Table 1-6
32 Degrees North

Normal Mean Temperatures, F

Month	Montgomery, Ala.	Tucson, Ariz.	Savannah, Ga.	Shreveport, La.	Jackson, Miss.	Abilene, Tex.	Dallas, Tex.	Average, °F
Jan	49.2	49.7	51.6	47.8	48.3	43.3	45.7	47.9
Feb	51.6	53.2	53.4	52.0	51.0	47.9	49.8	51.3
Mar	57.1	57.9	58.7	58.2	56.8	55.1	57.4	57.3
Apr	64.7	65.0	65.7	67.1	64.9	64.5	66.3	65.5
May	72.5	73.1	72.9	73.7	72.3	71.7	73.7	72.8
Jun	79.6	82.1	79.0	81.2	79.9	79.8	81.9	80.5
Jul	81.2	86.0	81.2	83.5	82.1	83.1	85.5	83.2
Aug	80.9	83.8	80.4	84.2	81.4	82.8	85.8	82.8
Sep	77.1	80.1	76.7	78.8	76.7	75.5	78.9	77.7
Oct	66.4	69.6	67.0	68.5	66.4	66.2	68.8	67.6
Nov	55.2	58.2	57.5	56.3	55.2	53.7	55.8	56.0
Dec	49.4	52.0	51.7	49.8	49.0	45.8	48.3	49.4
Average	65.4	67.6	66.3	66.8	65.3	64.1	66.5	66.0

Courtesy of Heating, Piping, and Air Conditioning magazine.

Table 1-7
40 Degrees North

Normal Mean Temperatures, F

Month	Spring-field, Ill.	Indian-apolis, Ind.	Good-land, Kan.	Balti-more, Md.	St. Joseph, Mo.	Lincoln, Nebr.	New York, N.Y.	Colum-bus, Ohio	Pitts-burgh, Pa.	Average °F
Jan	27.4	28.8	26.0	34.2	27.3	25.0	33.0	29.7	29.0	28.9
Feb	30.8	31.5	30.2	35.0	31.8	29.3	32.6	31.2	29.8	31.4
Mar	40.2	40.1	36.5	43.2	41.9	39.4	40.5	39.8	38.6	40.0
Apr	51.7	50.8	47.2	52.4	54.5	52.6	50.1	50.2	48.8	50.9
May	62.0	61.4	58.1	62.9	64.5	62.5	61.3	60.8	59.8	61.5
Jun	71.9	71.4	68.5	72.2	74.4	72.8	71.0	70.7	68.6	71.3
Jul	76.3	76.0	75.6	76.3	80.1	79.2	76.0	74.4	72.3	76.2
Aug	74.0	74.0	73.9	74.2	78.1	76.9	74.4	72.4	70.1	74.2
Sep	66.9	67.2	64.6	67.9	69.9	68.1	68.4	66.5	64.9	67.2
Oct	55.9	55.9	52.1	56.5	58.1	56.4	57.6	54.5	53.0	55.6
Nov	44.9	41.5	37.5	45.6	42.3	40.3	46.8	41.9	41.0	42.0
Dec	30.6	31.1	28.6	35.7	31.2	29.1	35.7	31.7	30.9	32.7
Average	52.4	52.5	49.9	54.7	54.5	52.6	53.9	52.0	50.6	52.6

Courtesy of Heating, Piping, and Air Conditioning magazine.

45 Degrees North

Normal Mean Temperatures, F

Month	Port-land, Me.	Esca-naba, Mich.	Minne-apolis, Minn.	Roches-ter, Minn.	St. Cloud, Minn.	Huron, S. Dak.	Sioux Falls, S. Dak.	Burling-ton, Vt.	Green Bay, Wisc.	La Crosse, Wisc.	Average, F
Jan	20.7	17.5	14.6	14.3	10.5	13.5	14.2	17.9	16.1	17.7	15.7
Feb	21.5	17.6	18.2	18.0	13.6	17.6	19.5	18.1	17.3	19.3	18.1
Mar	31.5	26.2	30.9	30.4	26.9	31.7	32.0	29.3	28.5	31.6	29.9
Apr	41.9	38.2	46.0	45.0	42.9	46.4	46.4	42.3	41.8	46.6	43.8
May	52.3	49.8	58.5	56.8	55.2	58.0	58.1	55.4	54.4	59.0	55.8
Jun	61.8	60.7	68.2	66.8	65.0	68.2	68.0	65.5	64.7	68.6	65.8
Jul	67.8	66.9	74.1	71.8	70.6	65.4	74.8	70.4	69.9	74.0	71.6
Aug	66.4	64.9	71.5	69.4	68.1	72.9	72.4	68.1	67.8	71.4	69.3
Sep	58.6	57.4	62.2	60.8	58.8	62.8	62.4	59.9	60.2	62.3	60.5
Oct	48.4	47.1	50.4	49.1	46.6	50.0	50.0	48.2	48.4	50.8	48.9
Nov	37.5	33.9	33.0	32.5	29.4	32.5	32.2	36.4	33.5	34.3	33.5
Dec	25.1	22.4	19.4	19.0	15.5	19.6	19.4	22.8	20.1	20.5	20.4
Average	44.5	41.9	45.5	44.5	41.9	45.7	45.8	44.5	43.6	46.3	44.4

Courtesy of Heating, Piping, and Air Conditioning magazine.

The following calculation illustrates this method by solving Example No. 1 (a) of Sec. 1.1, with z taken as 6110 by interpolation in Table 1-9, given 4871 degree-days for New York City.

$$F = \frac{24 \text{ hr/day} \times 500,000 \text{ Btu/hr} \times 6110 \text{ deg-days}}{.70 \times 100,000 \text{ Btu/therm} (70-11) \text{ F} (1,872,731 \text{ MJ})}$$

$$= 17,750 \frac{\text{therms}}{\text{year}}$$

The result, 17,750 therms per year, which is substantially higher than that of 13,840 obtained in the original solution of Example 1 (a), indicates the conservatism of this method and the caution that must be exercised in its use.

ASHRAE's Task Group TC 2.6 in 1967 came up with its own version of Equation 1 6.[12,13] The committee proposed some correction factors to compensate for different structural arrangements and indoor temperatures. The 65 F (18.3C) base was established before 1930. Due to subsequent increased use of insulation and greater internal heat gains, the committee noted that home heating may not be required until outdoor temperature is well below 65 F.

Now, according to ASHRAE, the house heating load equals Q, heat lost by heat transfer plus the air exchange with the outdoors. This relationship is expressed by:

$$Q = \frac{24 H (t-t_a) N}{(t_d-t_o)} \quad (1.7)$$

Where

t = average indoor temperature during heating season, F
N = number of heating hours in the heating season

Hence, if we equate Q_h = heat supplied by the heating system for Q (total heat loss) and $D = (t-t_a)$ for degree-days, as derived in Section 1.2, and make these substitutions in Equation 1.7, we have

$$Q_h = 24 \frac{H}{t_d-t_o} D$$

and if we further define h as the heat loss rate of the residence (Btu per hr per degF difference between indoor and outdoor temperature), then $h = H/(t_d-t_o)$ and Equation 1.7 now becomes

[12] ASHRAE TC 2.6, "New Methods for Estimating Fuel or Energy Consumption, *ASHRAE Journal*, October, 1967, pp. 45-50.
[13] Warren S. Harris *et al.*, "Estimating Energy Requirements for Residential Heating," *ASHRAE Journal*, October, 1965, pp. 50-55.

$$Q_h = 24 \, hD \quad (1.8)$$

Equation 1.8 would yield estimates that are too high because contemporary insulation practice and high internal heat loads, previously mentioned, have still to be reckoned with.

A modified equation is:

$$Q_h = Khd \quad (1.9)$$

Evaluations of factor K have varied between 11 and 20. The National Electrical Manufacturers Association (NEMA) uses 18.5 for electric heating. A study of 170 residences, located in sixty cities, indicated that K is a function of degree-days per season, D, heat loss rate, h, and average indoor temperature.

Equation 1 9 may be expressed thus:

$$Q_h = 24 \, [1 - K_d \, (65 - t + \Delta t_o)] \, hD \quad (1.10)$$

K_d, h and Δt_o may be selected from Tables 1-10 and 1-11.

Equation 1 10, although more accurate than the degree-day method, can still result in a substantial error, due to cumulative errors in heat loss calculations, assumption of temperatures, etc., by as much as 30%. This equation should be employed with great caution along the Pacific Coast, where a greater value for K_d in Table 1-10 is indicated. Also, Q_h should be reduced by 3.5% for every 1000 ft above sea level.

To illustrate use of Equation 1 10, the following example may be cited:

Example No. 2:
Estimate the seasonal heatin requirements of a residence located in Columbus, Ohio, whose hourly heat losses amount to 141,740 Btu/hr, given

D = 5211
t_d = 72 F (22.2C)
t_o = 2 F (−16.66C)

Then,

t = 71 F (21.66C) (average between 70-72 F)
h_d = 0.0459 (from Table 1-10)
Δt_o = 6.96 (from Table 1-11)

$$Q_h = 24 \, [1-0.0459 \, (65-71+6.96)] \, \frac{141,750}{72-2} \times 5211$$

$$Q_h = 242.1 \times 10^6 \text{ Btu per heating season}$$

Assuming 0.70 utilization efficiency and 100,000 Btu/therm gas fuel,

$$F = \frac{Q_h}{EC} = \frac{242.1 \times 10^6}{0.70 \times 100,000} = 3458 \text{ therms per heating season (364,839 MJ)}$$

Table 1-9
Values of Function z

Deg Days	z	Deg Days	z	Deg Days	z	Deg Days	z
100	459	2600	3522	5100	6377	7600	9025
200	585	2700	3640	5200	6487	7700	9127
300	712	2800	3758	5300	6597	7800	9228
400	837	2900	3875	5400	6706	7900	9329
500	963	3000	3993	5500	6815	8000	9430
600	1088	3100	4109	5600	6923	8100	9530
700	1213	3200	4226	5700	7032	8200	9605
800	1337	3300	4342	5800	7140	8300	9729
900	1462	3400	4458	5900	7247	8400	9829
1000	1585	3500	4574	6000	7354	8500	9928
1100	1709	3600	4689	6100	7461	8600	10026
1200	1832	3700	4804	6200	7568	8700	10125
1300	1955	3800	4918	6300	7674	8800	10222
1400	2077	3900	5033	6400	7780	8900	10320
1500	2200	4000	5146	6500	7880	9000	10417
1600	2322	4100	5260	6600	7990	9100	10514
1700	2443	4200	5373	6700	8096	9200	10611
1800	2564	4300	5486	6800	8200	9300	10707
1900	2685	4400	5599	6900	8305	9400	10803
2000	2806	4500	5711	7000	8409	9500	10899
2100	2926	4600	5823	7100	8512	9600	10994
2200	3046	4700	5934	7200	8615	9700	11089
2300	3165	4800	6046	7300	8718	9800	11184
2400	3284	4900	6156	7400	8821	9900	11268
2500	3403	5000	6267	7500	8923	10000	11372

Courtesy Building Systems Design magazine (formerly Air Conditioning, Heating and Ventilating).

Had we used the degree-day method, Equation 1 1, we would have obtained

F(gas) = 0.0049 × 141,750 × 5211 × 1.033
= 3738 therms per heating season

Using Equation 1.10 results in smaller energy requirements than by applying either the degree-day method or the z function.

The ASHRAE method embodied in Equation 1-10 is limited by the restraints of Table 1-12, with a maximum heat load in the neighborhood of 150,000 Btu/hr. The accuracy of this method depends on the exactitude of heat loss calculations.

1.4 Bin Method

The bin method is based on the temperature frequency concept. The temperature frequency bands are called bins and are usually in increments of 5 (2.8C) or 10F (12.2C). Bins can be further categorized by periods of

Table 1-10

D	K_d	D	K_d
1000	.1159	9000	.0335
1500	.0941	9500	.0324
2000	.0797	10000	.0315
2500	.0701	10500	.0306
3000	.0631	11000	.0298
3500	.0577	11500	.0290
4000	.0534	12000	.0283
4500	.0499	12500	.0277
5000	.0470	13000	.0271
5500	.0444	13500	.0265
6000	.0423	14000	.0259
6500	.0404	14500	.0254
7000	.0387	15000	.0249
7500	.0372	15500	.0244
8000	.0358	16000	.0240
8500	.0346	16500	.0236

Courtesy ASHRAE Journal.

Table 1-11

h	Δt_o	h	Δt_o
250	15.99	1150	7.93
275	15.05	1175	7.88
300	14.27	1200	7.83
325	13.61	1225	7.79
350	13.04	1250	7.75
375	12.55	1275	7.71
400	12.12	1300	7.67
425	11.75	1325	7.63
450	11.41	1350	7.60
475	11.11	1375	7.56
500	10.84	1400	7.53
525	10.59	1425	7.50
550	10.37	1450	7.47
575	10.17	1475	7.44
600	9.98	1500	7.41
625	9.81	1525	7.38
650	9.65	1550	7.35
675	9.50	1575	7.32
700	9.37	1600	7.30
725	9.24	1625	7.27
750	9.12	1650	7.25
775	9.01	1675	7.23
800	8.91	1700	7.20
825	8.81	1725	7.18
850	8.72	1750	7.16
875	8.63	1775	7.14
900	8.55	1800	7.12
925	8.47	1825	7.10
950	8.40	1850	7.08
975	8.33	1875	7.05
1000	8.26	1900	7.04
1025	8.20	1925	7.03
1050	8.14	1950	7.01
1075	8.08	1975	6.99
1100	8.03	2000	6.98
1125	7.93	2025	6.95

Courtesy ASHRAE Journal

the day or night or by virtue of occupancy or by whatever distinction the designer wants to subcategorize bins. The greater the breakdown, the greater the complications that ensue but, at the same time, the greater the accuracy. Advanced bin methods are able to account for various climatological conditions such as sunshine and cloudcover, as well as occupancy, internal heat loads, control modes, thermal storage, infiltration, zoning, etc. Most bin methods are computerized and some are referred to in the bibliography at the end of Chapter 6, under "Computers and Automation." The complexity and diversity of these programs render them immune to quick examination.

Carrier Air Conditioning Co. has developed a manual bin method of considerable sophistication, titled *Rational Energy Analysis Procedure* (REAP), which allows the engineer to insert his own inputs. Similarly L.M. Windingland and D.C. Hittle developed a manual method for Army installations, using cooling degree days. [14]

The method under consideration here assumes that internal loads and solar radiation are constant during the period under discussion. The disadvantage of this approximation can be easily overcome by eliminating sensible internal loads during unoccupied periods and calculating a solar load for each bin. This basic bin method applies to residential structures and cannot accommodate changes in occupancy or functions during the estimated period. This restriction precludes handling well enough such items as latent heat loads and thermal storage and release.

One assumption is that internal heat gain, Q_G, includes body heat, direct solar radiation, lighting heat, heat from motors as well as other sources such as heat from cooking and radiation from boiler jackets and piping. One further assumption is that this heat gain is constant throughout the heating season. The reasoning that may justify these assumptions is that solar radiation penetrates the windows during the day and that lighting at night tends to take the place of daytime solar gain. Certainly, a partial accounting of gains is preferable to ignoring them or assuming that they are identical for all buildings.

Table 1-12 applies only to New York City. It is constructed as a worksheet to illustrate the technique used in plotting the curves in Fig. 1-2, Annual Heating Requirement for (New York City) Residences, and Fig. 1-3, Heat Loss Rate, New York City, so that the technique can be applied to any locality.

Sources of bin weather data are few, but dependable.[15,16,17]

In Figs. 1-2 and 1-3, degree-day and NEMA curves are superimposed on the graphs, to make apparent that estimates calculated by these methods will be low for poorly insulated buildings with light internal heat gains, and high for well insulated, high internal heat gain structures.

---

[14] L. M. Windingland and D. C. Hittle, *Energy Utilization Index Method for Predicting Building Energy Use.* Document AD-A0 39913. Springfield, Va.: National Technical Information Service, 1977.

[15] Engineering Weather Data, *Air Force Manual*, AFM 88-29, U.S. Government Printing Office, Washington, D.C. 20402, July 1, 1978.

[16] "York Heat Pump Application," *Form 165.05-AD1*, York Corp., York, Pa, 1960.

[17] U.S. Department of Commerce, National Climatic Center, Federal Building, Asheville, N.C. 28801.

The recommended high insulation standard for houses heated by electric resistance equipment may account for the success of the degree-day method in predicting electric consumption for heating these buildings because greater insulation reduces h, building heat loss, while at the same time diminishing solar gain; the two factors of energy consumption thereby tending to cancel each other out.[18]

---

[18]Nathaniel E. Hager, Jr., "Estimation of Heating Requirements," *ASHRAE Journal*, August, 1962, p. 44.

Figure 1-2 was plotted from data derived in Table 1-12. The method of plotting Fig. 1-3 starts by substituting H, the calculated peak heat loss, for Q_h, heat supplied by the heating system, in Equation 1.9,

$$Q_h = 24\,KhD \qquad (1.11)$$

$$H = 24\,KhD \qquad (1.12)$$

$$K = \frac{H}{24\,hD} \qquad (1.13)$$

Table 1-12

Annual Heating Requirements for Residences (New York City)

①	②	③	④	⑤	⑥	⑦	⑧	⑨	⑩	⑪	⑫	⑬	⑭	⑮	⑯	Degree-Day Method	NEMA Method
Climatic data (New York City)				Internal heat gain, Q_G, Btu/hr													
				0		2000		4000		6000		8000		10,000			
				③ x h	④ x ⑤	⑤ − Q_G	⑥ x ⑦/⑤	⑤ − Q_G	⑥ x ⑨/⑤	⑤ − Q_G	⑥ x ⑪/⑤	⑤ − Q_G	⑥ x ⑬/⑤	⑤ − Q_G	⑥ x ⑮/⑤		
High-Low Temp. Limits, F	Med-ium Temp, F	Av. Indoor-Outdoor Δt, F	Time in Bin, hr/yr	Heat Loss, Q_L, Btu/hr	Ann'l Energy, H, Btu/yr X 10^6	Q_L −Q_G, Btu/hr	H	Q_L −Q_G	H	Q_L −Q_G	H	Q_L −Q_G	H	Q_L −Q_G	H		
70-60	65	5	1631	10,000	16.31	8,000	13.05	6,000	9.78	4000	6.54	2000	3.27	—	—	—	—
60-50	55	15	1500	30,000	45.00	28,000	42.30	26,000	39.20	24,000	36.00	22,000	33.02	20,000	29.90	—	—
50-40	45	25	1639	50,000	81.95	48,000	78.60	46,000	75.30	44,000	72.00	42,000	68.80	40,000	65.70	—	—
40-30	35	35	1111	70,000	77.77	68,000	75.70	66,000	73.00	64,000	71.20	62,000	69.30	60,000	66.70	—	—
30-20	25	45	401	90,000	36.09	88,000	35.20	86,000	34.50	84,000	33.60	82,000	32.90	80,000	32.10	—	—
20-10	15	55	91	110,000	10.01	108,000	9.82	106,000	9.63	104,000	9.47	102,000	9.27	100,000	9.09	—	—
10-0	5	65	7	130,000	9.10	128,000	8.97	126,000	8.81	124,000	8.68	122,000	8.52	120,000	8.38	—	—
TOTAL, h = 2000 Btu/hr°F			6380	—	276.23	—	263.64	—	250.22	—	237.49	—	225.08	—	211.87	234	180
TOTAL, h = 1000			6380	—	139.16	—	127.17	—	114.07	—	102.76	—	92.17	—	78.22	117	90
TOTAL, h = 500			6380	—	67.00	—	56.19	—	43.95	—	34.60	—	25.66	—	17.34	58.5	45
TOTAL, h = 200			6380	—	26.82	—	15.60	—	7.62	—	2.77	—	0.69	—	0.09	23.4	18
TOTAL, h = 100			6380	—	13.41	—	3.81	—	0.34	—	—	—	—	—	—	11.7	9

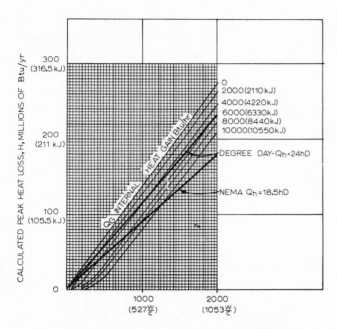

Figure 1-2. Annual heating requirements for residences.

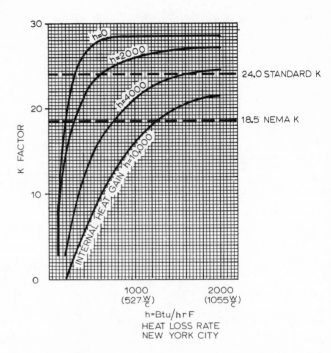

Figure 1-3. Heat loss rate for New York City.

By substituting values from Fig. 1-2 into Equation 1.11, Fig. 1-3 may be plotted. Figure 1-3 shows the correlation among degree-days, insulating qualities of a house and internal heat gain. (This technique may also be used to obtain energy estimates for types of buildings not appropriate to degree-day or K-factor analysis.)

Another rudimentary illustration of the bin method can be shown in the case of a ventilation system with 100% outside air, operating 24 hours per day and based on an assumed 1000 ton design cooling load. In this case, the refrigeration load is directly proportional to outdoor temperature. By referring to Table 1-13, Frequency of

Hourly Temperatures, for New York City (LaGuardia Airport), the following table may be quickly constructed.

The bin method, as examined above, besides being applicable only to small buildings, also does not account for changes in ventilation requirements, infiltration rates, storage and building response. Yet, even in this basic form, it is an advance over previous methods because it does take into consideration some of the aforementioned factors, if only partially. The bin approach estimating has contributed immeasurably toward sophisticated computerization of building load now taking place.

Temp. bin, F (C)	Hours (Table 1-13)	Percent load*	Load, tons (kW)	Energy, ton-hours (kW-hrs)
60-65 (15.6-18.3)	743	14.285	142.85 (502.4)	106,137 (373 283)
65-70 (18.3-21.1)	829	28.570	285.70 (1000.4)	236,845 (832 979)
70-75 (21.1-23.9)	876	42.855	428.55 (1505.2)	375,409 (1 320 307)
75-80 (23.9-26.7)	683	57.140	571.40 (2009.6)	390,266 (1 372 556)
80-85 (26.7-29.4)	383	71.425	714.25 (2512.0)	273,557 (962 096)
85-90 (29.4-32.2)	189	85.710	857.10 (3014.4)	161,991 (569 721)
90-95 (32.2-35.0)	559	100.00	1000.00 (3517.0)	59,000 (207 503)
(95-60 = 35)				1,603,205 (5 638 445)

$$\text{Equivalent full load hours} = \frac{1,603,205 \text{ ton-hrs}}{1000 \text{ tons}} = 1603.2$$

*For example, in line 1, 5÷35; in line 2, 10÷35; etc.

Table 1-13
Frequency of Hourly Temperatures for U.S. Cities
Hours per year by 5° class intervals [Based on 1951–1960, except stations marked (5)]

City	−30 to −25	−25 to −20	−20 to −15	−15 to −10	−10 to −5	−5 to 0	0 to 5	5 to 10	10 to 15	15 to 20	20 to 25	25 to 30	30 to 35	35 to 40	40 to 45	45 to 50	50 to 55	55 to 60	60 to 65	65 to 70	70 to 75	75 to 80	80 to 85	85 to 90	90 to 95	95 to 100	100 to 105	105 to 110	110 to 115
Birmingham, Ala.								3	6	17	69	143	292	433	528	614	668	742	805	908	1138	924	658	448	264	62	5	+	
Mobile, Ala										3	7	49	109	214	377	506	609	698	882	1038	1411	1341	750	536	208	30	2	+	
Montgomery, Ala.									2	5	33	88	172	346	464	598	679	729	784	882	1172	1101	721	537	363	85	8		
Anchorage, Alas. (5)		5	28	46	94	124	211	285	376	560	662	795	831	713	637	741	995	930	499	180	50	8							
Cold Bay, Alas. (5)							39	39	132	303	476	673	1245	1755	1327	1484	1042	211	31	6	2	+							
Fairbanks, Alas. (5)	126	159	206	270	332	401		379	401	447	457	455	495	409	429	513	637	658	515	361	202	118	54	14	1				
Kg. Salmon, Alas. (5)	22	43	87	144	221	222	250	237	259	323	435	628	994	1047	872	923	984	595	289	131	43	13	3						
Phoenix, Ariz.											2	8	57	182	391	540	659	769	767	776	762	774	798	698	611	507	324	123	15
Tucson, Ariz. (5)												25	98	248	417	598	716	800	763	781	870	959	777	656	520	357	152	31	
Little Rock, Ark.						+	+	1	5	23	50	172	363	509	605	669	638	672	725	803	940	947	695	505	313	107	24	1	
Bakersfield, Cal. (5)												7	77	247	541	746	908	977	966	898	831	742	613	474	371	254	100	18	+
Burbank, Cal.													10		792	661	1186	1562	1562	1163	808	565	431	257	129	48	9	3	
Fresno, Cal.													168	883	107	952	1036	1006	921	803	709	607	490	392	297	192	56	5	+
Los Angeles, Cal.															509	428	1054	1904	2193	1654	881	380	117	28	7	4	1	+	
Oakland, Cal.															701	971	1298	1329	1498	756	339	135		26	8	2	+		
Sacramento, Cal.													10	355	449	1049	843	1736	1071	773	630	486	375	276	192	93	34	+	
San Diego, Cal.																328	2341	2341	2244	1956	1016	99	126	29	7	2	+		
San Francisco, Cal.																1153			1264	665	285	397	40	15	5	1	1		
Colo. Spgs., Colo. (5)		1	+	1	7	18	43	75	136	262	448	626	755	673	678	725	760	805	784	600	477	437	276	176	46	10			
Denver, Colo.		1	+	1	6	22	36	78	119	216	359	553	721	717	692	704	678	731	684	684	549	419	332	236	103	3	2		
Hartford, Conn. (5)				2	3	11	33	77	153	233	370	552	825	807	683	575	649	752	751	755	617	630	274	118	29	16			
Wilmington, Del.						+	1	9	39	107	200	369	682	816	752	708	668	673	804	766	849	749	390	209	66	21			
Wash'tn, D.C. (Nat'l)								2	17	54	138	254	542	744	790	684	690	692	740	975	960	1581	524	306	114	57	1		
Jacksonville, Fla.											2	24	83	154	268	355	530	277	452	810	1334	2463	885	654	293	2			
Miami, Fla.														84	26	71	147	564	830	1123	1708	1732	1795	888	125	15			
Orlando, Fla. (5)												4	30	219	156	245	377	760	895	1061	1680	1275	1004	666	260	4			
Tallahassee, Fla. (5)											9	57	126	48	331	428	568	570	877	1187	1618	1910	739	530	149	6			
Tampa, Fla.												1	10	18	137	216	345	291	455	835	1387	2413	1126	752	195	3			
W. Palm Beach, Fla.													2		57	115	202	784	823	926	1672	883	1664	860	183	6			
Atlanta, Ga.								2	8	19	44	112	271	468	598	676	735	749	827	931	1185	909	625	401	177	29	2		
Augusta, Ga.									2	17	59	134	258	388	523	609	678	755	800	905	1137	995	644	500	306	85	11	5	
Macon, Ga. (5)									3	11	39	109	210	362	487	545	668	739	888	1003	1239	1193	665	538	354	85	3	+	
Savannah, Ga.										3	23	68	155	264	401	526	633				1344	2126	768	477	224	53	5		
Hilo, Hawaii (5)																				5	3001	3458							
Honolulu, Hawaii																		11	122	722	2789	3336	1552	24	1				
Wake Isl. (Pac.) (5)																		6						863					
Boise, Idaho					2	6	14	26	53	148	307	522	829	878	878	798	786	702	643	575	492	375	316	230	136	44	8	1	

Table 1-13 (Continued)

	−30 to −25	−25 to −20	−20 to −15	−15 to −10	−10 to −5	−5 to 0	0 to 5	5 to 10	10 to 15	15 to 20	20 to 25	25 to 30	30 to 35	35 to 40	40 to 45	45 to 50	50 to 55	55 to 60	60 to 65	65 to 70	70 to 75	75 to 80	80 to 85	85 to 90	90 to 95	95 to 100	100 to 105	105 to 110	110 to 115
Bingh'm, N.Y. (5)			1		5	23	67	156	263	350	515	658	873	666	598	595	673	694	808	781	548	318	145	31	2				
Buffalo, N.Y.			1	2	5	24	5	81	170	267	426	602	849	756	647	624	666	700	760	772	646	422	242	89	16	1			
New York (Int'l), N.Y.						1	1	10	26	2	188	330	603	858	838	796	722	745	754	877	926	604	263	96	28	5	1		
New York (LaGua), N.Y.				1	2	+		10	28	95	183	312	546	828	797	743	718	730	743	829	876	683	383	189	59	15	1		
Rochester, N.Y.					9	9	36	97	181	292	430	586	851	763	662	648	631	695	734	732	607	392	246	122	41	7			
Syracuse, N.Y.			2	2	5	23	55	102	190	282	392	547	830	720	651	641	656	717	723	735	627	442	269	122	32	3			
Charlotte, N.C.							2	4	5	23	64	166	360	611	634	684	730	734	839	908	1013	747	567	397	203	52	5		
Greensboro, N.C.							2	5	22	54	138	284	498	515	653	668	696	707	827	923	1087	660	498	337	125	20	1		
Raleigh, N.C.									11	38	103	236	410	527	638	672	707	762	848	937	1087	715	518	359	161	34	4		
Winston-Salem, N.C. (5)								1	18	43	116	245	456	622	708	704	710	694	801	907	1086	703	518	323	105	7		1	
Bismark, N.D.	9	20	47	77	131	208	278	292	338	371	474	550	653	604	518	520	563	606	614	566	454	357	252	151	78	28	6		
Akron-Canton Apt., Ohio				+	5	11	32	67	153	226	429	646	837	657	630	616	633	686	778	839	675	437	275	112	24	3	+		1
Cincinnati, Ohio			+		2	8	18	44	68	131	249	460	711	698	627	599	611	639	726	843	879	640	447	261	94	13	+		
Cleveland, Ohio				1	1	23	23	49	109	198	346	581	809	760	614	615	633	645	733	831	721	562	311	155	56	8	+		
Columbus, Ohio					4	10	20	40	94	169	280	502	772	730	658	603	622	648	720	820	774	556	405	238	90	15	1		
Dayton, Ohio				1	5	16	23	55	99	182	309	558	786	698	601	576	580	627	717	832	817	607	402	202	65	2			
Youngstown, Ohio			+	1	3	10	31	63	158	242	420	645	840	681	645	620	664	698	773	821	641	421	266	102	21	8			
Oklahoma City, Okla.							2	12	36	77	173	287	468	535	641	637	645	643	717	752	881	789	649	481	283	119	28	2	
Tulsa, Okla.							2	12	29	75	159	265	438	570	611	637	622	636	671	564	838	816	576	436	333	149	43	10	
Medford, Ore.									1	10	62	270	667	908	1119	1083	1012	898	713	564	432	339	281	211	132	55	12	1	
Portland, Ore.								+	4	20	40	123	343	772	1238	1271	1274	1163	1001	581	373	211	126	54	23	6	1		
Salem, Ore. (5)				+				+	3		46	172	435	707	1100	1398	1352	1316	833	540	405	261	169	100	47	14	5	+	
Harrisburg, Pa.							4	18	52	125	222	427	749	888	722	659	635	692	737	824	807	556	376	201	60	13			
Philadelphia, Pa.							30	9	32	100	189	335	654	818	758	701	663	710	735	809	863	655	420	225	74	17			
Pittsburgh, Pa. (5)						9	31	60	159	233	360	569	774	688	631	587	637	678	799	910	722	503	311	98	12	1			
Scranton, Pa. (5)					2	4	9	78	178	264	392	575	848	805	628	592	629	719	784	804	666	434	254	83	14				
Providence, R.I.					1			39	84	163	275	491	729	829	790	763	762	764	822	799	668	434	229	86	25	2			
Charleston, S.C.									2	5	27	79	192	321	434	576	651	787	889	1090	1267	1143	724	425	137	20	1		
Columbia, S.C.										19	56	138	293	411	523	623	673	722	808	895	1113	940	657	489	301	92	11	1	
Huron, S.D.	1	7	21	56	83	145	208	262	305	419	476	571	652	574	502	488	513	569	614	624	554	443	318	205	103	47	9		
Rapid City, S.D. (5)		1	5	10	46	79	128	194	246	273	420	594	742	694	675	621	632	666	655	590	488	370	318	198	108	52	10	+	
Chattanooga, Tenn.								4	14	45	113	228	414	553	642	679	713	722	775	895	1021	703	554	406	221	56	9	1	
Knoxville, Tenn.					+		2	7	21	41	101	217	456	590	648	689	672	675	746	889	977	703	595	384	153	28	3	+	
Memphis, Tenn.					1	1	1	4	10	25	74	196	374	532	614	633	618	690	715	798	1056	796	670	489	308	102	18	+	
Nashville, Tenn.					1	1	3	9	28	67	132	263	463	565	627	619	637	697	738	838	933	814	582	443	227	66	11	1	
Amarillo, Tex.			+	+	2	2	6	21	52	124	213	374	547	617	640	651	685	668	729	819	787	600	474	389	257	95	14	1	

Table 1-13 (Continued)

	−30 to −25	−25 to −20	−20 to −15	−15 to −10	−10 to −5	−5 to 0	0 to 5	5 to 10	10 to 15	15 to 20	20 to 25	25 to 30	30 to 35	35 to 40	40 to 45	45 to 50	50 to 55	55 to 60	60 to 65	65 to 70	70 to 75	75 to 80	80 to 85	85 to 90	90 to 95	95 to 100	100 to 105	105 to 110	110 to 115
Austin, Tex.									1	7	10	46	115	253	411	496	617	679	717	934	1016	1242	842	609	474	249	48	2	
Brownsville, Tex.										+	1	6	6	35	116	205	323	459	653	996	1351	1878	1444	833	438	19	+		+
Corpus Christi, Tex.								+	+	+	3	27	27	83	180	302	444	551	748	1041	1175	1538	1408	785	436	36	1	9	
Dallas, Tex.								1	4	17	34	91	231	371	504	576	629	656	693	795	831	942	880	659	493	273	79	4	
El Paso, Tex.								+	2	10	34	104	233	369	494	611	687	760	794	839	933	884	743	592	428	204	44	3	
Ft. Worth, Tex. (5)								1	3	12	44	132	294	422	538	591	648	622	689	774	889	982	788	596	440	246	54		
Galveston, Tex.											2	6	27	77	178	336	557	717	1000	1042	1006	1144	1763	849	62	1		1	
Houston, Tex.										1	2	4	18	64	141	291	452	570	681	772	980	1172	1611	949	677	327	56	6	
Laredo, Tex.											2	8	27	82	192	343	471	563	687	838	958	1346	1161	801	626	463	189	2	
Lubbock, Tex. (5)							5	7	33	86	180	346	490	546	620	618	642	700	688	829	833	708	544	447	322	109	14	1	
Midland, Tex. (5)								1	5	23	80	163	337	451	602	669	631	678	720	793	914	865	673	532	426	177	28	+	
San Antonio, Tex.								1	1	4	11	31	94	190	387	445	569	669	789	943	1086	1336	873	620	470	230	21	1	
Waco, Tex. (5)									1	3	24	84	216	354	501	558	651	622	701	830	909	1101	829	612	482	249	42	+	
Wichita Falls, Tex. (5)							+	4	10	27	114	228	388	497	581	636	627	606	677	714	784	825	741	560	406	260	83	2	
Salt Lake City, Utah				+	2	16	41	80	158	328	564	798	831	755	685	682	635	614	615	569	447	368	309	196	66	8		+	
Burl'tn, Vt. (5)	2	5	17	39	81	135	216	272	332	491	561	752	716	727	724	668	694	757	820	573	362	189	53	9	1				
Norfolk, Va.										1	61	175	371	558	637	603	668	757	703	850	953	722	481	297	136	19			
Richmond, Va.								2	19	67	138	285	478	632	727	724	668	694	820	920	1079	916	524	341	163	38	1		
Roanoke, Va. (5)								1	11	28	186	307	533	702	710	712	695	704	758	924	928	620	459	305	91	8	6		
Seattle-Takoma, Wash.									3	20	39	104	427	914	1408	1445	1462	1272	750	448	258	123	62	24	6	2			
Spokane, Wash.					8	16	29	52	91	153	302	625	1060	974	853	805	786	715	633	525	414	294	212	136	66	16	1		
San Juan, P.R. (W.I.)																			10	276	1717	3384	2440	920	19	3			
Charleston, W.Va. (5)					1	7	22	73	135	252	356	630	633	607	667	661	689	767	758	658	474	331	176	66	9				
Green Bay, Wisc. (5)	1		3	19	42	95	160	231	321	373	515	689	820	649	542	522	648	720	685	658	474	331	176	66	9				
Madison, Wisc.		1	4	9	34	68	106	163	241	323	469	688	843	641	521	526	535	648	715	685	597	459	290	143	50	8	+		
Milwaukee, Wisc.		1	2	4	18	47	83	116	176	285	421	659	913	774	611	591	634	634	749	753	597	390	226	96	32	6	+		
Casper, Wyo.		2	3	15	30	45	73	116	200	324	495	683	806	831	782	670	642	606	592	592	423	347	283	201	66	3			

Table 1-13 (Continued)

	-30 to -25	-25 to -20	-20 to -15	-15 to -10	-10 to -5	-5 to 0	0 to 5	5 to 10	10 to 15	15 to 20	20 to 25	25 to 30	30 to 35	35 to 40	40 to 45	45 to 50	50 to 55	55 to 60	60 to 65	65 to 70	70 to 75	75 to 80	80 to 85	85 to 90	90 to 95	95 to 100	100 to 105	105 to 110	110 to 115
Chic. (O'Hare), Ill. (5)		+		4	10	23	40	57	86	111	186	320	430	369	283	269	275	306	346	388	363	246	166	81	25	1	+		
Chic. (Midway), Ill.				3	12	25	59	85	117	196	335	551	822	800	591	543	569	592	653	769	762	563	370	220	105	27	2	1	
Moline, Ill.			1	8	22	45	81	177	168	247	389	603	784	665	529	530	541	576	692	737	725	584	397	215	94	19	1		+
Springfield, Ill.			+	1	1	3	12	29	57	168	272	531	775	693	588	557	545	576	652	745	802	671	491	289	137	27	1	1	
Evansville, Ind.			+	2	6	19	40	69	124	104	189	371	665	725	665	619	569	608	689	724	858	718	538	376	180	49	6	1	+
Ft. Wayne, Ind.			+	8	3	13	35	60	97	205	381	596	905	712	601	552	586	585	699	777	728	524	363	189	73	13	1		
Indianapolis, Ind.			+	14	7	26	47	81	166	152	449	661	870	694	564	526	567	608	722	815	821	619	413	220	80	12	7	1	1
South Bend, Ind. (5)				1	23	59	104	152	211	250	405	557	747	627	510	512	535	600	728	806	698	507	318	150	41	3			
Des Moines, Iowa		+	2		42	75	125	176	262	281	443	545	694	611	530	520	567	550	681	751	662	566	378	219	102	28	4		
Sioux City, Iowa		+	3	8	2	9	31	70	125	206	315	480	638	635	597	561	519	578	623	707	757	547	400	240	122	38	7	1	
Topeka, Kans.					1	3	14	45	85	161	273	426	607	611	530	592	589	603	641	679	758	677	509	338	212	99	23	4	
Wichita, Kans.				2	1	7	16	35	80	144	238	441	654	584	597	611	644	656	710	709	957	703	527	381	259	137	50	8	1
Lexington, Ky. (5)			+	1	+	3	8	25	45	97	169	332	631	627	611	634	619	654	693	898	869	630	464	263	62	43	7		
Louisville, Ky.						+			1	3	7	38	117	204	347	488	601	669	797	758	957	742	541	355	192	46	1		
Baton Rouge, La.										3	2	17	72	164	353	477	607	718	807	968	1366	1344	820	600	354	37	+		
Lake Charles, La.						+		+		+	2	9	47	128	282	449	621	692	850	984	1240	1439	893	644	310	12			
New Orleans, La.								2		6	23	72	200	361	516	609	619	679	772	987	1189	1671	979	620	229	1	+		
Shreveport, La.					4	60	109	190	293	6	408	599	820	839	772	748	760	808	780	886	1063	1076	758	583	394	123	25	4	
Portland, Me.				5	15	29	60	109	43	293	184	328	642	755	770	673	683	696	729	627	407	275	149	58	14	1			
Baltimore, Md.			+	1	1	2	2	7	89	89	256	429	674	848	828	757	766	781	804	794	883	655	438	263	109	23	2		
Boston, Mass.					1	9	4	35	74	151	256	674	848	808	595	566	592	633	695	819	676	433	245	127	39	10	+		
Detroit, Mich.					4	4	17	61	131	248	377	618	884	808	595	597	574	638	695	783	721	516	314	148	47	9	+		
Flint, Mich. (5)		+		2	11	34	75	142	208	347	487	707	863	675	565	565	571	647	712	745	588	411	253	88	11	+			
Grand Rapids, Mich.		+	1	1		10	31	78	172	293	469	690	938	742	554	617	705	733	739	739	634	451	286	137	41	5			
Duluth, Minn.	4	12	34	80	131	187	238	279	376	490	635	609	688	572	544	482	538	602	639	495	342	192	101	32	2				
Minneapolis, Minn.	2	4	10	31	62	119	186	246	311	383	514	632	632	560	500	605	618	677	695	690	621	468	295	147	54	8			
Jackson, Miss.							1	2	2	6	41	103	224	367	484	482	553	572	790	922	1169	1056	690	557	339	99	14		1
Kansas City, Mo.				+	4	4	21	53	99	175	265	407	591	625	623	562	553	575	601	723	761	742	615	404	236	103	25	1	
St. Louis, Mo				1	7	7	15	40	77	134	212	411	650	671	620	578	585	575	646	728	823	763	579	376	198	65	13	2	+
Springfield, Mo.				1		4	14	29	68	139	235	437	588	621	616	621	602	620	759	846	876	647	475	331	169	60	6	1	1
Great Falls, Mont.	4	15	43	51	68	101	118	136	167	218	355	533	698	813	832	830	822	754	636	520	407	296	187	113	46	5			+
Omaha, Neb.			3	15	15	40	93	135	189	287	390	511	663	655	543	543	539	558	606	721	726	610	445	288	149	44	13	6	
Las Vegas, Nev. (5)										1	7	44	194	396	591	716	769	786	699	644	651	669	602	474	474	431	301	101	16
Reno, Nev. (5)				1	1	4	15	37	101	227	387	530	733	829	890	909	845	690	572	477	418	371	333	243	120	35	2	1	1
Newark, N.J.						2	2	11	38	109	202	360	637	846	784	701	692	697	755	814	819	615	383	200	79	21	2	2	1
Alburquerque, N.M.					1	1	1	4	21	66	154	346	552	689	741	734	687	651	719	831	767	634	502	371	229	65	3	+	+
Albany, N.Y.	1	3	5		10	32	63	110	184	278	404	574	793	769	647	625	652	708	740	733	588	417	256	131	39	8			16

Other researchers have attempted to bring a greater degree of sophistication to the bin method itself,[19,20,21,22,23] developing time-consuming methods, some of which can provide results for larger projects using manual methods.

Large contemporary structures present many factors which affect their energy consumption: structural differences (glass vs poured concrete facade); massing and orientation (requiring heating in one zone at the same time another requires cooling); occupancy and use patterns that vary with the time of day and season; air conditioning systems with greatly different thermodynamic characteristics (from variable air volume to terminal reheat and a host of others); and the "heat island effect."

The heat island effect is characterized by higher temperatures in the microclimate of dense urban areas due to large heat release from buildings per unit of land area. It is said that a structure in the middle of the city behaves substantially differently than an identical building located in the countryside among trees and lakes, that center-city design temperatures are higher and that the factors increasing cooling load are more severe than in the country. This postulate is subject to verification but initial data appear to support such claims.

The more sophisticated bin methods allow for these individual variations and can incorporate a large number of conditions and variables to provide a fairly realistic estimate of energy requirements. Unfortunately the more complex formats yield more accurate information at the expense of greater effort and money.

Table 1-13 facilitates energy calculations and gives dry bulb temperatures for many large cities in the United States. It was prepared by the old U.S. Weather Bureau and represents averages over a ten year period. On the other hand, the Air Force Manual (see footnote 14, this chapter) breaks down weather data into eight-hour bins and gives coincident wet-bulb temperatures; unfortunately this manual is restricted to fewer locations, mostly Air Force bases. The National Climatic Center (see this chapter's bibliography, under "Climate Information") provides specialized weather data tabulations for a fee.

1.5 Cooling Degree-Day Method

In the middle 1950's as the air-conditioning industry began to expand rapidly, it was then still necessary in many cases to provide economic justification for an air-conditioning installation. Energy studies, usually, were part of an economic justification and, therefore, designers began to look for easy-to-use energy calculation methods. Since the concept of heating degree-days had been around for quite a long time, it was therefore natural to suppose that the equivalent concept of cooling degree-days might be a useful measure of energy consumption and comparison. Actually, the Weather Bureau started this work in 1930's with WPA assistance, using 75 F (29.3C) as a base. This work never passed the manuscript stage and the method never proved to be satisfactory.

Another approach was the combining of wet and dry bulb temperatures in different proportions and in inventing intriguing names such as "climatic factor" or in splitting the country into four "air-conditioning need" zones. Consolidated Edison of New York City used such a combination in the early years to predict its electric load by averaging the wet and dry bulb temperatures, thus approximating comfort zone temperatures.

Earl C. Thom,[24] of the U.S. Weather Bureau, proposed a cooling degree-day method based on the Discomfort Index (DI) Formula, over the base of 60 F (15.6C).

$$DI = 0.4 \text{ (dry bulb temperature} + \text{web bulb temperature)} + 15 \qquad (1.14)$$

The same investigator[25] provides tables of cooling degree-days over a 5-year period (1953-1957) for 55 selected stations in the United States over a base of 60F (20.0-21.1C), although at the same time posing the question as to weather a base of 68–70F (20.0-21.1C) would not be more appropriate, John J. Drummond explains to some extent the correlation between base selection and accuracy of load prediction.[26]

[19] E. E. Umlang, "Calculating Heating and Cooling Energy Using Building Load Characteristics and Temperature Data," *ASHRAE Journal*, December, 1965, pp. 46–51.

[20] Calvin Singman and Ephraim K. Cohen, "Use of Air Temperature Tables in Estimating Energy Requirements," *ASHRAE Journal*, January, 1966, pp. 43–48.

[21] H. G. Werden, "Weather Data Vs. Operating Costs," *ASHRAE Journal*, October, 1964, pp. 60–64.

[22] Trevor R. Tiller and Frederick H. Kohloss, "Energy Cost for Air Conditioning of Office Buildings in Mild Winter Climates," *ASHRAE Journal*, February, 1965, pp. 54–62.

[23] ASHRAE Handbook and Product Directory, 1976 Systems, Chapter 43, "Bin Method Procedure," pp. 43.10 to 43.15.

[24] Earl C. Thom, "A New Concept for Cooling Degree-Days," *Air Conditioning, Heating and Ventilating*, June, 1957, pp. 73–80.

[25] Earl C. Thom, "Cooling Degree-Days," *Air Conditioning, Heating and Ventilating*, July, 1958, pp. 65–70.

[26] John J. Drummond, "Development of a System of Cooling Degree-Days," *Proceedings, National District Heating Association* (now International), Vol. IL, 1958, pp. 108–116.

Another researcher, T.T. Porembski,[27] elaborated tables of heating and cooling degree-days, with some allowance for heating or cooling a fixed amount of outside air to a certain temperature, using a base of 0F (−17.8C) through 100F (37.8C), in 1-degree F (0.55C) spans.

The cooling degree-day method attempts to formulate a basis for comparing degree of discomfort at different times of the year or from year to year or between different locations. In addition, such a system could predict the magnitude of air conditioning need and the cost of operating an air-conditioning installation or it could define variations in energy requirements.

These objectives remain unfulfilled to this day. Although our techniques, especially computerized techniques, are on the verge of major achievements, our field measurement or field testing programs remain practically nonexistent. Of course, cooling-degree days depend a great deal on field test data to verify the correctness of its assumptions. Such data never became available and, therefore, this method remains a paper theory.

The cooling degree-day method depends on statistical weather averages in the same manner that the heating degree-day does, and it fails to account for such variable factors as solar intensity, wind velocity, lighting levels, human occupancy, hours of occupancy, percent fenestration and other important items.

The correlation between theoretical and field data was never established and it is almost impossible to gauge the accuracy of this method or to calculate the correction factors that must be applied under various circumstances.

1.6 Equivalent Full Load Hour (EFLH) Method

Estimating energy consumption by the EFLH method unfortunately depends on inadequate data, mostly collected and interpreted by utilities, with very wide fluctuations reflecting such variables as operating hours, type of structure, occupancy, lighting, use, etc. Three columns of data—low, medium and high—reflect these uncertainties so that one must be careful in utilizing this method. The EFLH method can be useful, however, for quick or preliminary estimates.

This method converts the energy used for the entire cooling season into a number of hours during which equipment preforms at rate load.

Let

T = Tons of refrigeration required at any given load level

h = Hours of operation during cooling season at the given load level

T_{fl} = Maximum design or rated tons of refrigeration

EFLH = Theoretically the number of hours during which the chillers would operate continuously at full load to supply cooling through the season.

Then,

$$T_1 h_1 + T_2 h_2 + T_3 h_3 + \cdots + T_n h_n = T_{fl}(\text{EFLH})$$

Hence

$$\text{EFLH} = \frac{\displaystyle\sum_{i=1}^{h}(Th)}{T_{fl}} \qquad (1.15)$$

The advantage of this conversion is that a criterion is established whereby energy consumption of buildings of diverse characteristics operating under a host of conditions can be compared.

The basis for the equivalent full load hours data will be found in the bibliography at the end of this chapter under the subheading "Cooling Statistical Data" and from data from various consulting engineers.

In order to find the energy consumed, it is necessary to multiply EFLH by the power requirement per ton plus the auxiliaries as reflected in Table 1-16. The cost is equal to EFLH × kW/ton × electric rate.

1.7 Prediction of Peak Demand

Prediction techniques yield valid projections for existing systems that are about to undergo expansion when the collection of consumption data is accomplished by dependable, if not metered, means for substantially long periods.

1.7.1 Demand Factor Method

This method of steam load prediction is empriical and depends on a systematic collection of flow data; weekly, in the opinion of those at Pennsylvania State University,[28] who originated this method. The condensate meters of various buildings or groups of buildings are read weekly at the same hour and a calculation of the heating factor

[27] T. T. Porembski, "Predicting Annual Heating-Cooling Loads," *Air Conditioning, Heating and Ventilating*, March, 1963, pp. 66–76.

[28] Thomas B. Kneen, "Steam Loads in Recent University Buildings," *Proceedings, International District Heating Association*, Vol. LIV, 1963, pp. 139–143.

Table 1-14
Equivalent Full Load Hours (Applicable to New York City only)[29]

Structure	Equivalent Full Load Hours (EFLH)[30]			Cu Ft / Ton			Hours of Operation Per Day
	Low	Ave.	High	Low	Ave.	High	
Office Buildings (curtain wall)	828	1625	2395	2206	2800	4760	10–14–16
Office Buildings (brick masonry)	571	868	1255	2850	4555	5640	10–14–16
Apt Houses	794	1487	1680	4614	5200	6300	24
Depart. Stores	540	682	1172	5000	—	5780	10
Hotels	—	1505	—	—	3475	—	24
Schools	—	800	—	2063	—	5750	10
Motels	—	1595	2090	3750	—	5150	24

for that week is made, expressed as lbs of steam per hour per cubic ft. or lbs of steam per degree day MCF. These points are plotted throughout the heating season as shown in Fig. 1-4. The curve must be extrapolated in the low temperature range because there are not seven consecutive days below a certain range of temperatures.

In order to find the demand factor at 0F, the following formula may be used:

$$D_O = f_O \frac{\text{cubic feet}}{1000} \times \frac{65}{24} \qquad (1.16)$$

Where

D_O = Demand in lbs of steam per (hour)(feet cube) at 0F outside temperature

f_O = Demand factor at 0 F in lbs of steam per deg-day per 1000 cu. ft.

65 = Degree-days per 24 hours at 0F

24 = Hours per day.

1.7.2 *Radiation Factor Method*

This method is used by some distinct steam utilities to determine system peak demand. Total connected radiation

is multiplied by a radiation factor. The radiation factor is defined as the amount of steam required per square foot of connected equivalent direct radiation. System losses, in the example below, were assumed to be 12%.

This method is satisfactory as a prediction tool of an on-going utility enterprise provided radiation factors undergo constant revision by comparing calculated versus actual peak loads for each building type. The method takes no account of building volume but is based on connected radiation surface area only. It is applied by the Detroit Edison Co. Table 1-17 shows calculations performed for its entire network in 1965.

1.8 Factors Affecting Heating Consumption and Demand

Changes in architectural styles have led to the use of new construction materials. The past decades have seen

[29] To find full load equivalent hours for any other city, multiply figures for New York City by multiplier in Table 1-15.
[30] For additional information on EFLH see:
a. ASHRAE 1966 Guide and Data Book, p. 1021.
b. W. H. Carrier, R. E. Cherne, W. A. Grant and W. H. Roberts, *Modern Air Conditioning, Heating and Ventilating* (New York: Pitman Publishing Corp., 1959), p. 516.
c. Edison Electric Institute, *Electric Heating and Cooling Handbook*, Section 4, 1965.
d. ASHRAE Handbook and Product Directory, 1976 Systems, Table 5, p. 43.9.

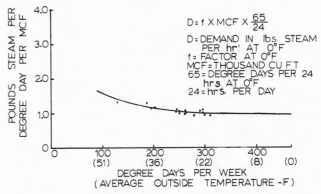

Courtesy International District Heating Association.

Figure 1-4. Heating factor *f* curve.

Table 1-15*
Equivalent Full Load Hours (EFLH) Multiplier
for Various U.S. Cities

Abliene,Tex.	1.93	Louisville, Ky.	1.31
Albany, N.Y.	0.55	Memphis, Tenn.	1.93
Albuquerque, N.W.	0.85	Miami, Fla.	3.72
Amarillo, Tex.	1.14	Minneapolis, Minn.	0.82
Atlanta, Ga.	1.75	Montgomery, Ala.	2.18
Baltimore, Md.	1.23	Nashville, Tenn.	1.70
Bakersfield, Calif.	1.11	New Orleans, La.	2.74
Billings, Montana	0.50	New York, N.Y.	1.00
Boston, Mass.	0.81	North Platte, Neb.	0.87
Brownsville, Tex.	3.51	Oklahoma City, Okla.	1.70
Casper, Wyo.	0.38	Philadelphia, Penna.	1.15
Charleston, S.C.	2.09	Phoenix, Ariz.	2.18
Chicago, Ill.	1.05	Pittsburgh, Penna.	0.71
Cleveland, Ohio	0.82	Providence, R.I.	0.99
Columbus, Ohio	1.18	Raleigh, N.C.	1.60
Denver, Colo.	0.50	Red Bluff, Calif.	1.14
Detroit, Mich.	0.76	Reno, Nev.	0.23
El Paso, Tex.	1.41	Rochester, N.Y.	0.71
Fargo, N.D.	0.64	Sacramento, Calif.	0.83
Fresno, Calif.	1.12	Salt Lake City, Utah	0.62
Fort Smith, Ark.	1.89	San Antonio, Tex.	2.55
Forth Worth, Tex.	2.53	San Francisco, Calif.	0.17
Hatteras, N.C.	1.97	Seattle, Wash.	0.16
Houston, Tex.	2.76	Shreveport, La.	2.35
Indianapolis, Ind.	1.15	St. Louis, Mo.	1.50
Jackson, Miss.	2.16	Tallahassee, Fla.	2.36
Jacksonville, Fla.	2.63	Tampa, Fla.	2.97
Kanas City, Mo.	1.58	Tuscon, Ariz.	1.69
Knoxville, Tenn.	1.60	Yuma, Ariz.	2.43
Laredo, Tex.	3.28	Washington,D.C.	1.34
Las Vegas, Nev.	1.44	Winslow, Ariz.	0.70
Los Angeles, Calif.	0.83		

*Table 1-15 based on the ratio of cooling degree-days for any city divided by the cooling degree-day for New York City. Data were obtained from an article by Earl C. Thom, "Cooling Degree-Days," *loc. cit*. New York City multiplier is (1) one.

the application of large-scale architectural glass, aluminum, outside wall insulation, double-glazed windows, and greater glass areas in order to bring in light and blend the function of the building with the outside surroundings. The demand for more efficient space utilization, in turn, induced builders to create larger structures with more interior spaces requiring year-round cooling. Office building occupancy periods increased because various departments of the tenant companies, such as computer operations, payroll, etc., had to work overtime. Schools extended the school year by expanding the curriculum or by offering evening adult education courses or opening

Table 1-16
Approximate Power Inputs[31]

Item	Compressor kW/Design Ton	Auxiliaries kW/Design Ton
Window Units	1.46	0.32
Through-Wall Units	1.64	0.30
Dwelling Unit, Central Air-Cooled	1.49	0.14
Central, Group, or Bldg. Cooling Plants		
(3 to 25 tons) Air-Cooled	1.20	0.20
(25 to 100 tons) Air-Cooled	1.18	0.21
(25 to 100 tons) Water-Cooled	0.94	0.17
(Over 100 tons) Water-Cooled	0.79	0.20

Reprinted by permission, from *ASHRAE Systems Handbook 1976*.

school spaces for community activities. Air conditioning with positive mechanical ventilation removed the need for operable windows, and winter infiltration through window cracks became a thing of the past. Here are some of the basic characteristics of the old versus the new buildings:

Old Buildings

1. No rigid temperature control. Radiation was continuous rather than intermittent, thus large consumption requirements.
2. Poor control of quantity and temperature of the outside air supply.
3. Small glass areas, hence, smaller installed radiation per unit volume.

New Buildings

1. Zone control, taking advantage of load diversity. Controls activated not only by indoor conditions but, by actual outside conditions or the rate change of outside conditions.
2. Control of outside air quantity to minimize intake during very low or very high outside temperatures, enlarging O.A. intake for "free cooling" when appropriate, for economy.
3. Increased installed radiation per unit volume due to increased glass area, often 45–70% of the outside wall.
4. Tighter construction, resulting in minimal infiltration and controlled stack draft; hence, reduced radiation load.
5. Increased lighting, use of business machines and other electrical loads, the heat output of which tends to lower heating requirements and increase cooling requirements.

---

[31] ASHRAE Systems Handbook 1976.

6. Sharply peaking steam loads. Some utilities have instituted steam demand penalties, inducing their customers to control load peaks.

Office buildings offer the most concrete examples of how design standards have changed radically over the years:

New Office Buildings: ceilings lowered to between 8–9 feet, more business machines, 3–6 watts per square foot lighting intensity, less area per person, more insulation, less infiltration through windows, off business temperatures of 60–62F, and curtain walls, 2680 Btu/yr ft^3 average annual consumption, increased solar radiation heat gain, advanced heating controls of the modulating type, licensed operators, personnel more cost conscious.

Old Office Buildings: 0.5-1.5 watts per square foot (5.38-16.14 W/m^2) lighting intensity, higher ceilings (10–15 feet) (3.04-4.57m), high infiltration, 3300 Btu/(yr) (ft^3) (123 MJ/yr m^3) average annual consumption, vacuum or gravity two-pipe steam systems, brick and masonry construction, off business temperatures of 55–57F (12.8–13.9C), operable windows, *intermittent* manual control settings.

In the tabulations that follow, three categories of energy consumption have been used: low, average and high. The tabulations are, of course, of general character and ought to be used with some caution, although all of them have been compiled from field reports. Here are some of the considerations that have been included under the three categories:

High: Irregular or long hours (75-90 per week), night work, owner-occupied, smaller building, certain areas work overtime, high infiltration, high stack effect, large glass area, special occupancy such as graduate student residences with lots of children, some insulation, special heavy usage such as laundry or restaurant water use.

Average: Steady hours of operation, usually 5 1/2 days per week, average type of building in its class, some insulation.

Low: Ideally or at least well-zoned, night setback, double glazing, fewer exposures, hot water perimeter heating, well-insulated, good operating personnel, large buildings, average fresh air ventilation 30% (range 25-40%), good plant efficiency of 70-75% (usually includes electric generation or heat recovery).

1.9 Factors Affecting Cooling Consumption and Demand

1. An important factor is the duration of the occupancy period. Staggered working hours, commuter schedules, stock house overtime operations, payroll schedules, research departments working into the night, community activities in the evening, adult education courses and numerous other activities have increased the length of the operating day. A 14-hour operating day for an office building is not at all unusual and a 16-hour campus operating day for a metropolitan university is routine.

2. A second important factor is the amount of outside air used on an overall basis for a building or a group of buildings. Cooling or heating this air from outside to room conditions is an appreciable refrigeration or heating load. The amount of outside air is determined by ventilation standards, the need for toilet exhaust, requirements for positive pressure by lab hood exhaust and odor control requirements, whichever is largest. Most designers take into the building huge quantities of outside air in order to dilute objectionable odors; 0.25 to 0.35 cubic feet of outside air per minute per square foot of floor area was fairly ordinary for New York high rise buildings. Although such a quantity may amount to only 25% of total air circulated in the interior space, it nevertheless may amount to 50% of the interior zone refrigeration load. Better and cheaper filtration equipment may permit reduction of outside air requirements by reconditioning return air. After the 1973 oil embargo, there has been a tendency to reduce outside air requirements to 0.1 cfm per sq ft, or even less.

3. Outside design conditions constitute an important, although not always readily recognizable, factor. What statistically prevailing combinations of outside wet-bulb and dry-bulb temperatures contribute the most to large blocks of energy consumption? The approach the designer assumes toward this statistical phenomenon will affect his sizing of refrigeration equipment and the estimation of energy consumption. If one considers 95Fdb (35C) and 30% rh, versus 85Fdb (29.4C) and 50% rh, the greatest air conditioning load will occur at 95Fdb (35C) and 30% rh. The higher 95Fdb governs the transmission load, and since the wetbulb temperature is the same for both conditions, the outside air refrigeration load remains the same for both conditions. On the other hand, if we compare 80Fdb (26.7C) and 40% rh, versus 80Fdb and 80% rh, the refrigeration load will be greatest at 80Fdb and 80% rh. Transmission remains identical for both conditions but the enthalpy at 80% rh is higher, making the refrigeration load for the outside air the governing factor.

Table 1-17
Peak-Load Calculations, February 3, 1965[32]

Code No.	Number of Customers	Classification	Radiation Connected, 1000's ft^2	Radiation Factor lb/hr/ft^2	Peak Load M lb/hr
1	111	Apartments and Apartment Hotels	686.6	.174	119.5
2	34	Bank and Loan Offices	405.9	.118	47.9
3	27	Churches	220.6	.076	16.8
4	9	Clubs	312.4	.090	28.1
5	24	Garages—Parking	168.5	.110	18.5
6	12	Hospitals	625.9	.162	101.4
7A	12	Hotels—over 1,500 sq ft	538.6	.133	71.6
7B	46	Hotels—under 1,500 sq ft	181.1	.231	41.8
8A	52	Office Buildings—over 1,000 sq ft	2,842.9	.108	307.0
8B	136	Office Buildings—under 1,000 sq ft	1,318.6	.112	147.7
9	32	Manufacturing	442.1	.290	128.2
10	11	Motels	117.0	.109	12.8
11	37	Printers and Newspapers	410.4	.136	55.8
12	139	Restaurants and Bars	182.1	.104	18.9
13	52	Schools	1,521.0	.091	138.4
14A	12	Stores—Department	664.5	.113	75.1
14B	289	Stores—Retail	610.2	.153	93.4
14C	21	Stores—Wholesale	58.0	.153	8.9
15	18	Theaters	292.0	.043	12.6
16	24	Warehouses	248.2	.070	17.4
17A	25	Miscellaneous—Government Buildings	1,139.4	.079	90.0
17B	167	Miscellaneous—Other Buildings	745.0	.144	107.3
	1,290		13,731.1		1,659.1
		Detroit Edison Buildings	200.6	.108	21.7
		Industrial Load (Stroh's)			65.7
					1,746.5
		12% System Losses			209.6
		Predicted Peak			1,956.1
		Actual Peak			1,969 M lb

*Courtesy of **International District Heating Assoc.***

4. The type of refrigeration equipment which is selected has an important bearing on the annual energy consumption depending on the energy demand of the equipment. The designer ought to be entirely familiar with all refrigeration machines and their driving mechanisms. Selecting machinery to do an effective job requires a thorough understanding of machine performance characteristics combined with an in-depth analysis of job economics.

5. A fifth factor is partial load operation. While most machinery and arrangements work satisfactorily at peak conditions, they fail to perform adequately at partial load conditions. Handling partial load at reasonable efficiencies depends a great deal on proper analysis and experience. Sizing of equipment, combination of components and adherence to good design practice are important elements for an acceptable part-load operation.

6. A sixth factor concerns the means of distributing energy. Terminal reheat or dual duct air distribution systems, for instance, are considered high energy consumers. The current energy crisis is imparting special significance to this aspect of engineering, and most designers are reexamining their concepts and are reevaluating system performance.

7. Construction materials and methods of wall construction are also a factor. Buildings constructed in the

[32] John V. Levergood and Roger L. Demumbrum, "What Has So Drastically Changed Load Forecasting?," *Proceedings, International District Heating Association*, Vol. LVII, 1966, p. 72.

twenties, thirties, and forties were of heavy masonry construction with small glass areas, light illumination intensities, sparser population and they had, in general, excellent storage capacity and slow thermal response. Refrigeration requirements were reduced accordingly. Contemporary buildings, with large glass areas and light curtain wall construction have very little storage capacity. Their thermal response is rapid. Individuals working within these buildings feel uncomfortable having to react almost continuously to high radiant space loads. In such systems, the storage capacity of the circulating central chilled water system assumes ever greater significance.

8. Another factor is the geographical orientation of all buildings in a complex. The orientation of the important buildings may be of crucial significance in limiting peak loads and achieving low diversity. A north-south orientation of the long sides is preferable to an east-west orientation, for instance. The mechanical designer ought to bring pertinent facts to the attention of the planners, pointing out savings in first cost and operating expenses.

9. Another factor is distribution, including the method of moving and controlling flow of air and water. An optimum objective is to move a minimum amount of fluid at minimum pressure at all times and still do an adequate job.

One must distinguish between old and new installations. Expanding or revamping an old installation offers a challenge in accounting for the criteria of the old installation while incorporating new features, which in time may be changed sufficiently to eliminate the old installation entirely. Making an effective working whole out of disparate elements calls for a full examination and understanding of the existing installation as well as selective and scheduled replacement.

Brand new installations afford the designer greater freedom of fulfilling the objectives of master planning and of making provisions for accommodating future expansion. The latest technical advances and most up-to-date thinking can be most easily incorporated into a new installation. A basic task is to weigh costs against advantages that may be obtained. Factors such as air pollution or the fuel crisis can be considered at this stage. Table 1-18 for annual cooling energy requirements gives three values: high, average and low; defined as follows:

High: 16-24 operating hours per day, 16 for office buildings, 24 for motels, hotels and some apartment houses, high glass area, high reheat, high full load equivalent hours, extensive partial load operation, equipment mismatching and inefficient thermodynamic operation,

maximun morning building precooling, high outside air requirements, low storage factor, lack of heat recovery, a use of drivers with high energy rates, curtain wall, absorption machines, smaller buildings.

Average: 12-14 operating hours, moderate climates, morning precooling, good storage factor, some insulation.

Low: 8-10 operating hours, well zoned and well-designed building, efficient and well-combined machinery, low heat rates, low ventilation rates, good storage factor, heavy masonry construction, efficient partial load operation, high design conditions such as are ordinarily used in department stores and other short occupancy structures, condensing steam turbine drive using high pressure steam, heat recovery, energy conservation, larger buildings.

The values given in Tables 1-18 and 1-19, for both heating and cooling, although authenticated, represent past practice. The new ASHRAE Standard 90-75, "Energy Conservation in New Building Design," undoubtedly will play a significant role in establishing new design concepts and standards of energy consumption.[33] However, data for the new designs will take years to verify. Therefore, Tables 1-18 and 1-19 can serve as guides and check figures for quick and preliminary estimates. Regretably, there are many gaps in the tabulations, but these data could not be obtained or verified.

1.10 Analytical Computers

The computer makes its appearance twice in central plant engineering, once during the initial stage of defining load magnitudes and annual energy costs: i.e., during the feasibility study period.

The computer is called into play again during the operating stage, as an operational and managerial tool. Here one must emphasize that a computer is an appendix to the entire automation process, a sort of super controller that overlooks and guides the entire operation. Historically, as the number of operations grew, the number of individual and separate control actions increased to the

---

[33] Energy consumption data for the years 1974, 1975, 1976, and 1977 will be found in AIA Research Corp. *Phase One/Data Base for the Development of Energy Performance Standards for New Buildings.* Sponsored by the U.S. Department of Housing and Urban Development in Cooperation with the U.S. Department of Energy. Document HUD-PDR-290-2, 1978.

Table 1-18

Annual Heating, Domestic Hot Water and Cooling Energy Requirements for Various Buildings—5000 Degree-Days

Structure	Heating — Annual Consumption MBtu/yr ft³ Gross Building Volume Above Grade—5000 Degree Days			Heating — Peak Demand, Btu/hr ft³ Gross Building Volume Above Grade—0F			Domestic Hot Water — Annual Consumption, MBtu/yr ft³ Gross Building Volume Above Grade			Domestic Hot Water — Peak Demand, Btu/hr ft³ Gross Building Volume Above Grade			Cooling — Annual Consumption, MBtu/yr ft³ Gross Building Volume Above Grade			Cooling — Peak Demand Btu/hr ft³ Gross Building Volume Above Grade		
	Low	Ave.	High	Low	Ave.	High	Low	Ave.	High	Low	Ave.	High	Low	Ave.	High	Low	Ave.	High
Office Buildings (curtain wall)	2.54	2.93	3.46	2.30	2.50	2.90	0.26	0.56	0.80				1.63	2.22	3.77	2.75	2.45	3.12
Office Buildings (brick-masonry)	2.50	3.09	3.78										2.36	9.75	13.85	2.99	5.77	8.10
Apt Houses	3.70	4.10	4.50	2.25	2.90	—	1.50	2.10	2.60	—	1.00	—	2.52	4.30	7.89	2.12	2.99	4.22
Churches	1.76	2.41	2.97				—	2.30	—				—	4.02	—	—	4.71	—
Clubs	5.36	7.10	9.85				1.67	2.42	3.70									
Department Stores	3.10	3.75	5.61				0.55	0.90	1.02				1.08	2.75	4.80	2.39	2.93	4.78
Factories	3.92	—	16.1				—	2.29	—									
Garages	—	1.98	—				—	0.44	—									
Hotels	5.23	8.25	9.61				2.60	3.03	4.80				—	8.13	11.5	—	5.6	—
Residences	3.98	—	4.40															
Restaurants	7.25	12.15	24.10				—	17.20	—									
Schools	2.64	3.87	5.43										—	4.67	—	2.69	—	5.87
Theaters	2.50	3.5	3.38				0.31	0.55	0.89									
Warehouses	1.34	1.94	7.12				—	—	—									
Hospitals	9.08	—	14.98				—	5.50	—									
Museums	3.06	3.67	6.07															
Taverns	—	4.07	—				—	1.0	—									
Printing Plant	—	18.10	—															
Nurses Residence	—	6.45	—															
Motel	—	13.50	—										5.90	7.05	8.42	2.81	3.53	5.06

Table 1-19

Annual Heating, Domestic Hot Water and Cooling Cooling Energy Requirements for Various Buildings—6000 Degree-Days

Structure	Heating						Domestic Hot Water						Cooling					
	Annual Consumption, MBtu/yr ft³, Gross Building Volume Above Grade—6000 Degree Days			Peak Demand, Btu/hr ft³, Gross Building Volume Above Grade—0F			Annual Consumption, MBtu/yr ft³, Gross Building Volume Above Grade			Peak Demand, Btu/hr ft³, Gross Building Volume Above Grade			Annual Consumption, MBtu/yr ft³, Gross Building Volume Above Grade			Peak Demand, Btu/hr ft³, Gross Building Volume Above Grade		
	Low	Ave.	High	Low	Ave.	High	Low	Ave.	High	Low	Ave.	High	Low	Ave.	High	Low	Ave.	High
Structures Built 1890–1950 Steam Heated																		
Residences with Dining Halls	7.35	10.25	12.75	–	2.97	–												
Residences (no food service)	7.15	10.55	16.58	–	2.82	–		*			*			*			*	
Class Rooms	3.87	5.50	9.20	–	1.91	–												
Miscel. Bldgs	3.57	8.17	16.95	–	3.11	–												
Research Labs.	16.35	29.00	35.70	7.65	8.90	13.63												
Structures Built in 1960's																		
Residences with Dining Halls (water heated)	8.40	9.35	10.35	2.17	2.98	3.41	1.12	1.47	1.60	0.96	1.23	1.28						
Class Rooms and Apts (water and steam heated)	5.11	8.65	11.70	2.03	2.93	3.11												
Entire Campus Complex—includes feedwater heater blowdown, line losses, heating dom. H.W., 10% makeup, no elec. generation 5500-6700 deg.-days	7.45	10.70	14.0	2.89	5.0	7.56	Included under heating											
Entire Campus—same as above 60-100% elec. generation 5-20% makeup 4800-6300 deg.-days	10.40	13.0	14.5	3.25	3.65	5.97	Included under heating											

*Figures not available.

point that one or more operators could not attend to all the functions, and the physical size of the control panel got out of hand, requiring several rooms to house the installation. To ease this situation, minicomputers, programmable controllers, and miniaturized electronic or solid state equipment was developed. This type of computer is examined in Chapter 6.

The computer as engineering tool, although it performs at unbelievable speed, depends for accuracy on the thoroughness of preparation and the principles that permeate a program. There is no initiative in the computer. Only programmed routines can take place. Since the computer executes programs with utter faithfulness, the programmer is faced with certain alternatives in the construction of the program in terms of weather data, system simulation, modeling techniques, assumptions, simplifications, detailing methods of load estimation, etc.

The existence of many commercially available programs poses even greater puzzles to the purchaser. Some programs have been initiated by consulting firms, others by trade associations and technical societies, and still many more by government agencies and manufacturing concerns. The accuracy of a program depends on the methods used, the available talent, and funds. The results can be excruciatingly different as Figs. 1-5 and 1-6 make evident.

An analytical computer program must not only account for the nature of the loads and the magnitude and rate of heat release, but must recognize the heat release and retention mechanisms and the relationship and reactions between these mechanisms. Such a program must be sensitive to effective heat removal and addition methods. In order to define the energy requirements of a structure or group of structures, it is not only necessary to establish an energy balance but also to understand the reaction of any related systems during all periods of operation at different load levels or different load combinations.

Most analytical computer programs have thus far been oriented towards calculating the requirements of a single building or, at best, a small number of buildings with similar characteristics. There are no satisfactory programs as yet that deal with variable and untypical building loads in a complex of tens or even hundreds of buildings. This computer capability is essential for large or even medium-sized central plants. Undoubtedly, future program development will correct this insufficiency.

Computers are capable not only of estimating building load but also energy requirements and costs for a structure that employs a variety of systems, cost differentials for different systems, costs for operating personnel, maintenance, scheduled repairs, and a financial analysis that demonstrates facts and figures of interest to investors, such as rate of return, cash flow or net present worth.

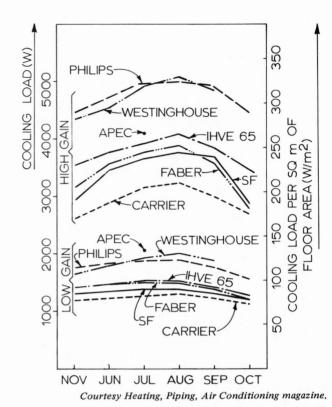

Courtesy Heating, Piping, Air Conditioning magazine.

Figure 1-5. Total monthly cooling loads of two rooms on opposite facades. Loads were calculated by six computer programs and Carrier manual method.

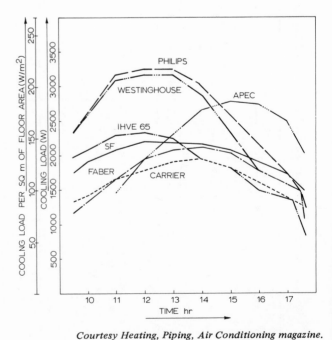

Courtesy Heating, Piping, Air Conditioning magazine.

Figure 1-6. Daily cooling load curve for July (south facade, high gain).

The allocation of energy costs to the plant and/or complex, including the costs of all central plant auxiliary services, is a specific arrangement for each project. An appropriate computer program should have the ability to categorize electrical demand charges, premiums, penalties or other special items. An adequate central plant computer program can account for simultaneous zoned heating and cooling loads of a complex, fully considering load diversity and helping the engineer to select prime components of the proper size to satisfy full and partial loads.

Thus, the analytical computer operation is one more engineering tool for energy computations. The degree of accuracy of a program is relative, since energy costs, although increasingly high, constitute only a portion of annual owning and operating costs. The expense of computer programs, the manpower necessary to fill hundreds of input cards and the finality of the program output makes them less attractive than they appear at first glance. Many clients are not willing to shoulder this cost and many engineering firms, in order to maintain their skills and profits, prefer manual programs, which at times reassure an engineer about the validity of his own input and his ability to change variables and procedures at will. Employment of computer programs is a matter of individual circumstances. However, medium and large central plants in the future will employ only computer programs for both preliminary and final design.

1.11 Systems Simulation

In a conceptual sense, simulation attempts to connect system design to system operation in order to facilitate energy analysis. It is a modeling method that brings together the dynamic interaction between three factors: 1. climatic conditions such as temperature, humidity, cloud cover, solar intensity, wind velocity, rain occurrence, etc.; 2. the nature of the structure, its response to outside and inside conditions and its role as a heat transfer barrier; 3. the interior electrical, pneumatic, or mechanical systems of the structure and their responsiveness to loads. Simulation is a modeling technique that takes into account varying indoor and outdoor conditions, the thermal and operating characteristics of a structure and the sensitivity or lack thereof of systems and components and their behavior. Simulation, then, results in a realistic approach to energy consumption.

Some of the difficulties that engineers face in using the simulation technique are the unlimited combinations of weather conditions, system types and building structural variations. It is nearly impossible to predict what combinations must be used in the program writing. Essentially, system simulation is nothing more than an economic optimization technique, based on mathematical modeling and a great deal of judgment on the part of the program author. Modeling depends for its representativeness not only on the author's mathematical dexterity but also on the manner in which all the variables are incorporated into the mathematical procedure and on the number of thermodynamic mechanisms and their sophistication which, as an integral part of the model, reinforce its accuracy.

Simulation is flexible in that it can take into account the owner's preferences and patterns for staffing, maintenance procedures, afterseason and afterhours operations and energy conservation programs, and also include part-load behavior of a structure under varying climatological conditions, hours of operation of a system in a particular operational mode, energy cost structure and selection of programs that prove effective.

The following are areas in which simulation is practiced with varying degrees of success:

Airside Simulation. Items which are ordinarily part of airside performance simulation include temperature, circulated air, performance of heat transfer components, such as heating, reheating and cooling coils and fan horsepower. The degree of sophistication involves a multitude of factors and its desirability depends on the overall importance of the systems under consideration and on whether accuracy in energy consumption predictions will ultimately contribute towards energy conservation. In some instances, if the results of airside simulation constitute the input to another program, accuracy is very important.

Refrigeration Plant Efficiency Simulation. A realistic program must consider compressor efficiency at part loads, the effect of condenser water temperature and compressor speed on motor horsepower. Rather than make simplistic assumptions about efficiencies—with predictably inaccurate results—it is preferable to obtain efficiency percentages from manufacturers. Overall horsepower predictions depend a great deal on an accurate load profile and modeling of the number of chillers in active operation to meet all loads at the predicted efficiency. Additional sophistication may be introduced, in connection with all auxiliary motors.

Heat Storage Simulation. A variety of methods are utilized in this area, including the Carrier building unit weight method, the ASHRAE method, etc. Due to the number of methods in existence and the diversity in accuracy, a predictable variation in results exists.

Control Mode Simulation. On-off or reset method of operation, constant leaving air temperature, minimum enthalpy control and a number of other control modes can add a degree of sophistication. This type of simulation is not ordinarily encountered in the average computer program.

Infiltration Simulation. Most residential programs assign an arbitrary number of air changes to infiltration and some use the window crack method. Both methods are valid for residences or smaller buildings. Industry or the technical societies never produced sufficient field-validated data to formulate a theoretical or experimental base. The current methods are undependable and inaccurate and will remain so until this technical void is filled.

Design Optimization Simulation. Most parameters in the subroutines remain as variables that are manipulated for design optimization. Leaving air temperature, air flow, chilled and condenser water flow are some of the manipulated variables.

In addition to the above, there are many other simulations, which include: economizer cycle, fixed outdoor air, outside air shutoff schedule, heating and cooling shutoff schedule, change over temperature. Of course not all of these simulations are available on every program.

BIBLIOGRAPHY

Cooling Loads

Anders, James M. "Energy and Equipment Selection Procedure for Post Office Buildings," *ASHRAE Journal*, (December, 1965), 52–56.

Mitalas, Gintas P. "Transfer Function Method of Calculating Cooling Loads, Heat Extraction and Space Temperature." *ASHRAE Journal*, (December, 1972), 54–56.

Nelson, Lorne W. and James R. Tobias. "Energy Savings in Residential Buildings," *ASHRAE Journal*, (February, 1974), 38–45.

Thom, Earl C. "Measuring the Need for Air Conditioning." *Air Conditioning, Heating and Ventilating*, (August, 1956), 65–70.

_____. "Cooling Degree-Days and Energy Consumption." *Air Conditioning, Heating and Ventilating*, (September, 1966), 53–54.

Cooling Statistical Data

Coad, William J. "Integrated Loop System for Campus Cooling." *Heating, Piping, Air Conditioning*, (May, 1973), 88.

_____. "Energy Economics: a Design Parameter." *Heating, Piping, Air Conditioning*, (June, 1973), 74.

Cohan, Harry A., Jr. "Choosing a Central Refrigeration Plant for an Apartment House." *Proceedings, National District Heating Association*, Vol. LVI, 1965, 38–43.

Hittman Associates, Inc. *Residential Energy Consumption–Multi-Family Housing Data Acquisition.* Document PB 216-440. Springfield, Va.: National Technical Information Service, 1972.

Imperator, Thomas. "Intermediate Cooling in the Thermocycle System," *Proceedings, International District Heating Association*, Vol. LVIII, 1967, 116–123.

Kight, Lucian T. "Steam Absorption Cooling Installations," *Proceedings, International District Heating Association*, Vol. LVIII, 1967, 123–128.

Heating Loads

ASHRAE Technical Committee (T.C.) 2.6. "New Methods for Estimating Fuel or Energy Consumption." *ASHRAE Journal*, (October, 1967), 45–50.

Billington, Neville S. "Estimation of Annual Fuel Consumption." *Journal of the Institute of Heating and Ventilating Engineers*, (November, 1966), 253–256.

Bossone, Edmund D. "Relation of Degree-Days to Temperature in Various Cities." *Proceedings, National District Heating Association*, Vol. XLII, 1952, 111–118.

Boyd, Donald W. "Converting Heating Degree Days from Below 65 F to Below 18 C." *Transactions, the American Society of Heating, Refrigerating and Air-Conditioning Engineers,* Vol. LXXXII, 1976, Part 2, 448.

Burggraaf, Robert. "Alaska Summer Weather Data." *Transactions, The American Society of Heating, Refrigerating and Air-Conditioning Engineers*, Vol. LXII, 1960, 240.

Cousley, John M. "Sizing of Service Water Equipment in Commercial and Institutional Buildings." *Proceedings, International District Heating Association*, Vol. LXI, 1970, 92.

Ewing, William H. "Domestic Water Heating." *Proceedings, International District Heating Association*, Vol. LVI, 1965, 102.

Parsons, Roger A. "How to Estimate Net Building Space Heating Requirements," *Heating, Piping, Air Conditioning*, (January, 1962), 155–159.

Segeler, George C. *et al.* "What Type of Heating Uses Least Energy?" *Heating, Piping, Air Conditioning*, (August, 1973), 61–64.

Skagerberg, Rutcher and J. E. Phifer. "Fuel Consumption Analysis for Multi-Family Housing Projects," *Transactions, American Society of Heating, Refrigerating and Air-Conditioning Engineers*, Vol. LIX, 1953, 113.

Strock, Clifford. "Short Cut Heat Loss Estimating." *Heating and Ventilating's Engineering Data Book*, 1st ed., New York: Industrial Press, 1948, section 5, 32–34.

Strock, Clifford and Richard L. Koral, eds. *Handbook of Air Conditioning, Heating and Ventilating*, New York: Industrial Press, 1965.

Thom, Herbert C. S. "Seasonal Degree-Day Statistics for the United States." *Proceedings, National District Heating Association*, Vol. XLIII, 1952, 111.

_____. "The Rational Relationship Between Heating Degree Days and Temperature," *Monthly Weather Review*, (January, 1954), 1–6.

_____. "Normal Degree Days Below Any Base," *Monthly Weather Review*, (May, 1964), 111–115,

Heating Statistical Data

Bohonan, Robert E. "Heating of Post-War Buildings," *Proceedings, National District Heating Association*, Vol. L, 1959, 77.

Burke, Robert E. "Steam Performance Data on All-Year Air Conditioning Systems Using Steam in New York City," *Proceedings, National District Heating Association*, Vol. XLV, 1954, 120.

Gordon, William J. "More on the Efficient Utilization of Purchased Steam Service," *Proceedings, International District Heating Association*, Vol. LVIII, 1967, 82.

Graham, Michael P. "Energy Consumption and Conservation in School Buildings," *Heating, Piping, Air Conditioning*, (July, 1977), 85–90.

Hittman Associates, Inc. *Residential Energy Consumption–Multi-Family Housing Data Acquisition*, Document PB 216-440. Springfield, Va.: National Technical Information Service, 1972.

Jamison, Grahaeme. "Hot Water Consumption in Two College Dormitories and a Public High School," *Transactions, The American Society of Heating, Refrigerating and Air Conditioning Engineers*, Vol. LXVIII, 171.

Jennings, Burgess H. and Samuel R. Lewis. "The Heating Load," *Ar Conditioning and Refrigeration*. Scranton, Pa.: International Textbooks, 1958, 160–165.

Jones, Jerold W. and Berkley J. Hendrix. "Residential Energy Requirements and Opportunities for Energy Conservation." *Transactions, The American Society of Heating, Refrigerating and Air Conditioning Engineers (ASHRAE)*, Vol. LXXXII, Part 1, 1976, 417.

McCarrick, Edwin F. "Selling Steam in New York City." *Proceedings, National District Heating Association*, Vol. LI, 1960, 134.

_____. "The Pan Am Story," *Proceedings, International District Heating Association*, Vol. LVI, 1965, 92. 1965, 91.

Metzger, Albert F. "A Method of Estimating Steam Consumption," *Proceedings, International District Heating Association*, Vol. LV, 1964, 126.

Milusich, John. "Heating Requirements—Glass and Curtain-Wall Versus Conventional Buildings," *Proceedings, International District Heating Association*, Vol. LVI, 1965, 96.

_____. "Steam Requirements for Process Equipment," *Proceedings, International District Heating Association*, Vol. LVII, 1966, 54.

Morrow, James H. "Operating Experience on 2000-Ton Steam-Operated Air Conditioning System in 36-Story Office Building, 100 Park Avenue, New York, N.Y.," *Proceedings, National District Heating Association*, Vol. XLIII, 1952, 90.

Rapson, F. Lee. "Steam Consumption of Buildings Based on Proposed Exposed-Area Method," *Proceedings, National District Heating Association*, Vol. XLII, 1951, 36.

Sigworth, Robert Y. "Steam Consumption in University Buildings," *Proceedings, National District Heating Association*, Vol, XLV, 1954, 125.

Sigworth, Robert Y. and Thomas B. Kneen. "Comparison of Campus Heating Systems," *Proceedings, National District Heating Association*, Vol. L, 1959, 139.

Sprenger, Carl B. "Steam Requirements of New-Type Buildings with Metal and Glass Walls," *Proceedings, National District Heating Association*, Vol. LIII, 1962, 146.

Venkata, S. Manian. "Off-Peak Domestic Hot Water Systems," *Transactions, The American Society of Heating, Refrigerating and Air Conditioning Engineers*, Vol. 80, 1974, 174.

Windingland, L. M., *et al.* (Hittman Associates). *Energy Utilization Index Method for Predicting Building Energy Use. Volume I: Method Development*, Prepared for the U.S. Army Construction Engineering Research Laboratory, NTIS document AD-A039 913/9WE. Springfield, Va.: National Technical Information Service, 1977.

Climatic Information

National Climatic Center, *Information Brochure*. (Asheville, NC: National Climatic Center), 1970.

Richard, Oscar E. *et al.* "Environmental Services Available to the Engineer." *ASHRAE Journal*, (November, 1966), 42–46.

CHAPTER 2

Plant Criteria

The significance of plant criteria is that a general framework is provided within which plant operation may be defined and rules are established that are helpful to the plant designer in making decisions. In addition, criteria help outline the limits of "good practice" so that the plant will fulfill its designer's objectives and operate within its design parameters. Criteria provide the basis for interrelating principles underlying the design of the various plant systems with a view toward achieving a plant that is a successful unit, technologically and economically.

2.I The Rationale for Central Plants

We are entering an era of central plant construction to serve urban areas. Central plants hold immense promise for the urban environment, consisting of living quarters which consume energy and industrial complexes which both use and produce energy.

The economic benefits of central heating and cooling plants for large installations have long been appreciated, most recently in connection with new town projects, where the latest technological advances in plant design and distribution can be applied to provide superior urban services.[1] Now, with rising energy and utility costs, these benefits can be multiplied where such plants incorporate electricity generation and heat recovery, and are interconnected with a regional energy grid.[2]

Central plants are attractive from a capital investment as well as from an operating point of view.

From a *capital investment* point of view, there are many advantages:

1. *The initial cost of equipment* is less than it is for a number of decentralized plants of equal capacity because cost per unit capacity is less for larger equipment.
2. *No capital investment* is necessary for the user, who does not have to construct a central plant but is obligated only to pay for all energy used. This may also be true for institutions, which can set up independent corporations to operate the central plant facilities.
3. *More usable or rentable space* becomes available in individual buildings by eliminating rooftop or basement machine rooms, and space for cooling towers, piping shafts, etc.
4. *Less mechanical space* is required in a central location for large refrigeration machines or boilers than for a greater number of smaller machines interspersed throughout the installation. In addition, considerable savings are effected by eliminating individual chimneys and avoiding duplication of auxiliary equipment, standby units, etc.
5. *Lower construction costs* will be incurred for the central plant than for several dispersed plants of the quality of the central plant and of an aggregate capacity equal to that of the central plant. Individual building plants require additional structural support for upper level mechanical spaces or more excavation for below-grade mechanical rooms. These lower construction costs plus the increased rentable spaces are usually of a sufficient magnitude to reduce, and in many cases eliminate, the cost of distribution piping associated with a central plant.

[1] Fred Smith, *A Manual of Specific Considerations for the Seventies and Beyond*, (New York:Man and His Urban Environment Project, 1972).
[2] George Polimeros, "Some Considerations in Industrial Heat Recovery," *Buildings Systems Design*, April/May, 1973, pp. 36–38.

6. *The total capacity of a central plant is invariably less* than the sum of multiple, smaller, dispersed plants because, in a central plant, a diversity factor may be used to reduce plant size, for not all buildings peak simultaneously due to differences in topographic location, occupancy, and construction.

From an operating point of view, the central plant offers the following advantages:

1. *Plant efficiency is maximized* simply because, even at partial load, those of the machines that are then in operation work near their optimum efficiency. This is one area where, from an energy conservation point of view, the central plant offers tremendous hope for profitable energy management. Machine operation can be computerized to optimize component selection and equipment performance for minimum energy input, maximum output and heat recovery.

2. *Labor costs are minimized* due to easier supervision and scheduled maintenance. A few high quality engineers are necessary in addition to a floating watch that does heavy maintenance and substitutes for personnel on vacation or on holidays. Centrally located manpower working in one place offers maximum utilization of the available labor force. The contrary is true for a multiple plant installation, where provision for 24-hour service to all individual plants would be prohibitive. (One common way to overcome these difficulties would be to install 15 psig steam boilers and absorption machines which do not require licensed operators in most localities. However, this approach cannot be considered seriously for a large central plant.)

3. *Fuel savings* are feasible because of mass fuel consumption and the possibility of negotiating more advantageous commercial terms with a number of fuel dealers through competitive buying.

4. *Central maintenance* has the advantage of closely located repair shops and of stored materials and spare parts, all essential factors for a well-kept plant. Thus, emergency breakdown service becomes swifter and more efficient. Responsible executive direction is available at all times and the existence of central maintenance leaves executives more time for planning and other business activities.

5. *Refinements in design, construction, and controls* are feasible and economically viable for the large central plant.

6. *Air pollution, noise, and vibration are reduced* because boilers are attended on a 24-hour basis and flue gas analysis is part of normal operating procedure. More efficient smoke control equipment is available. Because of the physical separation of most large plants from the buildings they serve, noise and vibration, if at all present, affect only the central plant.

7. *Plant safety* is improved by around-the-clock plant attendance.

8. *A complete plant shutdown* is improbable, especially if the plant generates its own electricity or if it is backed by an emergency diesel engine. A central plant possesses greater reserve capacity because there is greater flexibility and large equipment is better-constructed, hence, more reliable. The boiler plant, in particular, can meet weekend or other low loads very successfully and at near peak efficiency if equipped with part-load boilers.

9. *Individual building design flexibility* is enhanced because an architect can design without the constraints imposed by mechanical space for boilers or refrigeration machines, and without chimneys, cooling towers and their vapors. The physical separation of the central plant removes any interference between itself and the buildings it serves. The spatial logistics of a central plant cannot tolerate architectural restrictions, other than aesthetic, when incorporated in a larger building. It is almost mandatory that the central plant be a separate building, with its own architectural criteria based on its function and the need to relate creatively to its surroundings.

10. *Respect for community activities* is guaranteed by placing the plant in a relatively remote location. Coal silos, oil storage, fuel trucks, cooling tower noise and vapor, smoke stacks, machinery noise, maintenance activities, etc., are matters that can antagonize a community if they are placed where they interfere with day-to-day community life.

11. *Water treatment* is carried out at a central point, resulting in high quality water.

12. *Monitoring of the plant* can be accomplished by a central monitoring console, located in the control room, which supervises the status of the entire complex it serves. The monitoring console can be supplemented by trending capabilities, start

and stop features, alarms, and resetting of controllers. The control room can incorporate a computer for operating or safety reasons. If the control room is located near the central plant, the central plant can be monitored and operated by the same computer and control installation that serves the remainder of the complex. This coordinated operation provides an additional measure of safety and operating efficiency.

13. *Heat recovery* is a distinct possibility in every central plant. From an energy conservation point of view, the central plant presents great opportunities for maximum energy utilization. Heat recovery from boiler effluent gases and condensate is practicable. In process industries such as steel mills, copper smelters, coke kilns, etc., it is possible to recover huge amounts of energy by cooling process gases from 2000F (1093C) or more down to 500F (260C) or lower and generating steam, hot water, or electricity. Heat recovery is also possible in total energy plants. Of course, a commercial market must exist for the products of heat recovery in order to make a project financially viable.

2.2 Central Plant Heating and Cooling Loads

Central plant load depends on different conditions that exist in a number of buildings at the same moment. In defining the condition of a particular building, the following factors affecting thermal loads must be considered:

1. The constant physical characteristics of the structure,
2. The continuously changing internal conditions, and
3. The changing external environment.

Some of the variable items, which a procedure for establishing central plant load should be able to handle with a least some degree of detail, are:

1. Solar load as a function of building construction, including location of internal and external shading devices, orientation, time of day and rear, and cover,
2. Hourly variation in building occupation and related design temperatures, operation of lighting and other internal heat sources over the year,
3. Infiltration and changing outdoor air temperature,
4. Control temperatures as well as night setback or system stoppages during unoccuoied periods, and
5. Effects of lag and energy storage on instantaneous heat gain and carryover loads.

In the case of a number of buildings or a building complex, the following additional factors enter the picture:

1. Exfiltration, since many buildings are pressurized. (Exfiltration is constant, in contrast to infiltration which varies according to height or atmospheric conditions.)
2. Underground piping losses which result from ground conditions, mass flow in piping and related heat transfer to the ground.
3. The combination, in operation, of a number of buildings, each operating with distinctly different or similar occupancy patterns, lighting and other heat sources, orientations, wind velocities, solar loading, diversity factors, and load factors. The probability that the combination of any number of events will happen in a repeatable pattern is extremely slight. The accuracy of any load duration curve will be contingent upon the permutations and combinations that the engineer deems acceptable. Since computer techniques are the only ones indicated for solving such highly interrelated problems, it is simply a matter of the type of computer techniques adopted and the degree of accuracy pertinent to the problem at hand.
4. The type of occupancy is relevant to the type of energy pattern that each project develops. For instance, a complex of apartment houses will produce a more predictable energy utilization curve than a combination of office buildings, stores, and apartment houses.
5. Scheduled preoperation of various systems to meet design conditions at occupancy time. Preoperation is often programmed on a computer and is related to outdoor design conditions.

Contemporary computer programs make it possible to obtain daily and yearly load duration curves similar to those in Figs. 2-l and 2-2. The yearly load duration curve is affected by the length of.the heating or cooling season, which is dependent upon geographical location. The daily or yearly loads may be affected by use patterns, *e.g.*, in colleges, during summer vacations, when demand is less. The types of systems selected for various buildings affect the energy utilization pattern directly. Dual-duct, face and bypass dampers, terminal reheat or multizone systems require constant air flow and constant temperature chilled water supply. Pumping and energy costs will probably be high, although the systems may provide the best answer to a particular building design. Building system design selection cannot be independent of central

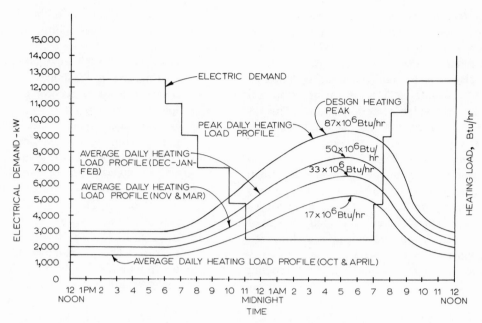

Figure 2-1. Building daily load profile—electrical and heating—for winter period (October–April).

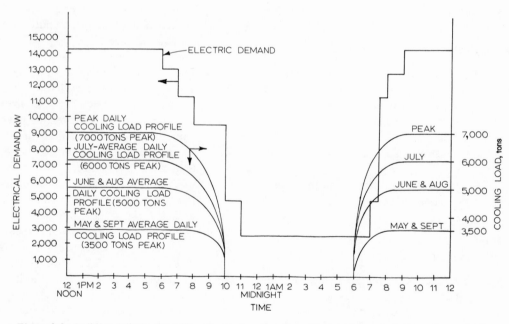

Figure 2-2. Building daily load profile—electrical and cooling—for summer period (May–September).

plant design considerations. Every design decision will be of consequence to the entire project.

2.3 Engineering and Design Responsibility

Establishing buildings systems selection procedures, obtaining approvals and coordinating the design of the central plant together constitute the most important responsibility of the designer in terms of obtaining low first costs, satisfactory systems operation and long term operating economy. This responsibility should remain in the hands of a single, most experienced engineering team. This procedure is essential in order to arrive at valid load duration curves, which can help define central plant component sizes and operating procedures. Parallel approval, at the very least, should be obtained from the central plant group for building service connections and secondary water systems. Primary chilled water, hot water, steam and electrical distribution must be the responsibility of the central plant team, and all work and decisions must remain within its jurisdiction.

Departure from these guidelines will result in chaotic conditions and an oversized, overflexible central plant constructed at premium cost.

A central heating and cooling plant must be housed in a separate structure. Heavy structural loads dictate specialized construction and heavy foundation work. Other items dictating separate housing include: separation criteria for chimney contaminants, high voltage yard equipment and transmission lines, dust control equipment, high chimneys, noisy cooling towers and their vapor plumes, noise caused by fans, boilers, compressors, diesel engines, space required for coal silos, oil storage tanks, truck-parking facilities and proximity to active neighborhoods in urban areas.

Above all, equipment arrangement must be guided solely by efficiency and economy criteria, not by the need to avoid an elevator shaft, a stairway egress, or an auditorium, which are bound to arise when a central plant is incorporated into a major building. A separate plant facilitates the moving in or out of equipment by allowing trucks to pull into the plant and to make deliveries feasible at an unloading dock. Ventilation air is available on all sides of the plant and power roof ventilators can be part of the plant ventilation system without competition from building shaft space required in some academic or medical building, for instance. Ordinarily, if the central plant is incorporated into a commercial building, indoor electrical transformers appropriate valuable exterior wall in order to be naturally ventilated through exterior wall louvers, leaving very little outside wall space for the outside air intakes of ventilating equipment. Such competition for outside wall accessibility shoves the main plant components into the lower depths of the structure. Incorporating the plant into academic, institutional or industrial areas makes it difficult to maintain steam pipe pitch in crowded ceilings, and power plant piping has to play second fiddle to important process pipes.

From one point of view, it would be ideal if all structures of an institutional, urban or industrial complex were completely designed before central plant design begins. On the contrary, practical considerations preclude such a possibility. For instance, the architects of the buildings served must know the location and size of the plant for sound landscape planning. From an overall project point of view, the distribution routes must be firmly established and coordinated with streets, roads and property access. All central plant components are long-lead delivery items and central plant construction follows a slower pace than the rest of the project. Besides, the central plant, or a good portion of it, must be ready long before the first building is up.

It is often found that in long-range, large projects the central plant itself must be phased simply because all this expensive equipment cannot lie idle while the project is being completed, let us say, over a 10-year period. Construction phasing and capital investment very often dictate central plant design and size of components. The engineer must select components on a modular basis while assuring himself that future loads will justify his decisions.

Thus, preliminary design may not follow computerized procedures, but other plausible, even rule-of-thumb criteria, until a firmer basis is established. At each juncture, the engineer must reevaluate his design and costs on the basis of the most recent, established data. The computerized processes may appear at the end, either simply confirming the engineer's previous assumptions or, if necessary, revising them.

One must not forget that building fenestration, outside wall construction and interior arrangements are not finished before the central plant is designed. In chemical process or nuclear plant refrigeration, this is not the case and loads can be readily established. The imponderables and gray areas are more prevalent in large commercial projects.

2.4 High Temperature Water (HTW) or Steam?

A basic, initial decision for a central plant designer is whether to generate steam or high temperature water (HTW). Aside from characteristics which are peculiar to each, the following are areas which can stand the test of comparison for either medium:

1. *Thermal expansion.* Introducing steam into a cold steam line can cause more shock than by gradually heating the water already in a pipe and allowing incurred stresses to be equally distributed. Steam tends to hug the upper part of the line, since it is lighter than air, expanding the upper portion of the pipe more and forcing it to bend upward. HTW piping heats up gradually, giving the ground time to absorb piping heat losses.

2. *Pipe pitch.* A portion of the steam condenses as it flows through piping, so that the steam line must be pitched to conduct the condensate to a

convenient collection point. Pitching is inflexible and leads to serious conflicts with other utility lines at crossover points. By contrast, water lines can dip under or can go over other pipes in difficult situations. HTW piping at times may present venting problems. Also, water velocities must be kept high, to carry along any entrapped air.

3. *Pumping requirements.* Water circulation requires continuous pumping. On the other hand, steam condensate, unless naturally pitched, must be pumped back to the boiler room. At first glance, there appears to be more pumping in a HTW system only if one ignores factors outside the distribution piping. This is discussed under operating costs.

4. *Operating costs, repairs, and maintenance.* For a valid comparison of operating costs of a steam and HTW plant, one must consider all items of operating cost, including the type of fuel used, boiler efficiency at all loads, and auxiliary motors and equipment. Where a HTW installation has boiler recirculation and primary HTW pumps, the analogous steam plant has a complicated steam cycle and associated equipment including pressure reducing stations, steam traps, condensate coolers, condensate return tanks, flash tanks, condensate pumps, deaerators, hotwell and feedwater pumps, drain pipes, etc. Steam boilers usually form scale: HTW boilers rarely. Steam plants, because of their abundance of components, ordinarily require more maintenance and more frequent repairs and replacements than an equivalent HTW plant.

5. *Plant auxiliaries.* Some of the larger auxiliaries are profitably powered by steam turbines in a steam plant while electric motors prevail in HTW installations. Steam-driven auxiliaries are favored by maintenance people for their economy, dependability and independence from loss of electric power, but are even more attractive now in the midst of the continuing energy crisis. Standby equipment in steam plants is electrically driven to provide fail-safe operation. Electric power may be financially attractive, if generated within the plant.

6. *Site Conditions.* HTW is more suitable for hilly terrain because piping can easily follow the contours of the land. Steam systems are more adaptable to fairly flat land because draining of condensate can become a major problem on steep slopes. Steam piping most often requires underground concrete expansion joint stations. It is common practice to employ expansion loops for underground water or overhead steam systems, which are run on trestles in industrial installations. Underground installations are used predominantly by city utilities or for aesthetic reasons at large institutional sites.

7. *Pipe sizing.* HTW pipe sizes are smaller than low pressure steam mains of the same capacity, but about the same size when compared to 125 psig (861 kPa) steam systems, and are larger when compared to 250 psig (1722 kPa) steam mains. However, items such as manholes, drain lines, trap assemblies, guides, restraints, etc., all contribute to the final cost comparison.

8. *Boiler efficiency.* One important but often overlooked factor is that HTW boilers, because of their recirculation pumps, are highly efficient at both full-load and part-load conditions. Steam boilers are progressively less efficient at lower loads.

9. *Load expansion.* Capacity of steam plants designed with adequate allowances can be substantially increased to serve additional clients by increasing line pressure or steam velocity or by absorbing larger pressure drops. HTW systems are distinctly less flexibile in this respect because water velocity can be increased only modestly, if at all, requiring greater horsepower, if not new pumps.

10. *Plant location.* A steam plant is best located in the center of the area it serves so that pressure at remote points will be sufficient. In a sense, pressure at remote points can be represented by the mean radius of a circle with the plant at the center of the circle. Most often, steam plants are relegated to the most undesirable and remote corner of the site. As a result, pressures are excessive near the plant and inadequate at the far end. A central location is not as important for a HTW plant, since piping and valving arrangements can lessen the significance of excessive pressure differentials.

11. *Operating pressure.* Setting the operating conditions for a HTW system is much simpler than for a steam system. The basic reason is that HTW is used mostly for heating purposes and that maximum temperature differentials for various pieces of equipment are well-known (see Tables 2-2 and 2-3) and water flow rates can be se-

lected that contribute to pumping economy.

Deciding on the final pressure level of a steam distribution system is a more complex matter. For instance, although 250 psig may be adequate for steam distribution, higher pressures may be necessary for turbogeneration, mechanical steam turbine or centrifugal compressor drives. At times, steam demand and consumption of two or three substantial clients may dictate distribution line pressures. If an industrial park and a residential community were to be served by a single central plant, for instance, what boiler steam pressure would serve the different pressure requirements of two users and the needs of the plant itself? High line pressure reduces pipe size but, on the other hand, increases costs by requiring additional, costly pressure-reducing stations. Multipressure plants can be justified more readily if reduction in steam pressure is used to produce work. Setting steam pressure is the final step in a long process of analytically predicting plant functions for the most compact and economical operation.

12. *System safety.* HTW systems, up until now, have operated at higher pressures than the average 250 psig industrial or institutional steam plant. Now, 600 psig (4134 kPa), 750 F (399 C) main steam is being used in a number of central plants. It must be borne in mind that HTW piping systems contain huge amounts of potential energy which, if released into a closed space because of a major pipe break, for example, may cause some localized physical damage. The question of whether steam or high-temperature water leaks are more dangerous is rather controversial and depends on the assumptions made. The risks of each system must be analyzed in terms of their operating environment.

13. *Distribution patterns and city services.* Large steam distribution systems, particularly those serving large urban areas, tend to grow in a tree-like fashion. A grid pattern, if found in the city, is rather of accidental origin. The reasons for this are historic and economic. A steam grid pattern must compete for underground street space with all the other utilities and is expensive to install and maintain without offering a distinctive advantage. The same is true of condensate return lines. Utility companies are resigned to non-returnable condensate; actually they discourage its return. A one-pipe steam system takes the least space and requires a low capital

investment. Customers ordinarily waste condensate into the sewer. Some cities, New York among them, mandate cooling of the condensate to 120F (49C) before discharging it into the sewer system.

Larger HTW installations tend to follow grid distribution patterns which are found even in smaller projects. Water flows to any point in the system from two directions. In the event that a pipe bursts, one side can be closed off for repairs, while the other remains in service. Strategically placed sectioning valves serve this function. In addition, flow from two directions means partial flow from each direction; hence smaller distribution piping can be employed. HTW systems have taken hold in the last two decades in new communities, campuses and medical complexes where adequate excavation space can be provided on a project basis for all utilities, usually underneath or adjacent to the street for easy access and service.

14. *Pipe rupture.* Interruption of service when a HTW line bursts poses more difficulties than a similar steam line break because the HTW disgorges a large quantity of extremely hot water, the task of disposing of which is formidable. Underground water leaks are difficult to locate; steam leaks, depending on excavation depth, often have telltale plumes on the surface. Sections of ruptured steam pipe can be replaced with minor interruption to service. A HTW system must be drained and refilled with treated water. The effects of a damaging break, whether steam or HTW, depend a great deal on the locale and the circumstances of the accident. No firm guidelines can be drawn for prevention of accidents except to use quality materials and insist on good workmanship, dependable engineering and meticulous maintenance.

15. *Pipeline thermal storage.* HTW systems have huge thermal storage capacity which can accommodate load peaks and fluctuations. The boiler has ample time to adjust its firing rate to load changes. On the contrary, storage in steam systems is insignificant, so that boilers and auxiliaries must adjust quickly to satisfy load demand. Peak loads cause uneven firing of steam boilers with attendant high stack losses.

16. *Capital investment.* HTW advocates claim lower costs because HTW generators are smaller than steam boilers of the same capacity and related

space requirements are also smaller; pipe sizes are reduced and less excavation is required because of the lack of pits or trenches to suit pipe pitch; there are fewer auxiliaries, less stringent water treatment, less corrosion and, therefore, fewer replacements. Although this is true, these factors may not be decisive because steam permits electric generation and serves process needs that HTW cannot. In the final analysis, it is the long-range, composite performance of all plant systems that will justify an investment rather than an optimum, partial advantage in a segment of the plant, such as the advantage enumerated above for HTW boilers.

17. *Corrosion and water treatment.* Steam systems, in particular their condensate lines, are subject to severe corrosion, under certain circumstances. Heat in condensate hastens corrosion, facilitating chemical reactions in the presence of carbon dioxide, oxygen and ammonia. Ammonia corrodes copper and zinc components; oxygen corrodes steel directly and, with carbon dioxide, forms carbonic acid, which causes pitting. The chief sources of carbon dioxide are carbonates and bicarbonates in make-up water, which decompose under high heat and give off carbon dioxide entrapped in the steam.

HTW systems have fewer internal corrosion problems and need less elaborate water treatment. Corrosion inhibitors work well in closed systems by forming a protective coating. Water does not come in contact with air, so the system, once effectively deaerated, stays oxygen-free. Chemical deaeration is adequate. Sometimes it is used exclusively, without a deaerator, for introducing make-up water into the system. Sodium sulfite, a deaerant, is fed internally, absorbing oxygen by forming sodium sulfate as in caustic soda or sodium hydroxide, to increase acidity (pH). A pH range of 9-9.5 is satisfactory for both ferrous and nonferrous metals.

Water softening is an external process for which resinous materials, such as sodium zeolites, are used in treating incoming water. To reduce residual hardness, small amounts of phosphates are added after the zeolite softener beds, because carbonates and bicarbonates must be eliminated from the make-up stream before being introduced into the system. If large quantities of oxygen have to be removed, hydrazine, a reducing agent, is used, combining with oxygen to form water and nitrogen.

Some say that deaeration is completely unnecessary for HTW. This is not true; deaerating may be required, depending on the quality of the make-up. The best corrosion protection is good maintenance that minimizes the need for make-up as much as possible.

Statistics are not available on steam boiler make-up quantities. As a rule, depending on the degree of water treatment, 3-5% blowdown may be required and, with receiver overflows, leaks, poorly packed valves and loose-fitting condensate systems, overall losses can amount to 5-10%. Industrial plants that serve steam processes can exceed 60% make-up. HTW water systems in the 30,000–50,000 gallon (113 700–189 500 l) range, should require no more than 1,000 gallons (3790 l) per week; for those systems in the 80,000–200,000 gallon (303 200–758 000 l) range, make-up should not exceed 2,000 gallons (7580 l).[3]

2.5 Diversity Factor, Load Factor, and Part Load Performance

Central plants are called upon to perform at peak load for a very small fraction of the year. Therefore, part load performance is the basic mode of operation and is of greater importance.

Satisfactory part load operation means acceptable efficiency throughout the load range. "Plant flexibility" is nothing more than the ability of the plant to "track" its load accurately and economically, which depends not only on proper boiler and chiller selections, but on how efficiently systems interact with each other.

The concept of *diversity* pertains to full load performance; *load factor* is indicative of part-load performance. The term diversity factor (DF) commonly is applied to maximum instantaneous refrigeration load, which can be equal to or less than chiller capacity, divided by the sum of all peak refrigeration loads. The same term has meaning when applied to boilers serving both heating and cooling loads:

$$DF = \frac{\text{Instantaneous load}}{\text{Sum of peak loads}} \qquad (2.1)$$

[3]Paul L. Geiringer, *High Temperature Water Heating*, (New York: John Wiley & Sons, 1963), p. 28.

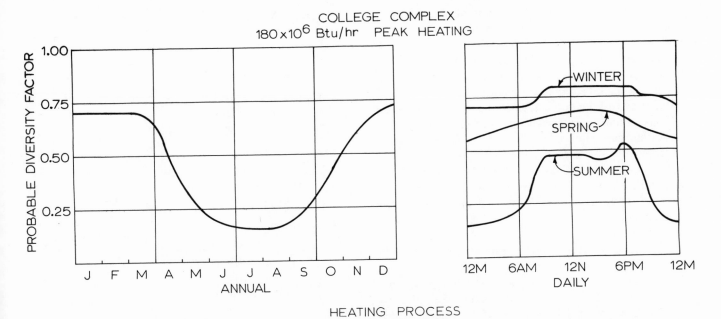

HEATING PROCESS

Figure 2-3. College complex—heating process.

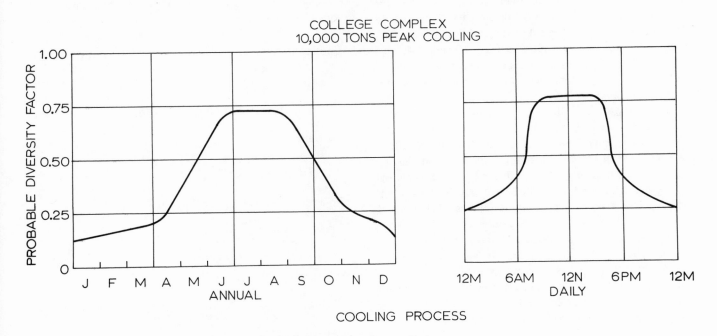

COOLING PROCESS

Figure 2-4. College complex—cooling process.

The load may be expressed in units of tons of refrigeration, Btu per hour, pounds of steam per hour, even in gallons per minute. The diversity concept is based on the fact that the loads of all pieces of equipment served by the plant do not peak simultaneously; hence, peak load is less than the sum of the connected loads.

Referring to Equation 2.1, since the sum of peak loads and maximum instantaneous demand vary from season to season, depending on any number of factors, one can conceive of a variable diversity factor and, in particular, a *minimum* diversity factor. Maximum diversity factor is useful for sizing pumps, chillers and boilers. Minimum diversity factor is an indication of minimum system requirements. Diversity factor is examined in more detail in Section 7.1.

The minimum diversity factor can be useful as a guide in specifying minimum machine size or ensuring that minimum load is within the operating range of machines on hand. Other intermediate value diversity factors, if they could be plotted, would help us understand system performance at intermediate loads and, therefore, would be of assistance in designing better systems. Currently, a load factor is estimated for the entire year and machines are selected to operate at top efficiency for the assumed load factor. In order for diversity factor to become a full design tool, actual installations can be instrumented to yield data from which we could plot curves similar to Figs. 2-3 and 2-4, or a series of simultaneous heat transfer equations could be solved and plotted to render theoretical diversity curves during the design period.[4]

Whereas diversity factor is an indication of power (the *rate* at which energy is consumed), load factor (LF) is an indication of energy consumption. Load factor is defined as actual energy (ton-hours) put out by a refrigeration plant per year divided by the potential energy (ton-hours) that could be put out by the plant if operated at maximum capacity for the same number of hours, *i.e.*, the hours in a cooling season. Another way of expressing it would be actual ton-hours per year divided by possible ton-hours per year. If we knew the operating hours and the load factor, we could predict operating expenses accurately, assuming constant compressor efficiency. However, knowledge of both of these items depends on statistical experience, like degree-days.

[4]Tamami Kusuda, "Procedure Employed by the ASHRAE Task Group for the Determination of Heating and Cooling Loads For Building Energy Analysis," Symposium Paper at the ASHRAE Semiannual Meeting in Dallas, TX, February 1, 1976.

Figure 2-5 graphically illustrates and interrelates the concepts of diversity factor, load factor, equivalent full load hours (EFLH) and chiller capacity. In Fig. 2-5,

Let

Ton_1 = Maximum chiller capacity = Maximum instantaneous load

Ton_2 = Average seasonal chiller capacity

Assume

area under actual operating curve

$$= \text{Rectangle ABIFGHA} = \text{ABCDIHA} \quad (2.2)$$

(thus, EFLH = equivalent full load hours.)

By definition

$$LF = \frac{\text{Area under actual operating curves}}{\text{ABCDEFGHA}}$$

$$= \frac{\text{ABIFGHA}}{\text{ABCDEFGHA}} \quad (2.3)$$

Since GHA is common to both rectangles,

$$LF = \frac{Ton_2}{Ton_1} \quad (2.4)$$

Using Equation 2.1 we can derive the relationship for equivalent full load hours:

$$(\text{Cooling season hours} \times Ton_2) = (\text{EFLH} \times Ton_1) \quad (2.5)$$

$$\text{EFLH} = \frac{Ton_2}{Ton_1} \times \text{Cooling season hours}$$

$$= \text{LF} \times \text{Cooling season hours} \quad (2.6)$$

Rearranging Equation 2.5,

$$Ton_2 = \frac{(\text{EFLH})(Ton_1)}{\text{Cooling season hours}}$$

Re-writing and rearranging Equation 2.1,

$$DF = \frac{Ton_1}{(\text{Sum of all peaks})} \text{ or,}$$

$$Ton_1 = DF (\text{Sum of all peaks})$$

Substituting in Equation 2 5,

$$Ton_2 = \frac{(\text{EFLH})}{\text{Cooling season hours}} (DF)(\text{Sum of all peaks})$$

Rearranging Equation 2.6,

$$LF = \frac{\text{EFLH}}{\text{Cooling season hours}}$$

thus,

$$Ton_2 = (LF)(DF)(\text{Sum of all peaks}) \quad (2.7)$$

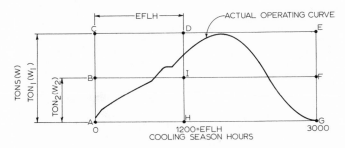

Figure 2-5. Load factor diagram.

There is very little empirical data on diversity and load factors for these two useful concepts to become truly instrumental for valid preliminary equipment selection and energy cost calculations. We hope that examination of current records and adequate instrumentation in new projects will provide these data.

If load factor can be pinpointed, refrigeration machine(s) can be selected for maximum efficiency at a loading that prevails for most of the cooling season. The seasonal cooling load factor for a district cooling plant of a typical university is 0.35–0.55.[5]

The diversity factor for hospital complexes with high make-up air quantities must be large, since it would reflect the wide fluctuation of outside air temperature relative to the diversity factor for an office building complex with regular working hours and low make-up air rates.

The load factor improves when both steam (or hot water) and chilled water are supplied. District heating plant load factor percentages fluctuate from the low twenties to the high thirties, depending on the type of business served, system losses and climatic conditions.[6]

A diversity factor of 0.60–0.80 is typical of a university complex. Where fresh air loads are part of the load, it would be near 0.80, and where air recirculation is expected it should be closer to 0.60.[7]

One investigator reports load factors in Table 2-1 for various applications in connection with incineration plants. Obviously, these factors should be applicable to central plants in general.[8]

2.6 Hot Water, Chilled Water Temperature Differentials, and Steam Pressures

2.6.1 *Hot Water Systems*

In medium and high temperature water systems, supply temperature should be the highest that the equipment can take and system economics can tolerate. High temperature differentials reduce water quantities and facilitate temperature control.

Tables 2-2 and 2-3 list some recommended differentials.

Table 2-1

Application	Load factor
1. Manufacturing, process use only	0.70–0.80
2. Industrial park	
Heating only	0.20–0.35
Heating and cooling	0.50–0.60
3. Urban area, heating and cooling	0.60–0.70
4. University campus, heating and cooling	0.40–0.50
5. Medical center, heating and cooling	0.47–0.57
6. Airport complex, heating and cooling	0.55–0.65
7. Shopping center, heating and cooling	0.35–0.45
8. Apartment complex, heating and cooling	0.37–0.47

Table 2-2
Recommended Temperature Differentials for Various Pieces of Equipment

Supply temp. F (C)	Possible temperature differentials F (C)[9]				Temperature differential in common use, F (C)
	Conventional radiation	Unit heaters	Fan coil units		
			Heating	Ventilating	
180 (82.2)	20 (11.1)	30 (16.7)	40– 80 (22.2– 44.4)	60–100 (33.3– 55.5)	20 (11.1)
240 (115.5)	30 (16.7)	40– 80 (22.2– 44.4)	80–120 (44.4– 66.7)	100–130 (55.5– 72.2)	20– 80 (11.1– 44.4)
320 (160.0)	40 (22.2)	60–120 (33.3– 66.7)	120–160 (66.7– 88.9)	140–180 (77.8–100.0)	80–100 (44.4– 55.5)
400 (204.4)	40 (22.2)	150–200 (83.3–111.1)	140–220 (77.7–122.2)	200–240 (111.1–133.3)	150–200 (83.3–111.1)

Courtesy Heating, Piping, Air Conditioning Magazine.

[5] Maurice J. Wilson, "Four Plans for Expanding District Chilled Water Systems," *Heating, Piping & Air Conditioning*, January, 1967, pp. 148–154.

[6] National District Heating Association, *District Heating Handbook*, 3rd ed., 1951, p. 36.

[7] David M. Hummel, "University Layout Forces Chilled Water System Design," *ASHRAE Journal*, April, 1969, p. 61.

[8] Maurice J. Wilson, "Heat Energy from Waste Incineration: Cash for Trash," *Heating, Piping & Air Conditioning*, April, 1974, p. 53.

[9] *Heating, Piping & Air Conditioning*, "MTHW Heating Reduces Costs, Requires Precise Engineering," April, 1961, 121–125.

Table 2-3
Recommended Temperature Differentials for Various Pieces of Equipment[10]

Service	Supply temp. F (C)	Return temp. F (C)	Temp. diff. F (C)
Domestic hot water heaters for 140–180 F (77.8–100.0 C) water	380 (193.3)	160–200 (71.1– 93.3)	180–220 (100.0–122.2)
Converters for 190–210 F (87.8–98.8 C) water	380	220–225 (104.5–107.2)	155–160 (86.1– 88.9)
Steam producer			
100 psig (689 kPa)	380	340–350 (173.9–176.7)	30– 35 (16.7– 19.5)
40 psig (276 kPa)	380	300–305 (148.9–151.7)	75– 80 (41.7– 44.4)
7–10 psig (48–69 kPa)	380	240–250 (115.6–121.1)	130–140 (72.2– 77.8)
Air heaters for 110–130 F (43.3–54.5 C)	380	150 (65.6)	230 (127.8)

Copyright © 1963 by John Wiley & Sons, Inc. Reprinted with permission of John Wiley & Sons, Inc.

Table 2-4

Supply water temp., F (C)	Saturation pressure corresponding to supply water temp., psig (kPa)	Generated steam pressure, psig (kPa)
300 (148.9)	52 (358)	10– 15 (69–103)
350 (176.7)	120 (827)	20– 40 (139–276)
380 (193.3)	181 (1247)	40– 60 (276–413)
400 (204.4)	233 (1605)	60– 80 (413–551)
425 (218.3)	311 (2143)	80–100 (551–689)
450 (232.2)	408 (2811)	100–125 (689–861)

Courtesy Heating, Piping, Air Conditioning Magazine.

Table 2-4 lists supply water temperatures and the corresponding temperatures and pressures of process steam generated from HTW which have proven economically feasible and which result in efficient and economic converter design.[11]

Rate of firing hot water boilers depends on the approach to project design. If steam is to be generated, then boiler outlet temperature must be constant. Otherwise, one can maintain a constant boiler outlet temperature and vary supply temperature to the district or keep the boiler outlet temperature and supply temperature constant, but vary flow to each distribution point according to demand. In the case of smaller, concentrated loads, one may vary the firing of the boiler sufficiently to maintain an outlet temperature capable of matching the heating load.

Steam-pressurized HTW systems demand constant temperature and pressure. In addition, constant water temperature is conducive to improved fuel consumption. Data in Tables 2-1-2-4 are guidelines only. Firm temperature differentials must be established for each project on the basis of an economic analysis of temperature differentials versus plant costs.

2.6.2 *Chilled Water Systems*

Selection of chilled water primary flow rates and temperature differentials is extremely important. If not established on the best available information, the result can be an improperly balanced system with excessive first or operating costs, or both. This is the area in which most misconceptions persist and where most arbitrary decisions are made. Few studies have been made and little actual operating data have been accumulated. Energy conservation strategies can be planned in this sector long before any plans are elaborated. Depending on the suppositions, the plant will be economical or wasteful. Selection of water flow rates and temperature

[10] Paul L. Geiringer, *High Temperature Water Heating* (New York: John Wiley & Sons, 1963), p. 126.

[11] J.S. Blossom, *et al.*, "Design and Control of High Temperature Hot Water Heat Consumers," *Heating, Piping & Air Conditioning*, June, 1960, pp. 169–182.

differences should indicate that all system variables and their effects on system operation, energy consumption and owning and operating costs are understood, and that the following items have received proper attention and all interrelated problems have been compromised or resolved: evaporator (series or parallel arrangement), condenser (series or parallel arrangement), condenser (water flow and related temperature differentials), type of refrigerant and related bhp/ton, route length and pipe sizes, incremental equipment sizes, pumping operation (either full flow or partial flow), mechanical space piping layout and related costs, distribution piping costs, pumping costs, compressor operational costs, and primary chilled water project demand and duration.

With so many variables to consider, it is almost impossible to consider all at the same time. Perhaps the condenser water cycle can be interrelated to the cooling tower and chiller operation and thus be considered as part of another cost sector. A nominal coordination check with the chilled water side decisions is sufficient to confirm that final decisions concerning the condenser water cycle meet the overall objectives of the chilled water plant.

A method has been developed that brings all these variables into play.[12] Figure 2-6 attempts to plot optimum flow against brake horsepower per installed ton. It is assumed that the pumps are operated at full flow during the entire cooling season and that pumping horsepower is constant for all flow rates. The shape of the pump curve will vary with pumping operation assumptions. The load factor curve in Fig. 2-6 for LF-1 can be drawn by obtaining the appropriate data from a manufacturer. The engineer can draw in all the remaining curves. It is assumed that return temperature is a constant 68F (20C). In this operation, the pumping horsepower becomes a greater percentage of the total horsepower at lower load factors. Furthermore, electrically driven pumps and centrifugal machines are assumed, costs are ignored and equal efficiencies at different load factors are taken for granted. The fact that the ordinate is bhp per installed ton aids in obtaining a common denominator for the optimum flow rate.

Obviously, an easily constructed curve similar to those in Fig. 2-6 can help the engineer select preliminary flow rates. Once firm preliminary piping plans and equipment selections are made, curves similar to those in Figs. 2-7

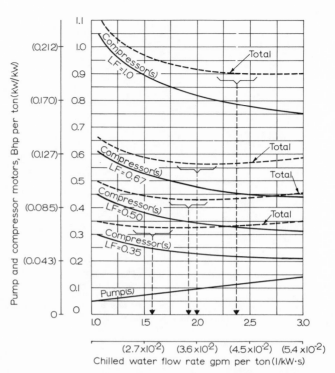

Figure 2-6. Optimum chilled water flow as determined by plotting pump and compressor horsepower at various load factors.

and 2–8 can be constructed.[13] These plans presume that the factors of constant versus variable flow, load factor, pipe friction, pump head and energy and labor costs have been dealt with. In an actual case, Fig. 2-8 would constitute a final check on Fig. 2-6. Here, Fig. 2-6 is simply an illustration and has no connection with Figs. 2-7 and 2-8.

2.6.3 Steam System Pressures

Central heating plant boilers fall into three general categories based on pressure: low (5-15 psig) (34-103 kPa), medium (15–350 psig) (103–2412 kPa), which is usually called high pressure in the HVAC field, and above 350 psig (2412 kPa), ordinarily associated with electric generation.

The main advantage of low pressure boilers is that they need not be attended by a licensed operator, which lowers operating costs; however, thermodynamic level is low for efficient central plant use. In waste heat recovery applications, where a corresponding refrigeration load exists, waste heat or auxiliary boilers generate 15-psig (103 kPa) steam which is fed at 12 psig (83 kPa) into a bank of absorption refrigeration machines. Although the steam

[12]For Fig. 2–6, see Maurice J. Wilson, "Four Plans for Expanding District Chilled Water Systems," *Heating, Piping & Air Conditioning*, January, 1967, p. 154:ASHRAE Guide, 1970, 174; *Campus Revolution* (Carrier Air Conditioning Co., 1967), p.11.

[13]*Campus Revolution*, pp. 13–14.

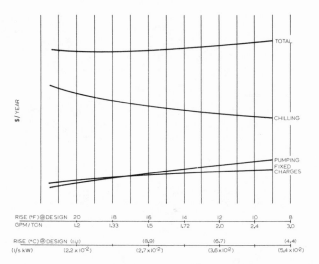

Figure 2-7. Hypothetical optimum flow rate based on economic considerations (includes primary loop fixed charges and pumping cost.)

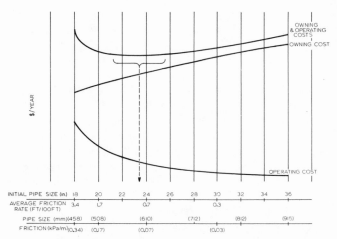

Figure 2-8. Hypothetical optimum friction drop based on economic considerations (includes piping fixed charges, pumping and chilling cost).

rate is high, on the order of 18-19 lb per ton-hour $(0.64-0.68) \times 10^{-3}$ kg/Ws, the operation can be a successful one because it effectively captures waste heat. The operation is limited by current absorption machine capacities, roughly 1500 tons (5280 kW). Therefore, we may be able to think of five to six machines used at a fairly high load factor.

A second range, 15-350 psig (103-2412 kPa), is most applicable to central heating applications not ordinarily

associated with electric generation. The latter, in general, utilizes high pressures with superheat. Superheat is not useful for heating purposes since condensation takes place at the saturation temperature. Central plant economics seldom justify steam distribution below 150 psig (1034 kPa). Furthermore, operation of feedwater pumps, soot blowers and other boiler room auxiliaries dictates steam pressures of not less than 125 psig (861 kPa) at which pressure use of cast iron fittings is permissible.

It is worth mentioning here that some small steam turbines operate normally at a back pressure of 10-15 psig (69-103 kPa) and cannot be used for condensing service. Unless demand for back pressure steam is constant, it is best to use electric motors. If a small amount of superheat is necessary for part of the steam, it is preferable to use an afterburner to reheat this portion of the steam rather than impose superheat on the entire boiler.

Most central plant heating applications utilize boilers in the 125-250 psig (861-1722 kPa) range and most manufacturers of watertube boilers have standardized on 250 psig (1722 kPa). Electric generation is sometimes attempted in this pressure range, depending on the needs of the project at hand, in which case, it is worth considering 50-100F (27.8-55.5C) of superheat, which causes no particular operating problems. The superheater improves turbine steam rate, eliminates moisture impingement in the high pressure turbine stages and minimizes condensation in the low pressure region. However, superheat adds to the expense, requires more controls, involves more care during low loads and the operator must make sure that feedwater is circulating through the superheater on startup.

The concept of varying steam pressure to suit load conditions runs counter to the practice of American stationary boiler operators who are accustomed to maintaining constant boiler pressure for all loads in the belief that fewer operating problems will be encountered and that all possible conditions are apt to be satisfied. Also, pressure is decreased through pressure reducing stations which are ordinarily designed to operate with constant inlet conditions.

Before examining high pressure boilers, let us first examine how pressure categories are established and, secondly, how boiler design pressure is specified. Although water tube boilers can be purchased at less than 250 psig (1722 kPa) rating, ordinarily there are, at best, only small saving in doing so. Note that water tube boilers operating at less than 100 psig (689 kPa) must be derated according to manufacturer's instructions.

Between 250 and 500 psig (1722–3445 kPa), boiler ratings increase in increments of 25 psig; between 500 and 1000 psig (3445-6890 kPa), in increments of 50 psig (344 kPa).

Working pressure is system pressure, which is less than design pressure. Design pressure is the pressure to which the various boiler parts must be fabricated in order to conform to code. These ASME and other codes establish minimum safety pressure requirements. Two items determine the ·minimum difference between working and design pressures: the excess pressure switch and the safety valves. The excess pressure switch may be set 5-10 psi above working pressure. The ASME code requires that there be a 4% difference between the pressure at which the safety valve unseats and seats itself. Most manufacturers exceed this, using 5% or 6%. when more than one safety valve is used, an additional 5 psig (34 kPa) must be added for each valve.

Example: What is the design pressure of a boiler feeding a turbine which requires a throttle pressure of 300 psig (2067 kPa)?

1. Throttle pressure	300 psig
2. Loss through piping between boiler and turbine and through stop-check valve at full flow	20 psig
3. Drop through super-heater at full flow	13 psig
4. Drum internals	2 psig
Working pressure	335 psig
5. Excess pressure switch	10 psig
6. First safety valve, 6% of 335 psig =	21 psig
7. Second safety valve	5 psig
Maximum drum pressure or minimum boiler design pressure	371 psig
Next higher standard boiler design pressure	375 psig (2584 kPa)

The high pressure steam range, *i.e.*, above 350 psig (2412 kPa), is ordinarily employed when central plants incorporate electric generation, but current concern for energy conservation has pushed steam pressures for centrifugal chiller turbine drives even higher. At the Dallas-Fort Worth Regional Airport, three 8500 ton (28 160 kW) centrifugal chillers are driven by turbines using steam at 600 psig (4134 kPa) and exhausting at 3 in. Hg abs (10.14 kPa). The temperature range is 60-36F

(15.6–2.2C) and the turbine steam rate is less than 7 lb per ton-hour (0.25×10^{-3} kg/Ws), by far the best steam rate known in a central refrigeration plant today. When total steam flow surpasses 150,000 lb per hr (68 100 kg/hr), it is realistic to step up steam pressures to 900 psig (200 kPa), when one considers today's fuel economics.

Pressure-temperature groupings are as follows:

125-250 psig (861–1 722 kPa); saturated steam to 150F (83.3C) superheat[14]
250-400 psig (1 722-2 756 kPa); saturated steam to 750F (399C) total temperature
400-800 psig (2 756-5 512 kPa); 750F to 825F (399–440C) total temperature
800-1000 psig (5 512-6 890 kPa); 800F to 900F (427–482C) total temperature.

The main advantage in combining electric generation with cooling and heating is that a straight condensing operation is avoided and the heat of steam condensation, which is usually transferred to circulating cooling water as rejected heat at an estimated loss of at least 60% of input energy, is instead transferred to the heated space at a plant loss of as little as 10%. In essence, the heated space takes the place of a steam condenser.

In general, it pays to produce electric power when purchased electric power is expensive, if waste fuel, such as blast furnace gas, bagasse, hogged wood, city refuse, etc., is available and if heating, cooling and power demands supplement each other. Depending on whether heating and cooling or electric generation is the byproduct, extraction or back pressure turbines may be selected. This subject will be dealt with more extensively in Section 4.2. It is pertinent here to discuss "Imponderables" which are central plant design requirement, the following ought to be specified: maximum flow, duration of maximum flow, constant load. However, if process steam is part of the central plant design requirement, the following ought to be specified: maximum flow, duration of maximum flow, minimum flow, and rate of change of flow.

The so-called turndown ratio or ratio of maximum to minimum firing rate, as translated into output, is an important factor in selecting the appropriate fuel and designing the furnace. Wild fluctuations in load and the required degree of boiler responsiveness can affect not only boiler design but all associated equipment of the steam cycle and, depending on the fuel that is ultimately available, automatic firing may be included or excluded. In the past, gas or oil were readily available and packaged

[14]Frederick T. Morse, *Power Plant Engineering*, 3rd ed. (New York:D. Van Nostrand Co., Inc., 1953), p. 263.

boilers increased in size and pressure range, fitting into the central plant picture quite comfortably. Coal may once more become the most readily available fuel for central plants, or we may have to rely on a multiplicity of fuels for continued operation, so that boiler size and equipment flexibility will play a primary role in conceptual engineering terms. Load factor or "capacity" factor, as it is known in boiler parlance, can be the basis for establishing boiler efficiencies in the most prevalent load range. Obviously, top efficiency throughout all loads is unnecessary and expensive. Knowledge of load factors and their significance and an understanding of load fluctuation must form the basis for sound judgment in equipment selection.

At about 400 psig (2756 kPa), water treatment begins to assume importance, and its importance increases as pressures become higher because at high temperatures and pressures impure water will result in the depositing of impurities. Water treatment together with deaeration will prevent the formation of scale, free oxygen and acids. Dissolved oxygen attacks steel, but modern deaerators are capable of limiting oxygen to 0.005 cc per liter. As a rule of thumb, 600 psig (4 134 kPa) steam should be the dividing line between base exchange and full demineralization.

Overall costs increase suddenly when 750 F (404C) in total steam temperature is exceeded. At this point, carbon steel pipe is no longer adequate and alloy carbon molybdenum pipe must be substituted. That is the basic reason why so many industrial installations use the popular 600 psig (4 134 kPa), 750F (404C) combination. At this point, a qualitative change takes place in boiler and steam turbine construction. Fittings and valves jump into the next pressure rating. Stepping the temperature from 750F to 825F (399-440C) poses no major problem. The cost of both boilers and turbines remains the same. The only penalty is that fewer manufacturers may be available and the increased cost of the carbon molybdenum pipe must be considered. The increased cost is for materials, not for labor.

Stepping the pressure up to the 850–900 psig (5856-6200 kPa), 900F (482C) range presupposes large flows and, in order to determine the lowest cost for the life of the plant, plant investment, fixed charges, fuel and operating costs for various alternatives must be thoroughly examined. Steam turbines, although of a higher pressure rating, require less material and, therefore, are not necessarily more expensive than those rated at 600 psig (4134 kPa), although the number of manufacturers will be drastically reduced. For very large cogeneration plants serving chemical and power utility complexes

steam pressures of 1800 psig, 1000F (12 402 kPa, 538C) are becoming the rule. Another classic pressure-temperature range for chemical-plant power stations is 1250 psig, 950 F (8621 kPa, 510C). On the other hand, 125 psig (861 kPa) mechanical drive steam turbines are standard, less expensive, and manufactured by a host of firms, thus providing a truly competitive situation. It is necessary to weigh current rising fuel costs against additional costly equipment or rearrangement necessary to employ the higher pressures.

2.7 Energy Conservation and Heat Recovery

The main equipment groupings in the central plant are refrigeration machines, prime mover combinations, cooling tower, condenser pumps, chilled water pumps, boilers, feedwater pumps, turbogenerator combinations and condensate or hotwell pumps. In addition, there are auxiliary and parasitic equipment groupings such as oil pumps, fuel pumps, make-up pumps, transfer pumps, condensate pumps, air compressors, lube oil coolers, intercoolers, sootblowers, etc.

Each piece of auxiliary equipment must be selected to operate in an efficient range, while pressure drops across heat exchangers must be minimized so that parasitic power remains minimal and auxiliary equipment operation does not lead to excessive sacrifice of plant capacity. The espousal of this philosophy promotes the use of steam turbine, gas turbine or internal combustion engine drives for the auxiliaries because, in addition to conserving plant production for sale, it also reduces electrical demand penalties if the plant uses purchased electricity.

Energy conservation is the result of intelligent use of machines based on thorough knowledge of the process. It is axiomatic that an efficient or conservation-oriented plant will revolve around the principle of executing the work for the least possible use of energy and the lowest possible charge. Numbers of the main components are selected to match the various load conditions during the design stage. It is important that each component be controlled both individually and in relation to the other functionally associated components. It may be necessary to fix some variables and make some simplifying assumptions without impairing the accuracy of the operation. Establishment of the main functions of the plant, its load conditions and the response of the component groupings is the engineer's most important job. Unless he resolves these matters successfully, plant operation and conservation in particular will not be served, for it is in these relationships that he can either "make or break" the plant.

To summarize, the economic and operational feasibility of a plant is established by the thermodynamic or operating levels of the equipment, the sequence of operations, the control mode, and the overall efficiency of each phase of the operation as well as the individual component efficiency, the rate at which work is being performed and the intricate and constantly changing relationship between components and their response, individually and en masse, to plant load. Major project operating economies are not accidental; they are engineered into a project by understanding and appraising the various design factors at initial and subsequent developmental stages.

Energy conservation can occur at any time, provided it is planned for. Heat recovery, on the other hand, can only take place when there is demand for the recovered energy. At other times, recaptured energy must be released unused to a heat sink, such as the atmosphere or a body of water. It stands to reason that heat recovery follows heat generation and that the two must parallel each other if the benefits are to be major.

The following are areas where practical conservation and heat recovery can take place without consideration of any engineering premises or concerns:

1. *Stack exhaust gases* can be used to:
   (a) Preheat boiler combustion air through an air preheater
   (b) Preheat boiler feedwater through an economizer, for steam boilers at high steam pressures
   (c) Preheat ventilation air or make-up air
   (d) Heat cooling tower exhaust air so as to minimize or eliminate hot vapor plume.
2. *Exhaust ventilation air* is useful in preheating the huge quantities of make-up air of the central plant itself by means of:
   (a) Runaround cycle, heat transfering coils
   (b) Heat pipes
   (c) Rotary regenerative heat exchangers of the thermal metallic wheel or dessicant variety.
3. *Condenser water* at as low temperature as possible enhances the performance of steam surface condensers, absorption and centrifugal machines. Temperatures as low as 50 F (10C) can be tolerated by both centrifugal and absorption machines; surface condensers can withstand even lower temperatures. As condenser water temperature drops, lift is reduced because of lowered pressure and temperature in the condenser; compressor brake horsepower per ton is lessened as a result. Low circulating water temperatures contribute towards lower vacua and heat rates in steam condenser

applications. Reciprocating and some screw machine chillers cannot fully benefit from low condenser water temperatures. For large central plant installations, the condenser water temperature level should be kept as low as cooling tower performance and chiller design will permit.
4. *Variable chilled water flow and minimum condenser water flow*, when in accordance with analyzed projects' needs, are two energy-conserving concepts dealt with in greater detail in Sections 4.10 and 7.5.
5. *Diesel engine and gas turbine exhaust gas heat recovery* constitutes a major application. The gases are piped to waste heat boilers which generate steam from 10 to 200 psig (69–1 378 kPa) or so to power absorption machines, centrifugal refrigeration compressors, mechanical turbine drives for chilled and condenser water pumps, induced and forced draft fan turbine drives for very large steam boilers and process steam needs for hospital complexes, campuses and industrial plants. In total energy plants, this type of heat recovery is most common. Successful capture of exhaust gas heat can boost plant thermal efficiency to above 70%. (See Section 4.1 for heat recovery at higher steam pressures.)
6. *Gas turbine exhaust gas*, due to its high excess air, can be directly fed to a boiler whose design allows the utilization of elevated temperature gas as combustion air, reducing the amount of fuel and increasing boiler efficiency. (See Sections 4.1, 4.3.1 and 4.3.2 for combined cycles using gas turbine exhaust gases.)
7. *Diesel engine jacket cooling* extracts heat from the engine block in the form of hot water in the range of 150–180 F (65.6–82.2C), which can be used for process or domestic hot water needs. In some specially constructed engines using ebullient cooling, steam at 250F (121.1C) saturated temperature or roughly 15 psig (103 kPa) can be generated for absorption machines or heating. Figures 2-9 and 2-10 indicate the performance and heat recovery capabilities of some medium size diesels.
8. *Lube oil cooler and aftercooler* heat recovery is accomplished by heating water to 150–160F (65.6–71.0C). The water is first passed through this equipment for preheat and, then, through the engine jacket. In industrial plants, tremendous quantities of water are used in cooling this equipment and the heated water is wasted. Certain cities, New York among them, require water conservation for such applications. Figures 2-9 and 2-10 indicate the extent of recovery possible for diesel engines.
9. *Auxiliary equipment* can be driven by gas turbines, internal combustion engines or steam tur-

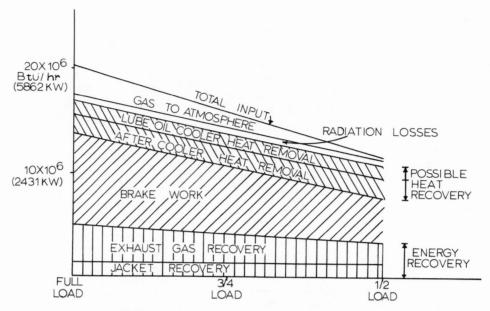

Figure 2-9. Heat balance of 2200 kW, 3088 hp diesel engine.

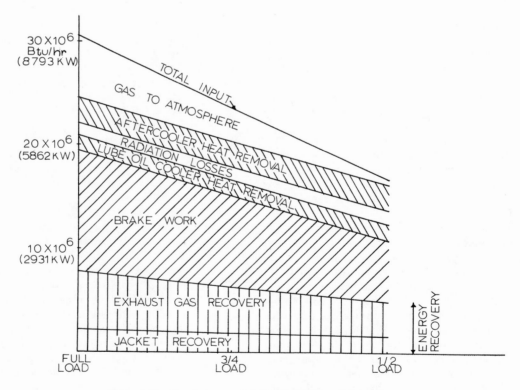

Figure 2-10. Heat balance of 3330 kW, 4655 hp diesel engine.

bines to increase availability of power for sale, to decrease dependence on purchased electricity or, if connected to the local utility network, to help keep electrical demand charges down. Standby units can be electric as an alternate and reliable source of power.

10. *Electrical demand* for cooling can be kept down by using absorption machines for peaking service, subject to analysis to prove such need.

11. *Back pressure turbines* are useful in promoting energy conservation, if a steady demand for back pressure steam can be established to parallel electric generation. (See Sections 4.2 and 5.1 for additional information on back pressure turbines.)

12. *Extraction turbines* conserve energy in that the extracted steam does useful work up to the point of extraction. Whether extraction or back pressure turbines are used, is a matter of project needs, pressure level and equipment performance. An engineering study will establish the criteria for a final selection. (See Sections 4.2 and 5.1 for additional information on extraction turbines.)

13. *Boiler blowdown* is a source of energy usually neglected. Boiler blowdown is a periodic letdown of hot boiler water into a receiver, the blowdown tank, in order to keep solids and impurities in the boiler at a desirable level. In industrial applications, make-up quantities sometimes amount to as much as 75% of boiler steam rate. Such voluminous make-up requires extensive water treatment. As a result, solids precipitate and require removal on a periodic basis. Then, boiler blowdown can be substantial and may be worthy as a source of heat recovery. An economic study may justify the required heat exchangers, storage tanks and pumps for a heat recovery system.

14. *Steam condensate* should be returned to the plant and should not be flashed to atmosphere in flash tanks. In the past, the cost of pumping and severe corrosion of underground condensate lines militated against condensate return to the plant. However, current economics justifies reconsideration. High pressure deaerators and chemical treatment may be used to maintain acceptable oxygen and purity levels. Flashing is a complete waste unless the flashed steam can be used in process work, thus contributing to energy conservation. Appropriately selected pumps can handle condensate at any temperature and pressure.

15. *Domestic hot water preheating* is within the scope of central plant tasks if the domestic hot water main for the project can be located near or inside the plant itself and if the resultant distribution network is sufficiently short and compact to make heat recovery economically attractive.

16. *The two-stage absorption machine* using steam at around 150 psig (1033 kPa) is a recent technological advance. Its steam rate, 11–12 lb per ton-hr $(0.39 \times 10^{-3}$–0.43×10^{-3} kg/kWs), is highly competitive with that of the centrifugal-absorption combination (piggyback) or high pressure steam-driven centrifugal, without technical complications or additional equipment such as a motor or turbine. The need for a licensed operator is within the manpower requirements of the central plant. The initial cost is higher than that of the single-stage absorption machine, probably less than the piggyback arrangement and competitive to that of the turbine-driven centrifugal. The machines are small in size, roughly 1100 tons (3 872 kW).

17. *Combined cycle* signifies the combination of gas turbines and steam turbines for electric generation. Ordinarily, a gas turbine drives an electric generator; the gases from the gas turbine are conducted to a waste heat boiler. The steam generated drives a turbogenerator. A variation of this cycle may substitute centrifugal machines in lieu of generators. The combined cycle is a total energy concept—*i.e.*, the waste of one machine becomes the feed of its companion. All these combinations are valid in a central plant, depending on an engineering study, and aim toward full utilization of available energy and the reduction of energy input per unit of work. Energy conservation lies at the heart of these concepts.

18. *Centrifugal-absorption machine combination or piggyback* is a form of combined cycle. The basic objective of combining centrifugal with absorption machines is a low steam rate. A load-sharing relationship exists between the centrifugal and absorption machine. The centrifugal machine acts as a perfect throttling steam control valve for the absorption machine; the absorption machine, in turn, fulfills the role of a perfect condenser for the centrifugal machine. Due to the varying efficiency of the steam turbine and the varying ratio of brake horsepower per ton at partial loads, the ratio of load shared by the two machines is not necessarily constant. Gravity condensate draining is not recommended. Rather a high temperature condensate pump is used which has the double advantage of allowing the absorption machine to operate below atmospheric pressure and of not flashing the condensate. Both of these advantages improve the heat rate. In combined cycles, the controls must be designed to maintain a proper balance between the two machines.

19. *Either water-ammonia absorption or steam jet refrigeration* may be used in industrial applications when waste steam is abundantly available. The

water-ammonia cycle can operate at low evaporator temperatures −50 to −40 F (−45.6 to −40C).

20. *Whether a parallel or series chiller arrangement* or a series-parallel combination is used is a matter for an early decision during the initial design stage, after the nature of the loads and their sequence is known. It is a basic project decision which will affect the operating costs of the plant throughout its life. Chiller arrangements and pumping configurations are examined in Chapter 3.

21. *"Thermocycle" cooling* is a method of operating a chiller without operating the compressor, only running the chilled and condenser water pumps. This is possible as long as the condenser water temperature is lower than that of the chilled water. The chiller has to be fitted specially to allow free circulation of refrigerant between the evaporator and condenser by creating a low pressure area in the condenser, drawing vapor from the evaporator to be liquified in the condenser. A necessary condition is the requirement for refrigeration at these low atmospheric temperatures. The benefit is economical refrigeration at partial project loads during part of the spring and fall and during the winter. In approximate terms, at 30 F (−0.6 C) atmospheric temperature, this mode can provide 30% of the refrigeration capacity; at 50 F (10C), an estimated 10–15%.

22. *Cooling tower water makeup*, compensating for drift, evaporation and bleed losses of water, is a serious waste. In areas lacking water resources, dry cooling towers may occupy a definite place in plant design, since make-up losses are entirely eliminated. In circumstances where steam turbine exhaust is condensed for want of applicable use, it is sometimes pumped to the cooling tower for make-up. However, this is low level conservation and should not be practiced until all other possibilities have been exhausted. If water charges are prohibitive, it may be worthwhile considering treatment of project effluent water for boiler and cooling tower make-up or for any other plant needs. Although these measures amount to economic recovery, they may nevertheless involve energy recovery as well.

23. *Large motors and transformers may be water-cooled.* Water-cooling of transformers saves structural costs by avoiding huge louvers required for air-cooling. Use of water-cooled motors, say 100 hp and above, is extremely economical when compared to air-cooling and is recommended. There is no inherent disadvantage to overcome except long-established air-cooling practice. Ductwork and air-handling are extremely expensive and large quantities of air are required to cool motors.

Outside design temperature is 95 F (35C) in most places and air temperature can be raised to 104 F or 40C, an ambient temperature at which motors are ordinarily rated. Some large motors or generators may be rated to operate at a higher ambient temperature than 40C, *i.e.*, 45 C and 50C. Assuming that the same insulation is used, the motor or generator must be derated. In order to maintain the same rating at the higher ambient, cable insulation must be upgraded from the usual type B to type F or H at some cost. It is therefore an economic balance between cable insulation and ventilation costs.

Power factor correction and the use of steam accumulators are both erroneously listed as energy conservation measures. Although power factor correction is highly desirable, it does not save energy; it only makes more power available.

Accumulators do not make more energy available; they help to make it available when it is needed. In addition, they help keep boiler size down as they smooth out demand peaks. Only in the sense of helping the boilers operate at full load during demand valleys—*i.e.*, at full efficiency—can they be considered energy conservers.

One realizes that energy conservation can be, and nowadays is, practiced within the various buildings served by a central plant. What is the relationship between conservation in an individual building and conservation and heat recovery in the central plant? Any conservation in the various buildings is bound to be reflected in the size of the central plant machinery and on the magnitude of the demand. In other words, effective building conservation results in a smaller plant and lower plant operating costs.

Central plant refrigeration limits the type of heat recovery system that is possible at distribution points. However, the run-around coil, heat pipe, rotary wheel, variable air volume, air distribution system, heat-of-light system, and others, do not involve the refrigeration system, and so can be employed regardless of plant location.

The double bundle condenser and, particularly, the heat pump, have come into play lately because they are ideal for recapturing low level energy from exhaust or return air and reintroducing it at a high energy level into the supply air, as needed for heating, via the chiller, compressor, and condenser portions of the refrigeration cycle. The heat pump becomes the heat transfer mechanism. These systems imply close coupling of plant and buildings served.

How can these various types of heat transfer machines be incorporated into the central plant concept? The main difficulty consists not so much in transferring the low level recovered energy from the buildings to the central plant, as the return chilled water system can serve this function very well, but in returning the reheated water at 105-125 F (41.6-51.6C) back to the buildings, since there is no distribution piping for low temperature water. Perhaps installation of such a system may be justified as a matter of economic balance. Let it be noted that current heat pump systems of the packaged type come in limited tonnage; however, custom-engineered heat pumps tied to large chillers come in an unlimited variety of designs and unrestricted tonnage. The costs of applying the heat pump locally or at the central plant must be weighed. There are not many precedents in connection with central plants because heat pumps as an energy-conserving device assumed immense popularity only recently, with the advent of the energy crisis. Application of the heat pump for energy conservation in buildings is more frequently encountered nowadays and there is a definite need for it, subject to economic analysis. The following brief description of the different heat pump systems will summarize their main operations and components.

1. *The double bundle condenser* is the oldest of the concepts, in use for at least a decade. The condenser consists of two shells or one shell with two compartments. One compartment is connected to the cooling tower. The other is engineered to provide water at 105-125 F (41.6-51.6C) which is pumped to reheat coils, perimeter coils, induction heating coils, domestic hot water preheat, multizone units, dual duct or any other use. The centrifugal machine has trouble maintaining 105 F as the load drops because the lift is simultaneously reduced. Unfortunately, the heat is unavailable when the machine stops. So the loads and hours of operation must be fairly well known. To maintain lift in large centrifugal chillers, one may use R-113 or employ a hot gas bypass, which also amounts to waste. Considerable investment is required for the premium priced double bundle, winterizing the cooling tower and providing controls and a fairly high flow in the heating circuit.

2. *Use of evaporative coolers* (*with closed coils*) is an alternate to the double bundle condenser. In this case, there is only one condenser in series with a heating coil and an evaporative cooler, which acts as a sink in dumping any heat that the heating coil cannot dispense. The system depends on the presence of a chiller load. If the heating load exceeds the capacity of the condenser, an auxiliary heat source is necessary. A dry cooler can serve in lieu of an evaporative condenser. If well water is available, the system can be arranged to use the well as both a source and a sink. Well water is ideally suited for this purpose since it is in the low fifties year-round. The evaporative condenser can also be piped appropriately to act as a heat source, but subject to maintaining minimum winter water temperatures.

3. *Heat storage* systems absorb excess heat during high cooling load periods to heat water in storage tanks. During unoccupied hours when heating is required, internal heat sources decrease or disappear and air systems are stopped entirely, the perimeter heating systems continue to operate by using hot water from the tank directly or by having hot water blended into the chiller to provide an artificial load so that condenser heat can be generated. Capital investment is high because, besides a huge tank or tanks, the double bundle condenser is still necessary as well as piping connections to the heating and chilled water systems. In addition to this cost, one must consider substantial, if not 100%, auxiliary heating sources to account for prolonged periods of no-storage.

4. *The cascade heat cycle* bears other names such as cascade heat reclaim cycle or cascade transfer cycle, etc. The cascade refrigeration machine cools condenser water from the base chiller. The cascade condenser generates all the heat for the heating system. There are several particular features in this cycle: two machines are required for heating, yet only one machine is available for cooling. The chiller of the cascade machine has cooling tower water circulating through it, and high hot water temperatures are feasible. There are several versions of this cycle, including double bundle condensers, series or parallel chiller design and a number of other features.

There are other variations: air-to-water and air-to-air, but these involve extraction of energy from an external source and are irrelevant to the central plant, unless the concept of a heat pump central plant proves feasible in the future.

As mentioned previously, when chilled water is available year-round, recapture of energy from exhaust, return, or even from outside air, if necessary, becomes easy and, since a piping system exists, it is also easy to carry the recaptured energy back to the chillers, as illustrated in Fig. 2-11.

Any of the heat pump systems may be applied in the central plant, the double bundle condenser leading the

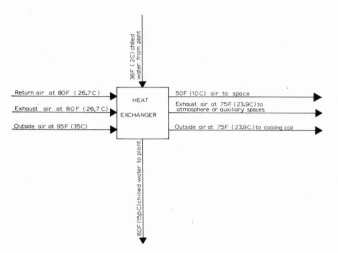

Figure 2-11. Method of transporting recaptured heat to central plant.

group. The question of returning the heat to the buildings is not subject to quick and easy solutions. If the complex is concentrated, the solution is liable to prove economically feasible. Project dispersion, in general, is costly and detracts from the economy of a central plant.

BIBLIOGRAPHY

Plant Criteria

Application Engineering Bulletin EB-APR-6, *Design Considerations for Central Chilled Water Plant Installations*, La Crosse, Wis.: The Trane Company, no date.

ASHRAE Guide and Data Book, 1970, *Heating and Cooling from a Central Plant*, Chapter 11, 167–180.

ASHRAE Handbook and Product Directory, 1974 Applications, *Refrigeration in the Chemical Industry*, Chapters 54, 58. 1–58. 16.

Central Chilled Water Conference Proceedings-1976. Held at Purdue University, November 9–11, 1976; Ed. J.T. Pearson and D.L. Weast, West Lafayette, Ind.: Purdue University, Department of Physical Plant, 1976, 1–265 and *Central Chilled Water Conference*

Central Chilled Water and Heating Plant Conference Proceedings-1979. Held at Syracuse University, August 21-23, 1979; Ed. Eugene E. Drucker, Syracuse, NY: The Syracuse University Institute for Energy Research, 1979, 1–255 and Central Chilled Water and Heating Plant Conference Addendum, 1979.

"Chilled Water as a Utility," Carrier Air Conditioning Company, no date.

Hoglund, M.L. "Guidelines for Central Plant Cooling," *Power*, (June, 1969), 84–87.

Hoglund, M.L. "Central Refrigeration," *Central Chilled*

Water Plants Technical Manual TM-104, La Crosse, Wis.: The Trane Company, 1966.

Johnson, George N. "Spare Capacity Requirements for District Heating Systems," *Proceedings, National District Heating Association* (now International), Vol. XLVIII, 1957, 171.

Miller, Jr., Samuel W. "Hot Water Vs. Steam for Space Heating Campus Buildings," *Proceedings, National District Heating Association* (now International), Vol. LI, 1960, 81.

Morabito, B.P. "How Higher Cooling Coil Differentials Effect System Economics," *ASHRAE Journal*, (August, 1974), 54–55.

Schoenberger, Paul K. "HVAC System Design Criteria for Energy Conservation," *ASHRAE Journal*, (August, 1974), 54–55.

Sinoski, Donald A. "Design Criteria for a New District-Heating System," *Proceedings, International District Heating Association*, Vol. LVI, 1965, 49.

Smith, B.H. "Selection of Boiler Equipment in Central Heating and Cooling Plants," *Proceedings, International District Heating Association*, Vol. LV, 1964, 54.

"Suggested Guide for Architects and Engineers in Designing District-Heating Installations," *Proceedings, International District Heating Associations*, Vol. LV, 1964, 131.

Hot Water, Chilled Water and Refrigeration Storage

Altman, Mansfield. *Conservation and Better Utilization of Electric Power by Means of Thermal Energy Storage and Solar Heating*, Document PB- 210-359, Springfield, Va.: National Technical Information Service, 1971.

ASHRAE Handbook & Product Directory, 1974 Applications, *Solar Energy Utilization of Heating and Cooling*, Chapter 59, 16–20.

Boester, C.F. "The Application of Storage Refrigeration to Air Conditioning," ASHVE Research paper 1142, *ASHVE Transactions*, 1939, 675.

Chada, A. and W.R.W. Read. *The Performance of a Solar Air Heater and a Rockpile Thermal Storage System*, (Paper No. 4/48, International Solar Energy Society Conference, Melbourne, 1970.)

Close, D.J., R.V. Dunkle and K.A. Robison. "Design and Performance of a Thermal Storage Air Conditioning System," *Mechanical and Chemical Transaction*, Institute of Engineers, Australia, Vol. MC-4, No. 1, May 1968, 45.

Cuplinskas, E.L. "A Simplified Heating-Cooling Thermal Storage System," *ASHRAE Journal*, (April, 1976), 29–30.

Dudley, J.C. and S.I. Freedman. "Power Reduction in Air Conditioning by means of Off-Peak Operation and

Coolness Storage," *Transactions of The American Society of Mechanical Engineers* (ASME), Vol. 96, 1974, 165.

Dudley, J.C. "Thermal Energy Storage in Air Conditioning: A Means for Reducing Peak Electrical Power Demand," PhD dissertation, University of Pennsylvania, 1973.

Dudley, J.C. and S. I. Freedman. "Economics of Operating Energy-Storage Air Conditioning Systems with Reduced Peak Power Demand," *Transactions, The American Society of Heating, Refrigerating and Air Conditioning Engineers* (ASHRAE), Vol. 81, Part 1, 1975, 436.

Freedman, S.I. and J.C. Dudley. "Off Peak Air Conditioning Using Thermal Energy Storage," *Proceedings of 7th Annual Intersociety Conversion Engineering Conference*, American Chemical Society, 1972, 1256.

Hirschberg, H.G. "Ways of Accumulating Refrigeration," *Sulzer Technical Review*, (March, 1964), 146-154.

Landman, William S. "Chilled Water Storage," *Heating, Piping, Air Conditioning*, (November, 1976), 73-75.

Linton, Karl J. "Chilled Water Storage," *Heating, Piping Air Conditioning*, (May, 1978), 53-56.

Meyer, Charles F. and David K. Todd. "Heat Storage Wells," *Water Well Journal*, (October, 1973), 35-41.

Meyers, Earl. "Chilled Water Storage," *Heating, Piping, Air Conditioning*, (March, 1965), 113-116.

Patel, Krishna N. and William J. Murphy. "Storage Water Application and Service After Supply Systems to Heat Recovery." *ASHRAE Journal* (December, 1974), 52-55.

Geiringer, P.L. and B.M. Venable. "How We Design a High Temperature Water System," *Power*, (February, 1957), 77-78, 204.

Hansen, E.G. "High Temperature Hot Water District Heating at the U.S. Air Force Academy, Colorado Springs, CO." *Proceedings, National District Heating Association* (now International), Vol. L, 1959, 129.

Harmon, Ray M. "Some Suggestions Regarding MTW Equipment Selection," *ASHRAE Journal*, (July, 1961), 44-48.

Hartman, Walter L. "Hot Water Campus Heating System at the Ohio State University," *Proceedings, National District Heating Association* (now International), Vol. L, 1959, 124.

Lindberg, Ralph E. "Medium Temperature Water: Its Advantages and Disadvantages," *ASHRAE Journal*, (June, 1966), 54-56.

Mumford, S.F. "Controlled Circulation Boilers for High Temperature," *Proceedings, National District Heating Association* (now International), Vol. XLVII, 1957, 158.

Strempel, E. "Problems of Heat-Power Economy and Town Heating," *Energie*, (November 28, 1976), 312-317. (In German.)

Teller, William M. and Joel Friedman. "High-temperature Water Heating: Will it Yield the Savings Needed Now?", *The 1976 Energy Management Guidebook*, ed. by *Power* magazine, 89-92.

Ziel, Perry H. and John S. Blossom. "Pressurizing High Temperature Water Systems," *ASHRAE Journal*, (May, 1965), 73-78.

High Temperature Water

American Society of Heating and Air-Conditioning Engineers, Inc. *Bibliography of High-Pressure, High-Temperature, Hot-Water Heating*, Prepared from the Engineering Index, 1935 through July 1954.

Blossom, J.S. and P.H. Ziel. "Pressurizing High-Temperature Water Systems," *ASHRAE Journal*, (November, 1959), 47-54.

Bord, Homer M. "More Economical MTW Systems," *ASHRAE Journal*, (April, 1961), 60-65, 88.

Carofano, D.A. "District MTW System is Pump-Pressurized, Has Open Tank Storage," *Heating, Piping, Air Conditioning*, (July, 1970), 68-70.

Diskant, William. "A High Temperature Water District Heating System for Sapporo, Japan," *Record of the Tenth Intersociety Energy Conversion Conference, 1975*, New York, N.Y.: Institute of Electrical and Electronics Engineers (IEEE), Cat. No. 15CHO 983-7 TAB, 473.

Ecabert, R. and W. Helbling. "Study for the Federal Bureau of Energy: Technical and Economic Possibilities of District Heating in Switzerland," *Sulzer Technical Review*, (2/1974), 96-104.

Steam Plants

Johnson, Robert E. "Combined-Cycle Operation and Use of Secondary Energy," *Proceedings of the American Power Conference*, Vol. 36, 1974, 657.

Miller, A.I., *et al. Use of Steam-Electric Power Plants to Provide Thermal Energy To Urban Areas.* ORNL-HUD-14, Springfield, Va.: National Technical Information Service, 1971.

Mockeridge, E.H. "Utilization of Package Boilers for the Combustion of Waste Fuels," *Proceedings of the American Power Conference*, Vol. 36, 1974, 649.

Whirl, S.F. "District Heating Using Electric Boilers and Steam Accumulators," *Proceedings, International District Heating Association*, Vol. LVIII, 1967, 62.

Power Plants

Christian, Jeffrey E. *Gas-Steam Turbine Combined Cycle Power Plants*, Prepared for Argonne National Laboratory, NTIS Document ANL/CES/TE 78-4, Springfield, Va.: National Technical Information Service, 1978.

Diamant, R.M.E. *Total Energy*, New York : Pergamon Press, 1970.

Diamant, R.M.E. and J. McGarry. *Space and District Heating*, London: Iliffe Books Ltd., 1968.

Modern Power Station Practice, 2nd Ed., 5 vols, London: Central Electricity Generating Board, 1963.

Morse, Frederick T. *Power Plant Engineering*, 3rd Ed., revised, New York: D. Van Nostrand Company, Inc., 1953.

Salisbury, J. Kenneth. *Steam Turbines and Their Cycles*, New York: John Wiley & Sons, 1950.

Skrotzki, Bernhardt G.A. and William A. Vopart. *Steam and Gas Turbines*, New York: McGraw-Hill Book Company, 1950.

Standard Practices for Stationary Diesel and Gas Engines, Sixth Ed., Washington, D.C.: Diesel Engine Manufacturers Association, 1972.

Steam, Its Generation and Use, 38th Ed., New York: The Babcock and Wilcox Company, 1972.

Power Plant Auxiliaries and Accessories

Brown, Robert, V. Ganaparthy and John Glass. "Design of Air-cooled Exchangers," *Chemical Engineering*, (March 27, 1978), 106–124.

Cato, G.A., K.L. Maloney and J.G. Sotter (KVB, Inc.) *Reference Guideline for Industrial Boiler Manufacturers to Control Pollution with Combusion Modification*, Prepared for U.S. Environmental Protection Agency, Industrial Environmental Research Lab. NTIS Document PB-276 715, Springfield, Va.: National Technical Information Service, 1977.

Cooling Tower Fundamentals and Application Principles, Kansas City: The Marley Company, 1967.

Christian, J.E. *Central Cooling-Absorptive Chillers*, Prepared for Argonne National Laboratory, NTIS Doctument ANL/CES/TE 77-78, Springfield, Va.: National Technical Information Service, 1977.

———. *Central Cooling-Compressive Chillers*, Prepared for Argonne National Laboratory, NTIS Document ANL/CES/TE 78-2, Springfield, Va.: National Techincal Information Service, 1977.

Design Manual, Mechanical Engineering, NAVDOCKS DM-3, Washington, D.C.: Department of the Navy, Bureau of Yards and Docks, 1962.

Dyer, David F. and Glennon Maples. *Measuring and Improving the Efficiency of Boilers*, Auburn, Ala.: Auburn University, 1979.

Elonka, Steve. "A Special Report, Cooling Towers," *Power*, (March, 1963), 1–16.

Evans, R.K. "A Special Report, Combustion Controls," *Power*, (December, 1967), 1–32.

Farahan, Ebrahim. *Central Heating-Package Boilers*, Prepared for Argonne National Laboratory, NTIS Document ANL/CES/TM-77-6, Springfield, Va.: National Technical Information Service, 1977.

Finsel, Edward E. "Boiler Blow-off Valves," *Air Conditioning, Heating and Ventilating*, (November, 1959), 53–55.

Fisher, P., J.W. Suitor and R.B. Ritter. "Fouling Measurement Techniques," *Chemical Engineering Progress*, (July, 1975), 66–72.

France, Howard B. "Tailor Boiler Continuous-Blowdown System to Suit Plant Heat Cycle," *Power*, (December, 1960), 232–233.

France, Howard B. "The Case for Continuous Blow-off," *Power*, (November, 1960), 202–204.

Fundamentals of Combustion, Caltex Fuel Oils, Application, Specification, Handling, The California Texas Oil Corp., 1963.

Hall, Frank. "How to Pick an IC Engine," *Product Engineering*, (December 10, 1962), 107–126.

Hot Water Systems, Components, Controls and Layouts, Milwaukee: Cleaver-Brooks Company, 1960.

Karassik, Igor J. "How to Select Boiler-Feed Pumps for a Modern Industrial Power Plant," *Power*, (February, 1962), 76–78 and (April, 1962), 78–80.

Karassik, Igor J. *Engineers' Guide to Centrifugal Pumps*, Milano: Hoeple Publisher, 1973.

Kent's Mechanical Engineers' Handbook, Power Volume, 12th Edition, Ed. J. Kenneth Salisbury, New York: John Wiley & Sons, Inc., 1950.

KVB, Inc. *Assessment of the Potential for Energy Conservation through Improved Industrial Boiler Efficiency, Volume I*, Prepared for the Federal Energy Administration, NTIS Document PB-262 576, Springfield, Va.: National Technical Information Service, 1977.

Liao, G.S. and P. Leung. "Analysis of Feedwater Pump Suction Pressure Decay Under Instant Turbine Load Rejection," *Transactions of the Society of Mechanical Engineers* (ASME), Series A, Vol. 94, 1972, 83.

Mason, Paul M., Robert E. Betts and Wesley W. Smith. *Engineers' Bibliography on the Design and Operation of Cooling Towers*, Bulletin 138, College Station, Tex.: Texas Engineering Experiment Station, 1956.

McElroy, Michael W. and Dale E. Shore (KVB, Inc.) *Guidelines for Industrial Boiler Performance Improvement (Boiler Adjustment Procedures to Minimize Air Pollution and to Achieve Efficient Use of Fuel)*, Prepared for the Federal Energy Administration, NTIS Document PB-264 543, Springfield, Va.: National Technical Information Service, 1977.

Perez, E.R. "Design Fuel-oil Systems for Better Combustion," *The Energy Management Guidebook*, Ed. by *Power* Magazine, 1976, 101–113.

Purtell, C.J. and Tom Timm. "How to Plan a Boiler Layout," *Air Conditioning, Heating and Ventilating*, (September, 1961), 59–62.

Stefani, R.M. "How to Choose Your Feed-Water Heaters," *Power*, (June, 1957), 82–84.

Stepanoff, A.J. *Centrifugal and Axial Flow Pumps*, 2nd Ed., New York: John Wiley & Sons, Inc., 1967.

Standards for Closed Heaters. 1st. Ed., New York: Heat Exchange Institute, 1968.

Swift, D.C. "Burners, Stokers and Combustion Air Affect your Boiler's Rating," *Power,* (October, 1955), 106–107, 210, 214, 216, 218.

Use of Fuel Oil for Heat or Power Generation in Commercial and Industrial Installations, Training Unit Text T-F-21, Socony Oil Company, 1951.

"Using Natural Gas Engines as Air Conditioning Prime Movers," *Heating, Piping, Air Conditioning,* (November, 1961), 152-156.

Power Generation, Central Plant and Process Heat Overview *(See Chapter 4 Bibliography on System Criteria and Design).*

Bos, Peter *et al.* (Resource Planning Associates), *The Potential for Cogeneration Development in Six Major Industries by 1985.* Prepared for the Department of Energy, NTIS Document HCP/M 60172-01/2. Springfield, Va.: National Technical Information Service, 1977.

Coxe, Edwin F. (Reynolds, Smith and Hills), *The Marketability of Integrated Energy/Utility Systems,* Sponsored by the National Bureau of Standards and the Department of Health, Education and Welfare, NTIS Document PB-266 042, Springfield, Va.: National Technical Information Service, 1976.

Cogeneration and Integrated Energy/Utility Systems, Proceedings of the Energy Research and Development Administration Conference, Washington, D.C. NTIS Document CONF-770 632, Springfield, Va.: National Technical Information Service, 1977.

"Cogeneration," *Power Engineering,* (March, 1978), 34–42.

Croke, K.G. *et al.* (Argonne National Laboratory), *Financial Overview of Integrated Community Energy Systems,* Sponsored by the U.S. Energy Research and Development Administration (ERDA), NTIS Document ANL-77-XX-71, Springfield, Va.: National Technical Information Service, 1977.

Decker, G.L., Robert E. Sampson, *et al.* (The Dow Chemical Company and Environmental Research Institute of Michigan), *Evaluation of New Energy Sources for Process Heat,* Prepared for the Office of Energy R & D Policy, National Science Foundation, NTIS Document PB-245 604, Springfield, Va.: National Technical Information Service, 1975.

Dow Chemical Company. *Energy Industrial Center Study,* NTIS Document PB-243 823, Springfield, Va.: National Technical Information Service, 1977.

Fulbright, Ben E. *MIUS Community Conceptual Design Study,* Sponsored by the National Aeronautics and Space Administration (NASA), NASA Document TM X-58174 and NTIS Document N 76-33085, Springfield, Va.: National Technical Information Service, 1976.

Gamze, Maurice M. "A Critical Look at Total Energy Systems and Equipment," *Heating, Piping, Air Conditioning,* (August, 1975), 43–46.

Goen, Richard L. *Assessment of Total Energy Systems For the Department of Defense,* Stanford Research Institute, NTIS Document AD-781 816, Springfield, Va.: National Technical Information Service, 1973.

Hebrank, John, *et al.* (National Bureau of Standards). *Performance Analysis of the Jersey City Total Energy Site: Interim Report,* Sponsored by the Department of Housing and Urban Development, NTIS Document PB-269 517, Springfield, Va.: National Technical Information Service, 1977.

Holness, Gordon V.R. "Designing a Central Chilled Water System," *Heating, Piping, Air Conditioning,* (September, 1978), 111–122.

Javetski, John. "Cogeneration," *Power,* (April, 1978), 34–42.

Karkheck, J. and J. Powell (Brookhaven National Laboratory). *Prospects for the Utilization of Waste Heat in Large Scale District Heating Systems,* Prepared for the Energy Research and Development Administration (ERDA), NTIS Document BNL-22550, Springfield, Va.: National Technical Information Service, 1977.

Karkheck, J. *et al.* "Prospects for District Heating in the United States," *Science,* (March 11, 1977), 948–955.

Karkheck, J. *et al. Technical and Economic Aspects of Potential U.S. District Heating Systems,* Document BNL 21287, Upton, N.Y.: Dept. of Applied Science, Brookhaven National Laboratory, 1976.

Kirsme, Dale W. and Steve B. Manyimo (Reynolds, Smith and Hills). *Integrated Utility Systems Feasibility Study and Conceptual Design at the University of Florida,* Sponsored by the National Bureau of Standards and the Department of Health, Education and Welfare, NTIS Document PB-266 043, Springfield, Va.: National Technical Information Service, 1976.

Kirsme, Dale W. and Carl Bronn, Jr. (Reynolds, Smith and Hills). *Integrated Utility Systems Feasibility Study and Conceptual Design at Central Michigan University,* Sponsored by the National Bureau of Standards and the Department of Health, Education and Welfare, NTIS Document PB-266 044, Springfield, Va.: National Technical Information Service, 1976.

Lackey, M.E. *Thermal Effects of Various Air Conditioning Methods Used in Conjunction with an Electric Power Production and District Heating Facility.* NTIS Document ORNL-HUD-13, Springfield, Va.: National Technical Information Service, 1971.

Levine and McCann. *Total Integrated Energy Systems (TIES) Feasibility Analysis for the Downtown Redevelopment Project, Pasadena, California,* Sponsored by the Energy Research and Development Administration (ERDA), NTIS Document SAN/1151-1, Springfield, Va.: National Technical Information Service, 1976.

Meckler, Milton. "Options for On-Site Power," *Power*, (March, 1976), 33–35.

Miller, A.J., *et al.* (Oak Ridge National Laboratory). *Technology Assessment of Modular Integrated Utility Systems*, Sponsored by the Department of Housing and Urban Development, NTIS Document ORNL-HUD-MIUS-25, Springfield, Va.: National Technical Information Service, 1976.

Mixon, W.R., *et al.* (Oak Ridge National Laboratory). *Technology Assessment of Modular Integrated Utility Systems*, Sponsored by the Department of Housing and Urban Development, NTIS Document ORNL-HUD-MIUS-25, Springfield, Va.: National Technical Information Service, 1976.

Nydick, S.E., *et al. A Study of Inplant Electric Power Generation in the Chemical, Petroleum Refining and Paper and Pulp Industries*, Prepared for the Federal Energy Administration, Report No. TE 5429-9796, Waltham, Mass.: Thermo Electron Corp., 1976.

O'Keefe, William. "A Special Report, In-plant Electric Generation," *Power* (April, 1975), 1–24.

Olszewski, M. *Preliminary Investigation of the Thermal Energy Grid Concept*, Publication ORNL/TM-5786, Oak Ridge, Tenn.- Oak Ridge National Laboratory, 1977.

Pickel, Frederick H. *Congeneration in the U.S.: Economic and Technical Analysis* (MIT Energy Laboratory Report). Cambridge, MA: Massachusetts Institute of Technology, Energy Laboratory, 1978.

Potential for Scarce-Fuel Savings in the Residential/Commercial Sector through the Application of District Heating Systems, Argonne, Ill.: Argonne National Laboratory, 1977.

Reale, Frank N. and Tharam S. Dillon. "Investigation into the Use of Large-Scale Total-Energy Systems in Mild and Warm Climates," *Fuel*, (July, 1977), 257–266.

Sakhuja, R.K. *et al. Technical and Economic Feasibility of Solar Augmented Process Steam Generation*, Prepared for the Energy Research and Development Administration (ERDA), Report No. TE4205-99-76, Waltham, Mass.: Thermo Electron Corp., 1976.

Sharp, A. "Optimizing Gas Turbine Power System Operation with Product Support Services," *Gas Turbine International*, (January-February, 1977).

Solt, J.C. *Fuel Savings Through the Use of Cogeneration*, San Diego, Calif.: Solar Turbines International, no date.

Wolfer, Barry M., *et al. Preliminary Design Study of Baseline MIUS*, Sponsored by the National Aeronautics and Space Administration (NASA), NASA document TM X-58193 and NTIS document N 77-34050, Springfield, Va.: National Technical Information Service, 1977.

Environmental

Breistein, L. and R.E. Grant (Oak Ridge National Laboratory). *MIUS Systems Analysis-Effects on Unfavorable Meteorological Conditions and Building Configurations on Air Quality*, NTIS Document ORNL/HUD/MIUS-29 (Addendum II of ORNL/HUD/MIUS-6) UC-38, Springfield, Va.: National Technical Information Service, 1976.

Hoffman, David R. "Spray Cooling for Power Plants," *Proceedings of the American Power Conference*, Vol. 35, 1973, 702.

IGGI/ABMA Joint Technical Committee. "Criteria for the Application of Dust Collectors to Coal-fired Boilers," *Proceedings of the American Power Conference*, Vol. 29, 1967, 572.

Landon, Richard D. and James R. Houx, Jr. "Plume Abatement and Water Conservation with the Wet-dry Cooling Tower," *Proceedings of the American Power Conference*, Vol. 35, 1973, 726.

Leung, Paul and R.E. Moore. "Thermal Cycle Arrangements for Power Plants Employing Dry Cooling Towers," Series A, *Transactions of the American Society of Mechanical Engineers* (ASME), Vol. 93, 1971, 257.

Mathews, Ralph T. "Some Air Cooling Design Considerations," *Proceedings of the American Power Conference*, Vol. 32, 1970, 612.

Senew, Michael J., *et al.* (Argonne National Laboratory). *Impacts of Environmental and Utility Siting Laws on Community Energy Systems*, NTIS Document ANL/ICES-TM-7, Springfield, Va.: National Technical Information Service, 1978.

"Special Report," Pollutant Control Equipment:Gases... Liquids...Solids," *Power*, (June, 1971), S.l–S.26.

Standards Support and Environmental Impact Statement Volume 1: Proposed Standards of Performance for Stationary Gas Turbines, Publication EPA-450/2-77-017a, Research Triangle Park, N.C.:U.S. Environmental Protection Agency (EPA), September, 1977.

Stationary Internal Combustion Engines, Background Information: Proposed Standards (Draft), Research Triangle Park, NC: U.S. Environmental Protection Agency (EPA), Office of Air Quality, Planning and Standards, 1978.

Turner, D. Bruce. *Workbook of Atmospheric Dispersion Estimates*, Research Triangle Park, N.C.: U.S. Environmental Protection Agency (EPA), MD-35, 1970.

Heat Recovery and Energy Conservation

Air Force Energy Conservation Handbook. Document NBSIR 77-1204, Vol. 1, 1977. Prepared by the National Bureau of Standards for USAF Civil Engineering Command, Superintendent of Documents Publications SD Cat. N. C13. Superintendent of Documents, U.S. Printing Office, Washington, D.C. 20402.

Applications Engineering Manual AM-FND6-174. *Conserve Energy by Design*, La Crosse, Wis.: The Trane Company, 1974.

"Heat Recovery." ASHRAE Handbook and Product Directory, 1976, Systems, Chapter 7.

Barnes, R.W., *et al.* (The Dow Chemical Company). *Evaluation of New Energy Sources for Process Heat*, Prepared for National Science Foundation, NTIS Document Pb-245 604, Springfield, Va.: National Technical Information Service, 1975.

Bell, W.E. "Heat Recovery from Diesel Engines," *Air Conditioning, Heating and Ventilating*, (March, 1959), 57–60.

Bridgers, Frank H. "How New Technology May Save Energy in Existing Buildings," *Heating, Piping, Air Conditioning*, (August, 1975), 50–55.

Clay, Paul E. "Design and Installation Consideration for Heat Recovery Systems," ASME paper 65-OGP-la.

Energy Research and Development Administration. *Building Energy Handbook, Volume 1, Methodology for Energy Survey and Appraisal*, NTIS document ERDA-76/163/1, Springfield, Va.: National Technical Information Service, 1976.

Fanaritis, J.P. and H.J. Streich. "Heat Recovery in Process Plants," *Chemical Engineering*, (May, 1973), 80–88.

Fleming, J.B., J.R. Lambrix and M.R. Smith. "Energy Conservation in New Plant Design," *Chemical Engineering*, (January 21, 1974), 112–122.

Gatts, Robert R., Robert G. Massey and John C. Robertson. *Energy Conservation Program Guide for Industry and Commerce*, NBS Handbook, publication 115 SD Cat No. C13 11:115, 1974 and Supplement 1, Publication SD Cat No. C13 11: 115, Supl. 1, 1975. Superintendent of Documents, U.S. Printing Office, Washington D.C 20402.

Gordian Associates, Inc. *Cooling Tower Energy Studies, Conservation Techniques Applicable to Existing Installations plus Comparative Economics and Energy Requirements of Mechanical and Natural Draft Towers*. Document PB-264 028, Springfield, Va.: National Technical Information Service, 1977.

Japhet, R.A. "Considering Heat Reclamation? Here Are Your Options," *Heating, Piping, Air Conditioning*, (April, 1969), 88–95.

National Bureau of Standards. *Waste Heat Management Guidebook*, NBS Handbook 121, Ed. K.G. Kreider and M.B. McNeil. Sponsored by the Federal Energy Administration. NTIS Document PB-264 959, Springfield, Va.: National Technical Information Service, 1977.

Ostrander, W.S. "Cascade Heat Reclaim," Centrifugal Heat Pump Systems, ASHRAE Symposium Papers of 1969 Annual Meeting.

Reay, David A. *Industrial Energy Conservation*, New York: Pergamon Press Inc., 1977.

Rex, Harland E. "Four Pipe Heat Recovery Systems," *ASHRAE Journal*, (August, 1971), 43–48.

Robertson, J.C. "Energy Conservation in Existing Plants," *Chemical Engineering*, (January 21, 1974), 104–111.

Sinnamohideen, K. and N.C. Olmstead. "Chiller Plant Energy Conservation Opportunities," *Energy Conservation*, ASHRAE Symposium Bulletin No. 73-5, 16–27.

Stamm, Richard H. "Getting the Most out of Industrial Refrigeration," *Specifying Engineer*, (December, 1975), 50–53.

Sumerell, Howard. "Recovering Energy from Stacks," *Chemical Engineering*, (March 29, 1976), 147–148.

Technical Guidelines for Energy Conservation. Document NBSIR 77-1238, 1977. Prepared by the National Bureau of Standards for USAF Civil Engineering Command. Superintendent of Documents publication SD Cat. No. C13. Superintendent of Documents, U.S. Printing Office, Washington, D.C. 20402.

CHAPTER 3

Chiller Arrangements

Air conditioning accounts for a substantial portion of the annual energy consumption in most buildings. Refrigeration constitutes a large sector of industrial energy applications. Indubitably, the efficiency of these operations is of immense importance.

Since chillers are energy-consuming machines, their many sizes, numbers and arrangements in various applications are not only evidence of design flexibility in handling air conditioning or refrigeration loads, but are proof that design engineers consider seriously the question of energy use.

This chapter addresses the factors that influence chiller selection and arrangement. There are many possible arrangements; those examined here are the four basic configurations exemplified in Fig. 3-1, as well as others shown in Figs. 3-8 through 3-16. These are somewhat complex, but appear often enough in actual practice to merit comment.

3.1 Basic Arrangement

Two or more chillers may be employed, either in series or in parallel, in the four basic arrangements *a*, *b*, *c*, and *d* of Fig. 3-1. A fifth, a series-parallel arrangement, is illustrated in Fig. 3-9, Section 3.6. Each arrangement presents the designer with specific choices, requiring a grasp of design objectives. Understanding the significance of the four basic arrangements will facilitate examination and analysis of the more involved cycles. These four arrangements deal with water side control and are predicated on full water flow through both evaporator and condenser and a fixed leaving chilled water temperature.

Central cooling plants invariably have more than one chiller. The justification for more than one chiller and the determination of the final number of chillers depend on a variety of factors.

At the present time, machine size is limited by shipping dimensions and a lack of technical and operating experience for machines in excess of 7000–8000 tons. However, 13,000-ton (45 760 kW) machines can be ordered. Large machines are shipped as components to be field-assembled. Undoubtedly, for plants in excess of 50,000-tons (176 000 kW) machines of greater than 13,000-ton capacity can be used; the huge refrigeration

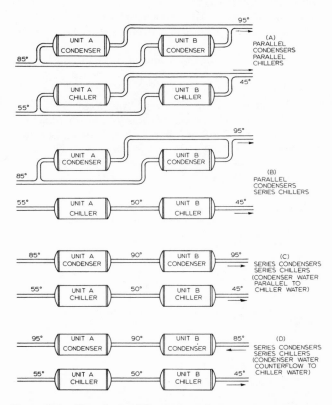

Figure 3-1. Four basic chiller and condenser arrangements employing conventional temperatures [temperatures given are F: 95 F (35 C); 90 F (32.2 C); 85 F (29.4 C); 55 F (12.8 C); 50 F (10 C); 45 F (7.2 C)].

plants of the future will require machines of increased tonnage. Commercial demand is bound to determine ultimate size development. Open centrifugal compressors, rather than hermetic, are used in large central plants, when the need is for centrifugals, since hermetic centrifugal machines are currently limited to approximately 1200 tons (4224 kW).

Spare or standby capacity is another important factor. A fully redundant unit is ordinarily too expensive to tolerate in commercial applications. In industrial work, it becomes a matter of process needs and weighing of acceptable alternatives. With no standby capability, a machine breakdown is liable to deprive the plant of its services for a few hours or many months. Decreasing the impact of the loss of one machine, not taking into consideration the size and total number of chillers, will insure minimal inconvenience except at times of top load conditions. Machine breakdowns are associated with the concept of reliability, which for centrifugal chillers is more than 99%.

In general, repair of large machines may be too costly and the downtime prohibitive. Although even a small hermetic machine may be out of service for a long time, dismantling of large chillers is very time-consuming and spare parts for them may be harder to obtain and the need for factory assistance more probable.

Efficiency and surging are two aspects that must be examined carefully. Large machines perform best at full load and become increasingly inefficient at diminishing part loads. At low-load ranges, centrifugal compressors are subject to surging and other operating difficulties. Thus, it is practical to select centrifugal machines for peak efficiency at prevailing part loads, perhaps in the neighborhood of 75% load. Efficiency selection at part load makes surging a little less likely or, at least, shortens the periods of occurrence, since most machines operate at partial loads most of the time.

If a substantial number of chillers are involved, the question of series versus parallel arrangement emerges from the beginning. A series arrangement is limited to three or four chillers, bearing in mind that a minimum six degrees of chilled water temperature difference is needed for each chiller. On the other hand, an unlimited number of chillers can be piped in parallel.

The designer must strike an economic and operational balance in arriving at the total number of machines, considering power consumption, piping and valving economy, maintainability, total installed cost, operating costs, space limitations, the extent of automatic operation of the refrigerant and water cycles and availability and capability of operating personnel.

3.2 Arrangement *a* (Fig. 3-1), Chillers in Parallel, Condensers in Parallel

Both evaporators and condensers are piped in parallel, an arrangement with widespread acceptance and one that designers are most familiar with, offering extensive flexibility and the ability to accommodate phased additions.

A cardinal advantage of parallel flow is that it lends itself extremely well to variable flow, especially in primary-secondary loops, and is the main reason for its popularity. As load decreases, condenser and chilled water pumps can be disengaged, saving horsepower and contributing to a more economical operation.

Based on chiller heat transfer area of 6–8 ft^2 per ton (0.15–0.21 m^2/kW) and maximum water velocity of 12 ft per sec (3.64 m/s), units connected in parallel require two- or three-pass evaporators and, therefore, offer excellent heat transfer performance for a large chilled water range. In the last two decades, economics have tended to increase water velocities both in the chiller and condenser sections from 8 to approximately 10 ft per sec (2.4–3 m/s), at times, even higher.

A great deal is known about this arrangement and many designers feel comfortable in using it. Nevertheless, there are some disadvantages of a rather serious nature, especially in the case of full system flow. To maintain full flow at partial loads, return chilled water must be circulated through all units, both active and inactive, with the non-refrigerated water being mixed with water that has passed through the evaporators and chilled to design temperature. The mixture, of course, is at a higher temperature than design. If design temperature water is desired, the active chiller must overchill the water passing through it. The problem of overchilling is examined a little further on in this section. In general, overchilling increases the risk of an evaporator freeze-up because of lower suction temperatures and tends to increase horsepower requirements, and therefore operating costs. Unpleasant alternatives to overchilling include a variable chilled water supply temperature above design levels, if one machine is operated, or operation of both machines at very low loads, which requires more horsepower per ton and is entirely uneconomical.

As for condensers in parallel, they present few difficulties. Water temperature problems do not exist, and lift is identical for all units. Whether condenser water is allowed to circulate through an inactive unit or not, does not affect the performance of the cooling tower. Reduced condenser water flow is desirable at times because it reduces pumping requirements (see Section

4.10 "Optimization of Condenser and Chilled Water Flow"). Various techniques can be used to minimize low flow problems. The subjects of reduced water flow and variable primary-secondary flow are discussed in Chapter 7.

Undoubtedly, a great deal more piping, valving, pumps, and instrumentation are needed for the parallel arrangement than for the series. However, pipe sizes may be reduced due to smaller flows through each branch. Piping in parallel is quite involved spatially and there is a greater abundance so that, in an overall sense, costs are high. On the other hand, series piping throughout the system handles full flow at all times and is full size, including the bypasses around the chillers. Costs may exceed those of the parallel arrangement. A piping layout, a flow diagram, and an accurate estimate will establish relative costs.

When examining the question of overchilling more closely one comes to the conclusion that the problem is caused perhaps by an unrealistic requirement for constant flow in the primary circuit at all times, regardless of load size. Specification of constant flow is a holdover from past practices, when constant flow for small systems was acceptable economically and offered the hydraulic stability that most engineers prefer. Full flow in large systems is wasteful of pumping horsepower and, in terms of wear and tear, of the machines and system in general.

Overchilling may be eliminated by adhering to one of four main curative methods. First, all units may be operated together down to a small portion of their capacity in order to handle part loads. In doing so, the chillers

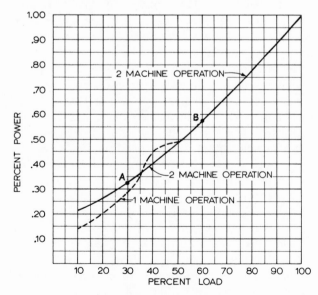

Figure 3-2. Typical reduced part load performance for centrifugal compressor system.

sacrifice operating economy. The following example illustrates the point:

Assume a 1000-ton (3520 kW) chiller requiring about 1000 Bhp (746 kW) at design conditions. If two identical 1000-ton (3520 kW) units are installed in parallel and at 30% system load, each chiller must operate at 300 tons (1056 kW) and 330 Bhp (246 kW) (Point A, Fig. 3-2) a total of 660 Bhp (492 kW). If the 30% system load were to be satisfied by operating only one machine, then it would operate at 60% capacity and 57% power (Point B) or 570 Bhp (425 kW) a clear saving of 90 Bhp (67 kW).

This approach introduces some complications. It requires automatic valves for each chiller if one pump is used, or a pump for each chiller with a check valve in its discharge. Furthermore, it is assumed that the pumps are non-overloading; if not, a pump bypass or a back pressure valve may be necessary to control the pump operating point.

A second alternative to overchilling is reduction of overall system flow by maintaining full flow only through the active units, if building flow requirements permit it and if pump transients are manageable.

A third solution to accept chilled water temperatures above design conditions, if it does not present control problems and can satisfy load conditions throughout the project. Ordinarily, the supply chilled water temperature is constant in large projects because buildings peak at different times and variable temperature control would hamper design flexibility. On the other hand, advocates of this solution maintain that higher design temperatures at part load can be accommodated if cooling coils are oversized when full load is anticipated in one building at the same time. Oversizing coils can be considered a viable design alternative, if it is applied to a limited number of coils.

A fourth method is a series arrangement of evaporators, provided economic as well as hydraulic considerations and project design philosophy are favorable. Overchilling is not possible since, at part load, chillers may be deactivated, while through flow is maintained.

Up to this point, overchilling and methods to avoid it have been defined. The question remains as to whether some degree of overchilling may be usefully employed and what may be its practical limits in application. Tables 3-1 through 3-4 examine two and three chillers piped in parallel, using 10 F (5.55 C) and 18 F (10 C) temperature differentials between supply and return water.

Table 3-1 indicates that two chillers in parallel with a 10 F (5.55 C) temperature difference constitute a usable combination. However, the low 37 F temperature is

Table 3-1
Ideal Performance of Two Chillers in Parallel
[Design Chilled Water Supply 42 F (5.6 C), Design Chilled Water Return 52 F (11.1 C)–10 F Temperature Differential]

System			Unit No. 1				Unit No. 2			
% Load	Supply Temp., F	Return Temp., F	% Load of System	% Load of Unit	Supply Temp., F	Return Temp., F	% Load of System	% Load of Unit	Supply Temp., F	Return Temp., F
Both Chillers Operating										
100	42	52	50	100	42	52	50	100	42	52
75	42	49.5	37.5	75	42	49.5	37.5	75	42	49.5
50	42	47	25	50	42	47	25	50	42	47
37.5	42	45.75	18.75	37.5	42	45.75	18.75	37.5	42	45.75
25	42	44.5	12.5	25	42	44.5	12.5	25	42	44.5
12.5	42	43.25	6.25	12.5	42	43.25	6.25	12.5	42	43.25
First Chiller Operating, Second Chiller Bypassing (First chiller providing 42 F water; mixture temperature increasing above 42 F)										
50	47	52	50	100	42	52	0	0	52	52
37.5	45.75	49.5	37.5	75	42	49.5	0	0	49.5	49.5
25	44.5	47	25	50	42	47	0	0	47	47
12.5	43.25	44.5	12.5	25	42	44.5	0	0	44.5	44.5
First Chiller Operating, Second Chiller Bypassing (First chiller overchilling, mixture temperature at 42 F)										
50	42	47	50	100	37	47	0	0	47	47
37.5	42	45.75	37.5	75	38.25	45.75	0	0	45.75	45.75
25	42	44.5	25	50	39.5	44.5	0	0	44.5	44.5
12.5	42	43.25	12.5	25	40.75	43.25	0	0	43.25	43.25

problematic and may need resetting to 38–39 F to prevent evaporator freeze-up. Two and three chillers in parallel using an 18 F (l0C) temperature differential are unfeasible, since they require overchilling to 31–28 F.

In general, a l0 F (5.55C) chilled water range may prove practical under limited conditions for smaller central installations of a few thousand tons with short primary piping networks, and not more than two chillers able to tolerate overchilling and not requiring close temperature control. If the number of chillers or the temperature differential increases to an appreciable extent, overchilling is not a useful design tool. Again, we conclude that an arrangement based on constant chilled water flow throughout the system at all times is not ordinarily attractive because constant flow at partial loads contributes to higher operating costs.

3.3 Arrangement *b* (Fig. 3-1), Chillers in Series, Condensers in Parallel

Arrangement *b* is next in popularity to arrangement *a*. The condensers are piped in parallel, since no con-

denser water temperature problem exists. Chillers in series deliver design temperature chilled water at all loads, entirely eliminating the overchilling problem of the parallel arrangement.

The designer must decide during the early stages of a project whether series or parallel chiller flow is best-suited to his project, taking the special conditions of a particular job into consideration. If constant flow is envisaged for the primary chilled water circuit, the series arrangement is preferable. Under special circumstances, some variation in flow can be tolerated even in a series arrangement, although this violates the old taboo against varying flow through the chiller for fear of a freeze-up. Constant flow may be the result of staggered demand, when diversity remains fairly constant. Large temperature differentials can dovetail well into a constant flow situation.

Although constant refrigeration loads do not necessarily accompany constant flow, they very often do. The series arrangement is strongly indicated when constant refrigeration loads are present.

The point can be made that, at full-load maximum horsepower, savings can be effected by using series flow

Table 3-2
Ideal Performance of Three Chillers in Parallel
[Design Chilled Water Supply 42F (5.6C), Design Chilled Water Return 52F (11.1C)—10F Temperature Differential]

	System			Unit No. 1				Unit No. 2				Unit No. 3			
% Load	Supply Temp., F	Return Temp., F	% Load of System	% Load of Unit	Supply Temp., F	Return Temp., F	% Load of System	% Load of Unit	Supply Temp., F	Return Temp., F	% Load of System	% Load of Unit	Supply Temp., F	Return Temp., F	

% Load	Supply Temp., F	Return Temp., F	% Load of System	% Load of Unit	Supply Temp., F	Return Temp., F	% Load of System	% Load of Unit	Supply Temp., F	Return Temp., F	% Load of System	% Load of Unit	Supply Temp., F	Return Temp., F
All Chillers Operating														
100	42	52	33.33	100	42	52	33.33	100	42	52	33.33	100	42	52
75	42	49.5	25	75	42	49.5	25	75	42	49.5	25	75	42	49.5
50	42	47	16.66	50	42	47	16.6	50	42	47	16.66	50	42	47
37.5	42	45.75	12.5	37.5	42	45.75	12.5	37.5	42	45.75	12.5	37.5	42	45.75
25	42	44.5	8.33	25	42	44.5	8.33	25	42	44.5	8.33	25	42	44.5
12.5	42	42.25	4.16	12.5	42	43.25	4.16	12.5	42	43.25	4.16	12.5	42	43.25
First Two Chillers Operating, Third Chiller Bypassing (First two chillers providing 42F water; mixture temperature increasing above 42F)														
66.66	45.33	52	33.33	100	42	52	33.33	100	42	52	0	0	52	52
50	44.5	49.5	25	75	42	49.5	25	75	42	49.5	0	0	49.5	49.5
37.5	44.08	47.62	18.75	56.25	42	47.62	18.75	56.25	42	47.62	0	0	47.62	47.62
25	43.25	45.75	12.5	37.5	42	45.75	12.5	37.5	42	45.75	0	0	45.75	45.75
12.5	42.62	43.88	6.25	18.75	42	43.88	6.25	18.75	42	43.88	0	0	43.88	43.88
First Two Chillers Operating, Third Chiller Bypassing (First two chillers overchilling; mixture temperature at 42F)														
66.66	42	48.66	33.33	100	38.66	48.66	33.33	100	38.66	48.66	0	0	48.66	48.66
50	42	47	25	75	39.5	47	25	75	39.5	47	0	0	47	47
37.5	42	45.75	18.75	56.25	40.13	45.75	18.75	56.25	40.13	45.75	0	0	45.75	45.75
25	42	44.5	12.5	37.5	40.85	44.5	12.5	37.5	40.85	44.5	0	0	44.5	44.5
12.5	42	43.25	6.25	18.75	41.38	43.25	6.25	18.75	41.38	43.25	0	0	43.25	43.25
First Chiller Operating, Second Two Chillers Bypassing (First chiller overchilling; mixture temperature at 42F)														
33.33	42	45.33	33.33	100	35.33	45.33	0	0	45.33	45.33	0	0	45.33	45.33
25	42	44.5	25	75	37.83	44.5	0	0	44.5	44.5	0	0	44.5	44.5
12.5	42	43.25	12.5	37.5	39.5	43.25	0	0	43.25	43.25	0	0	43.25	43.25

but that, at part-load conditions with all compressors operative, the difference between series and parallel flow diminishes substantially. Selection of the proper refrigerant on the part of the manufacturer is important because at a particular evaporating condition, the load carried by one unit versus the other can be greater, depending on which refrigerant is being used. However, the engineer is not in a position to select the refrigerant and often the manufacturer chooses the refrigerant which will make his machine most competitive.

Besides series arrangement and selection of proper refrigerant, there are two additional factors that affect energy consumption and power requirements in hp/ton, *i.e.*, suction temperature and the number of passes. At the higher temperature differentials of 15–20 F, (8.3-ll.lC) the series arrangement effects operating costs savings because less chilled water need be circulated. The leading chiller (unit A, Fig. 3-l) operates at a very high suction temperature and hp/ton is reduced to a minimum. On the other hand, lagging chiller (unit B) operates at low suction temperature and high horsepower requirements. Nevertheless, the average hp/ton for the two machines is below that of an equal tonnage arrangement in parallel. A chilled water range of l0 F (5.55C) or less for a series arrangement penalizes the application because of increased flow rate, velocity through the chiller becoming the limiting factor. The greater the number of machines in series, the greater the flow limitation. Ordinarily, more than three machines in series is disadvantageous. A minimum 6F (3.33C) chilled water temperature drop across each chiller is the rule of thumb. Conceivably, four machines in series could be considered with a 24F (l3.55C) differential. However, not all machines will be the same size. Most probably, the first two will be of one size and the last two of a different size.

A series arrangement most often results in a single-pass evaporator, due to chiller flow and piping considerations. Single-pass evaporators are less efficient in heat transfer

Table 3-3
Ideal Performance of Two Chillers in Parallel
[Design Chilled Water Supply 40 F (4.4 C), Design Chilled Water Return 58 F (14.4 C)–18 F Temperature Differential]

System			Unit No. 1				Unit No. 2			
% Load	Supply Temp., F	Return Temp., F	% Load of System	% Load of Unit	Supply Temp., F	Return Temp., F	% Load of System	% Load of Unit	Supply Temp., F	Return Temp., F
Both Chillers Operating										
100	40	58	50	100	40	58	50	100	40	58
75	40	53.5	37.5	75	40	53.5	37.5	75	40	53.5
50	40	49	25	50	40	49	25	50	40	49
37.5	40	46.75	18.75	37.5	40	46.75	18.75	37.5	40	46.75
25	40	44.5	12.5	25	40	44.5	12.5	25	40	44.5
12.5	40	42.25	6.25	12.5	40	42.5	6.25	12.5	40	42.5
First Chiller Operating, Second Chiller Bypassing (First chiller providing 40 F water; mixture temperature increasing above 40 F)										
50	49	58	50	100	40	58	0	0	58	58
37.5	46.75	53.5	37.5	75	40	53.5	0	0	53.5	53.5
25	44.5	49	25	50	40	49	0	0	49	49
12.5	42.25	44.5	12.5	25	40	44.5	0	0	44.5	44.5
First Chiller Operating, Second Chiller Bypassing (First chiller overchilling, mixture temperature at 40 F)										
50	40	49	50	100	31	49	0	0	49	49
37.5	40	48.75	37.5	75	33.25	46.75	0	0	46.75	46.25
25	40	44.5	25	50	35.5	44.5	0	0	44.25	44.25
12.5	40	42.25	12.5	25	37.75	42.25	0	0	42.25	42.25

than multiple-pass. Thus, single-pass evaporators for a particular leaving chilled water temperature require either lower evaporator temperature and increased horsepower or larger heat transfer surface and, therefore, greater initial investment.

In general, the greater the chilled water range, the more economically attractive a two-pass chiller becomes by reducing the horsepower gap between parallel and series chillers.

The designer of a series arrangement faces the problem of selecting identical or unequal chillers for the leading and follower unit. Equally sized chillers are desirable from a maintenance and logistical point of view, but, either case involves particular considerations in terms of performance and convenience.

Figures 3-3 and 3-4 indicate some of the parameters that affect chiller selection.

Loading the leading and follower unit properly in a series arrangement is of primary importance. Initial load balancing is required for optimum performance of chillers in series. When refrigeraion load is reduced or when one

of the machines breaks down, the other continues to handle its share of the load. Ordinarily, the leading machine assumes more than 50% of the load because of

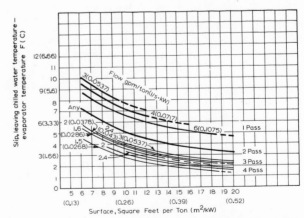

Courtesy York Division of Borg-Warner Corp.

Figure 3-3. Typical chiller performance. (Curves based on 16 ft length, 0.56″ I.D. pipes; 10 ft/sec velocity. Dashed curves indicate a velocity of less than 200 ft/min.)

Table 3-4
Ideal Performance of Three Chillers in Parallel
[Design Chilled Water Supply 40 F (4.4 C), Design Chilled Water Return 58 F (14.4 C)—18 F Temperature Differential]

	System			Unit No. 1				Unit No. 2				Unit No. 3			
% Load	Supply Temp., F	Return Temp., F	% Load of System	% Load of Unit	Supply Temp., F	Return Temp., F	% Load of System	% Load of Unit	Supply Temp., F	Return Temp., F	% Load of System	% Load of Unit	Supply Temp., F	Return Temp., F
							**All Chillers Operating**								
100	40	58	33.33	100	40	58	33.33	100	40	58	33.33	100	40	58
75	40	53.5	25	75	40	53.5	25	75	40	53.5	25	75	40	53.5
50	40	49	16.66	50	40	49	16.6	50	40	49	16.66	50	40	49
37.5	40	46.75	12.5	37.5	40	46.75	12.5	37.5	40	46.75	12.5	37.5	40	46.75
25	40	44.50	8.33	25	40	44.50	8.33	25	40	44.50	8.33	25	40	44.50
12.5	40	42.25	4.16	12.5	40	42.25	4.16	12.5	40	42.25	4.16	12.5	40	42.25
First Two Chillers Operating, Third Chiller Bypassing (First two chillers providing 40 F water; mixture temperature increasing above 40 F)															
66.66	46	58	33.33	100	40	58	33.33	100	40	58	0	0	58	58
50	44.5	53.5	25	75	40	53.5	25	75	40	53.5	0	0	53.5	53.5
37.5	43.37	50.12	18.75	56.25	40	50.12	18.75	56.25	40	50.12	0	0	50.12	50.12
25	42.25	46.75	12.5	37.5	40	46.75	12.5	37.5	40	46.75	0	0	46.75	46.75
12.5	41.12	43.37	6.25	18.75	40	43.37	6.25	18.75	40	43.37	0	0	43.37	43.37
First Two Chillers Operating, Third Chiller Bypassing (First two chillers overchilling; mixture temperature at 40 F)															
66.66	40	52	33.33	100	34.02	52	33.33	100	34	52	0	0	52	52
50	40	49	25	75	35.52	49	25	75	35.52	49	0	0	49	49
37.5	40	46.75	18.75	56.25	36.63	46.75	18.75	56.25	36.63	46.75	0	0	46.75	46.75
25	40	44.5	12.5	37.5	37.75	44.5	12.5	37.5	37.75	44.5	0	0	44.5	44.5
12.5	40	42.25	6.25	18.75	38.90	42.25	6.25	18.75	38.9	42.25	0	0	42.25	42.25
First Chiller Operating, Second Two Chillers Bypassing (First chiller overchilling; mixture temperature at 40 F)															
33.33	40		33.33	100	28	46	0	0	46	46	0	0	46	46
25	40	44.5	25	75	31	44.5	0	0	44.5	44.5	0	0	44.5	44.5
12.5	40	42.25	12.5	37.5	35.5	42.25	0	0	42.25	52.25	0	0	42.25	42.25

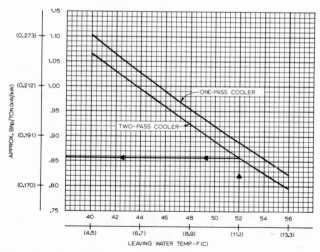

Courtesy Carrier Corp. From Handbook of Air Conditioning System Design, © Mc Graw-Hill Book Co. Used with permission of McGraw-Hill Book Co.

Figure 3-4. Centrifugal compressor power—refrigerant 11.

its higher suction temperature. This implies that the follower machine can be smaller, resulting in cost reductions. From a practical point of view, two identical machines at increased cost are preferred because this presents fewer maintenance problems. In selecting identical machines, and in order to produce identical speeds, it is necessary to divide the load at 55% for the leading machine and 45% for the follower machine as a first approximation, or at percentages that are not appreciably different. Centrifugal compressors are selected at a definite speed to satisfy a fixed design head. Since the leading and follower compressors have different heads, it is possible that different speeds are required if the 55%/45% ratio is exceeded appreciably.

It follows that balancing both chillers at 50% unbalances loads and requires a different head and speed for each machine, precluding the possibility of sequencing the two machines as leading or follower units alternately. Two or more machines may be placed in series provided that a minimum of 6F (3.33C) chilled water temperature drop is maintained across each chiller.

Automatic speed resetting by a controller is difficult, and more so for motor than for turbine drive. It is preferable that any seasonal speed adjustment be manual and that vane control be automatic so that there is less chance of the machines "hunting."

Constant flow through the machines promotes pumping smoothness and control stability, two distinctly desirable qualities, which are discussed in Chapter 7 as constant flow in the primary circuit.

Unfortunately, friction of the two chillers in series is additive, most often resulting in higher heads, although not necessarily so for single pass evaporators. Relatively speaking, there are fewer pipes than in the parallel arrangement, although of a larger diameter since they must handle full flow. The designer can compare costs of piping arrangements, and he must weigh the cost of a back-up unit in either parallel or series arrangement.

Series arrangement piping costs may be reduced to some extent by using four-way valves as Figs. 3-5, 3-6, 3-7 and 3-9 indicate.

Figure 3-5 shows single-pass and two-pass chillers arranged in opposed or side-by-side positions. Four-way valves eliminate two gate valves in each case. The star-delta arrangement of Fig. 3-6 allows either unit A, B or C to become the spare but does not permit unit B in the leading position. Bypassing around each unit is provided. The star-delta arrangement is capable of three combinations using ten gate valves. Four 4-way valves (Fig. 3-7) permit six possible combinations at considerably less cost. However, all six combinations may not be desirable and present difficult control problems.

The arrangement presented in Fig. 3-8 has both chilled water and condenser water in series in counterflow fashion and possesses much flexibility in refrigeration load handling and some variation in chilled water flow by employing more or fewer pumps as the circumstances demand. One chiller is used for low loads, two chillers for normal loads and three chillers for high loads. The third chiller serves as a spare when it is not used for high loads. Two or three pumps can be used, depending on load magnitude and system heat transfer characteristics.

3.4 Arrangement *c* (Fig. 3-I), Chillers in Series, Condensers in Series (condenser water parallel to chilled water)

This arrangement is of theoretical interest only. The leading machine (Unit A) has the lowest condensing

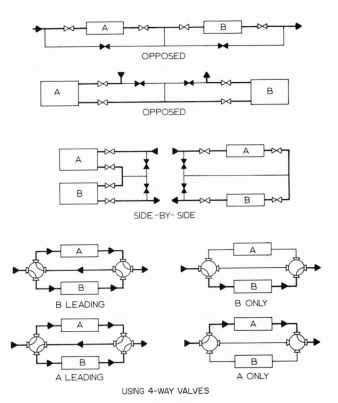

Figure 3-5. Two chillers in series without a spare unit.

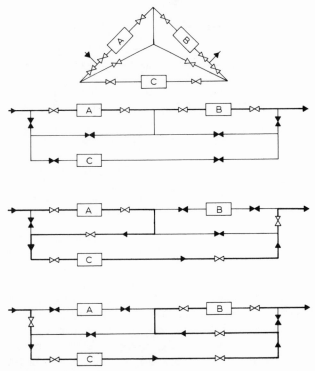

Figure 3-6. Two chillers in series with a spare unit in star-delta arrangement.

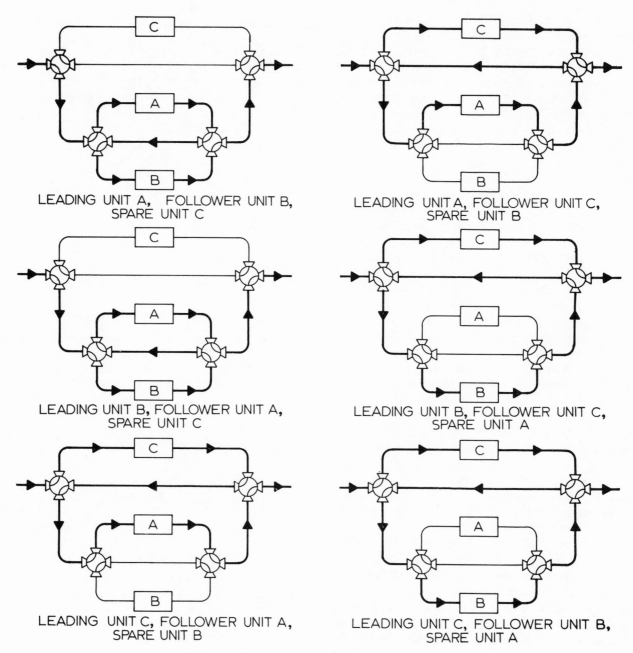

LEADING UNIT A, FOLLOWER UNIT B,
SPARE UNIT C

LEADING UNIT A, FOLLOWER UNIT C,
SPARE UNIT B

LEADING UNIT B, FOLLOWER UNIT A,
SPARE UNIT C

LEADING UNIT B, FOLLOWER UNIT C,
SPARE UNIT A

LEADING UNIT C, FOLLOWER UNIT A,
SPARE UNIT B

LEADING UNIT C, FOLLOWER UNIT B,
SPARE UNIT A

Figure 3-7. Two chillers in series with a spare unit using 4-way valves.

temperature and highest evaporator temperature. The follower (Unit B) has the highest condensing and the lowest evaporator temperatures. The lift (difference between condensing and evaporating temperatures) of the two units is distinctly diverse and, therefore, the heads of the two compressors will be quite opposed, creating problems of speed and control, and generally serious application problems. This arrangement cannot be recommended because it imposes severe design limitations.

3.5 Arrangement *d* (Fig. 3-l), Chillers in Series, Condensers in Series (condenser water in counterflow to chilled water)

By following the counterflow principle, this arrangement creates roughly equal lift (condensing temp. —

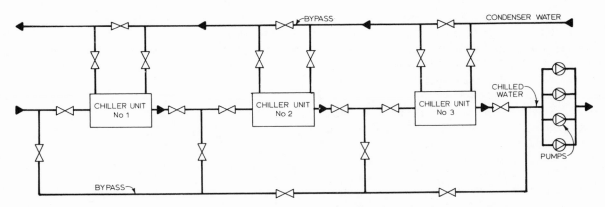

Figure 3-8. System load with flexibility in load handling and variation in chilled water flow possible because of use of multiple pumps.

evaporator temp.) for both units, matching the highest condensing and evaporator temperatures in the leading machine (Unit A) to the lowest condensing and evaporator temperatures in the follower machine (Unit B). Compressor head requirements are equalized at an equal load split.

Condenser series flow can be utilized at great advantage where there is a dearth of condenser water and its temperature is low, as usually is the case with well water. A large condenser water temperature rise is practically mandatory.

A valved bypass around each chiller and condenser is required for repairs and cleaning. As in any series arrangement, larger valves and pipes are required. The follower condenser and evaporator must be sized to do a leading unit job if exchange of duties between units is contemplated.

3.6 Arrangement *e*, Series-Parallel

The classic series-parallel arrangement is represented by the diagrams of Fig. 3-9. A tonnage of 25,000 is being handled in this fashion for the Capitol Building in Washington, D.C.

This arrangement is used extensively because it combines the most desirable features of both series and parallel arrangements, flexibility and economy, plus its own advantage of a multiplicity of components and the implied potential of combining these components to match loads and obtain maximum efficiency.

Some disadvantages of the parallel arrangement also appear here. Constant system flow at partial system refrigeration load signifies circulating water through inactive chillers, requiring either overchilling or acceptance of water temperatures higher than design.

The overchilling operation means that at 50% load the deactivation sequence is $A_1 - B_1$ or $A_2 - B_2$. Overchil-

ling may be eliminated by deactivating $A_1 - A_2$ first with B_1 and B_2 still on the line, handling their share of the load.

An alternate solution, reducing flow through the chillers at partial load, may cause pump head and water distribution problems. Controlling the operation of the units becomes more complex in the sense that ideal performance would require identical performance from corresponding units in each series branch. Whereas one single branch may not be particularly difficult to control, several series branches acting in concert demand more precision on the part of control and instrumentation systems. Introducing or withdrawing load from

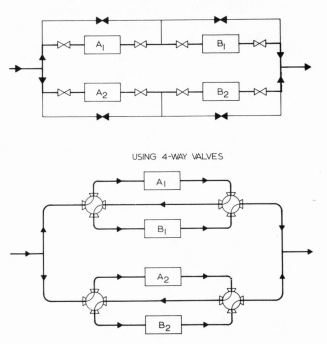

USING 4-WAY VALVES

Figure 3-9. Four chillers in series-parallel arrangement without a spare.

the system requires engagement or disengagement of refrigeration units in a particular pattern to maintain system design parameters with the least disturbance or inconvenience possible. A thermodynamic and operational analysis should provide the answer to the deactivation procedure for different control schemes. These are some of the questions that the designer must consider.

Like the parallel, the parallel-series arrangement can accommodate additional series branches to match any capacity requirement, if the main pipes are sized adequately for the extra capacity. Oversizing of the main pipes must be provided if future growth is anticipated. The series-parallel arrangement can fit almost any type of refrigeration machine, be it a leading or follower unit, and can accommodate all practical drives.

The designer is in a position to use a number of design tools to obtain the best chiller selection, including load factor, size and type of machine that will do high or low level cooling, primary pump arrangement, machine increments, flow quantities, water temperature rise, unit operational sequence, condenser water flow and condenser pump arrangement, cycle efficiency throughout the load range, component matching, etc.

Figures 3-10 through 3-15 illustrate some arrangements for a hypothetical 10,750-ton (37 840 kW) refrigeration plant. The location of the pumps as shown is entirely incidental and is only for illustrative purposes. Although the arrangements are possible, they are not necessarily advantageous unless shown to be so by analysis. No attempt has been made to discover the superiority or inferiority of any of these arrangements. Figure 3-10 shows four 2670-ton (9398 kW), electrically driven centrifugal machines installed in three phases. Phase 1 consists of two 2670-ton machines in parallel, representing 50% of total capacity, and two chilled water pumps at 6650 gpm (420 l/s), or each pump handling 50% of total flow. Distribution network flow is variable and the mechanical room return loop has constant flow. The purpose of the automatic bypass is to return the water

not needed in the distribution network to the machines. Loads below 50% are carried by Phase 1 machines working at part load. Loads above 50% are carried sequentially

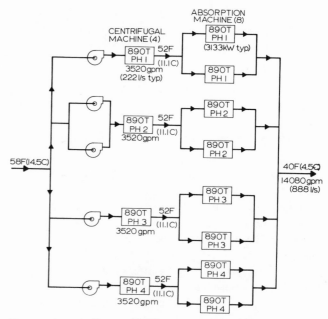

Figure 3-11. Combination centrifugal-absorption "piggyback" arrangement. Noncondensing steam turbine drive.

by phase 2 and 3 machines. This arrangement is responsive to load changes and is simple to operate. Nonelectric drives are possible, provided they prove economically feasible and compatible with the cycle.

In figure 3-11 each of four sets is composed of one 890-ton (3122 kW) centrifugal machine driven by a noncondensing turbine and two 890-ton (3122 kW) absorption machines. The exhaust from each turbine, with a steam rate of approximately 40 lb/ton-hr (1.43 × 10⁻³ kg/Ws) is the supply steam to the corresponding absorption machines operating at a steam rate approximately half that of the turbine. The steam rate of the turbine matches that of the two absorption machines fairly constantly throughout the entire load range down to fairly low loads. Each set represents 25% of capacity and installation can be completed in four phases. One of the sets shown has two pumps to illustrate the possibility of installing two pumps per set to match absorption machine capacity. Chilled water flow in the system is variable and condenser water flow must correspond proportionately to the number of machines in operation. If full chilled water flow in the system is mandated by project requirements, then water going through the active sets must be overchilled. Exhaust steam from the turbines can be mani-

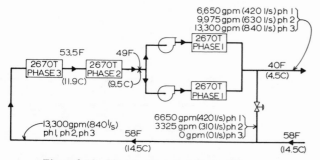

Figure 3-10. Electric drive centrifugal chillers.

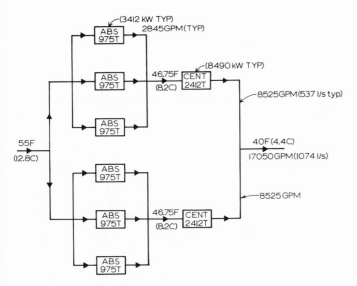

Figure 3-12. Combination centrifugal-absorption "piggyback" arrangement. Noncondensing steam turbine drive (leading units: absorption machines).

folded so that manual controls can combine any centrifugal with any absorption machine. Ordinarily automatic controls would operate each set as an entity.

Figures 3-12 and 3-13 are similar except that Fig. 3-13 represents a three-phase expansion and the leading machines are centrifugal. Both are piggyback arrangements; however, the criteria are different. In Fig. 3-12 the unit sizes are based on an attempt to match the steam output of the turbine to the steam rate of the absorption machines so that the latter act like condensers. At the same time the steam rate of the centrifugal was selected judiciously in order to match the output of the heating

plant boilers, which are sized to satisfy the heating load of the project. A balance must be struck in the steam rates of boilers, centrifugal, and absorption machines so that each matches the other two. A steam rate of 21.6 lb/ton-hr (0.77×10^{-3} kg/Ws) for the centrifugals and 18.02 lb/ton-hr (0.64×10^{-3} kg/Ws) for the absorption machines will bring about this balance. Boiler steam, turbine exhaust steam and condenser water conditions are important parameters in achieving a trial and error balance among components. Each set is composed of one centrifugal and three absorption machines.

Figure 3-13, on the other hand, is similar to Fig. 3-11 in principle but is composed of larger units and can be accomplished in three phases, rather than the four shown in Fig. 3-11. Steam headers can be provided in all arrangements to facilitate interchangeability and manual operation in case of emergency. Condenser pump arrangement can parallel that of the chilled water pumps because it renders each set an operational entity.

In Fig. 3-14, the exhaust from the gas turbines, because of its high excess air, is used as combustion air for high temperature water boilers 400F (204C) which serve project needs, including medium temperature water for powering the absorption machines. It is important to dovetail the performance of the gas turbines to that of the HTW boilers and the absorption machines for maximum heat recovery in order to make this arrangement economically viable, bearing in mind that the absorption machines have a high steam rate and that gas turbines have low thermal efficiency. In Fig. 3-15, the four 2150-ton (7568 kW) centrifugals are driven by gas turbines whose exhaust gases pass through a waste heat boiler which produces enough steam to drive the absorption

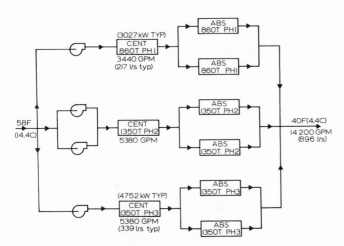

Figure 3-13. Combination centrifugal-absorption "piggyback" arrangement. Noncondensing steam turbine drive (leading units: centrifugal machines).

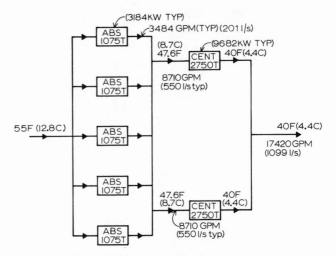

Figure 3-14. Combination absorption-centrifugal "piggyback" arrangement, gas turbine drive.

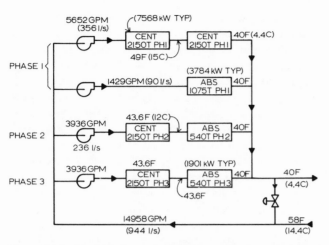

Figure 3-15. Combination centrifugal-absorption "piggyback" arrangement, gas turbine drive.

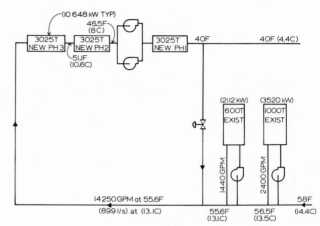

Figure 3-16. Series arrangement of new units incorporating existing units in parallel.

machines. In phase 1, two centrifugals at 2150 tons (7568 kW) each correspond to one 1075-ton (3784 kW) absorption machine using 12 psig (83 kPa) steam. Phases 2 and 3 consist of one 2150-ton (7568 kW) centrifugal and one 540-ton (1901 kW) absorption each. For every absorption machine there is a waste heat boiler. It is worth noting that the 1075-ton (3784 kW) absorption machine is in parallel with its companion centrifugals whereas the two 540-ton (1901 kW) absorption machines are in series and follower units to the 2150-ton (7568 kW) centrifugal. The condenser water pump arrangement parallels that of the chilled water pumps.

Occasionally, existing refrigeration machines may be incorporated into a new plant if they are worth saving and can fit well into the proposed expansion. In Fig. 3-16 one 600-ton and one 1000-ton existing machines are used as precoolers of the returned chilled water by simply diverting a portion of the return flow through them and cooling this flow 10 degrees according to the rating of the machine. Phases 1, 2 and 3 involve the installation in series of three 3025-ton (10 648 kW) centrifugals which come on the line as the load increases and vice versa. System flow is variable and mechanical room loop flow is constant. Each chiller has its own condenser water pump.

3.7 Gas-Side Interconnections

Up to this point only water cycle arrangements were examined. A new concept, that of interconnecting the gas side of centrifugal chillers, might lead to greater flexibility and other operating advantages.[1]

Figure 3-17 indicates a two-unit operation with suction and hot gas interconnections, circulating gas in either direction in order to balance the two units. Arrangement 3-18 represents one machine shut down below 50% plant capacity and the other compressor operating at the added advantage of having two evaporators and two condensers at its disposal. The idle condenser receives 50% of the gas from the operating compressor. Condenser

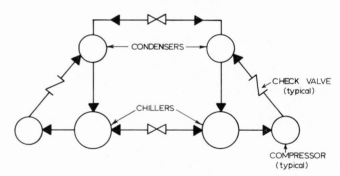

Figure 3-17. Arrangement of two operating, interconnected centrifugals.

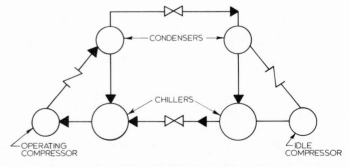

Figure 3-18. Arrangement of interconnected chillers and one idle compressor.

[1] R. E. Japhet, "New Theory Promises Hydraulic Breakthrough," *Actual Specifying Engineer*, January, 1966, pp. 66–69.

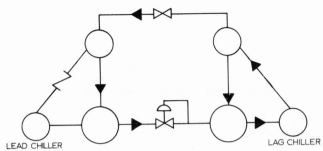

Figure 3-19. Arrangement of two compressors in series with interconnections.

water continues flowing to the idle condenser and condensed liquid drains to the "idle" evaporator, which is subject to suction pressure. The system, although operating on one compressor, possesses twice the normal evaporator and condenser surface, thus increasing the refrigeration effect. Both chillers deliver design temperature chilled water, avoiding the overchilling of parallel chillers. The excess surface has the tendency to decrease horsepower requirements and increase refrigeration capacity.

Figure 3-l9 illustrates two compressors operating in series through an automatic valve which maintains a difference in suction pressures between the two compressors. Undoubtedly, of all three arrangements, this must be the most difficult to apply and control without offering adequate compensatory benefits.

Generally speaking, the main advantage of gas side connections is the elimination of overchilling and better operating characteristics at part load. One difficulty that might be encountered is that in case of loss of refrigerant, both machines would be lost. Also, hydraulic, control, and service problems might occur that are not predictable because there is no operating experience with these types of systems. These arrangements remain theoretical possibilities. As of this moment, no manufacturer has packaged any of the arrangements because all research and development has been predicated on water side connections. A consultant who specifies such an arrangement would have to field-erect, control, and instrument a prototype.

3.8 High and Low Level Cooling in a Centrifugal-Absorption "Piggyback" Arrangement

What positions should the centrifugal and absorption machines occupy in a series arrangement? Which should be in the leading position? This question is of some practical importance.

Until very recently, the piggyback arrangement constituted a principal means of reducing steam consumption and operating costs throughout the load range. The combined steam rate is lower than the individual steam rates of the absorption and steam turbine-driven centrifugal.

There is still a place for the series arrangement of absorption and centrifugal machines under advantageous circumstances, although the question of placing the machines in the proper position is not as meaningful because the attractiveness of the piggyback arrangement has been offset to a great extent by significant advances in absorption machine design. The oncoming of solution control managed to reduce the steam rate at partial loads. However, one manufacturer still adheres to steam throttling control, which has a rising steam rate characteristic at low loads. Another fairly recent development is the two-stage absorption machine, capable of using steam pressures up to 144 psig (992 kPa) and an average steam rate of roughly 11 lb/ton-hr (0.39×10^{-3} kg/Ws). Both of these improvements stiffen the competition against the series arrangement. The distinctly lower steam rate of the piggyback arrangement as practiced in the past is no longer superior to the steam rates of some of the present day absorption machines either in the entire load range or in the 25% and lower range. The piggyback system, in general, has a good steam rate down to about 25% partial load. Depending on selection of components, at 25% the steam rate begins to rise until at l0% load it can vary between 23–30 lb/ton-hr (0.82-l.07 kg/Ws). Even though the new absorption machines exhibit a favorable steam rate at lower loads, this in itself may not be an important criterion, if the nature of the load is such that partial load operation is expected to be minimal.

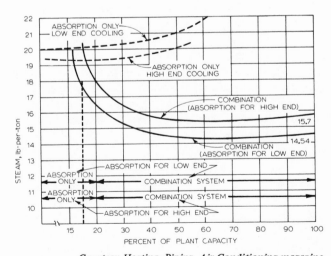

Courtesy Heating, Piping, Air Conditioning magazine.

Figure 3-20. Comparative steam rates for a combination absorption-centrifugal arrangement, with absorption machines doing high- and low-level cooling.

Ideal proportioning of total load between centrifugal and absorption machines results in no waste of steam at any load and a steam rate that is competitive at low loads. The controls must be able to reassign load as required to maintain an operation where steam is not wanted. The fact that the newer absorption machines have a lower steam rate requires that the load balance be shifted so that the absorption machine tonnage is increased while that of the centrifugal machines is decreased. Also, a specific load balance is required depending on the location of each machine in the cycle. When centrifugal machines do high-level cooling, they can assume 5-10% more of system capacity than is the case when they do low-level cooling.

Figure 3-20 indicates that it is far more desirable to have the absorption machines do low-level cooling as far as steam consumption is concerned.[2] The difference in steam rate is about 1.1 lb/ton-hr (0.04 $\times$ 10^{-3} kg/Ws) as number substantial enough to receive consideration. On the other hand, absorption machines doing low-level cooling require additional evaporator heat transfer surface at additional cost.

There is really no set rule governing the position of each machine, although practice usually has the centrifugal doing high-level duty. Each situation must be evaluated on an economic basis and on its own merits, both from the initial as well as the operating cost point of view.

In order to optimize central plant design, the difference between supply and return chilled water temperatures is constantly being increased, so that today differentials of 22-24F (12.2-13.3C) are not uncommon. In order to make such high differential temperatures practical, a 38F (3.3C) leaving chilled water temperature is highly desirable, thus indicating a centrifugal for low-level cooling. The lowest practical water temperature for an absorption unit is 40F (4.5C).

Occasionally, the nature of the refrigeration load may highlight the position of the centrifugal. A refrigeration load which is not expected to dip below 45% could dictate that the centrifugal do low-level cooling at constant load and speed, while the load modulation be carried out by the absorption machine.

Besides initial and operating costs and load factor, other considerations may have to be taken into account such as length of cooling season, complexity of controls for a particular arrangement and the possibility that any wasted steam which cannot be absorbed in the cycle can be effectively used in process or for heating domestic water.

---

[2]Maurice J. Wilson, "Combination Centrifugal-Absorption Systems: Which Units Should Do High, Low End Cooling?" *Heating, Piping & Air Conditioning,* August, 1968, pp. 78-80.

BIBLIOGRAPHY

Chiller Arrangement

Carrier Corp. special study, prepared by Randal J. Scheib and Harland E. Rex, *"Multiple Centrifugal Water Chillers,"* 1971.

Chun, Victor. "Combination Centrifugal/Absorption Cooling May Save Dollars," *Heating, Piping & Air Conditioning,* (March, 1961), 108-109.

Combination Centrifugal-Absorption Systems. York Corp., Application Data, Form 155.20-AP, 1964.

Dorwart, Richard A. "For 400 Tons and Up—Consider Multiple Chilling Units," *Air Conditioning, Heating and Ventilating,* (February, 1962), 60-64.

Multiple Chiller Applications. York Corp, Sales Application Manual, 1-6.

Ramsey, Melvin A. "Chillers in Series," *Building Systems Design,* (January, 1971), 21-24.

Stewart, L. E. "Pickaback Cooling Trims System Costs," *Power,* (June, 1960), 67-69.

Stevenson, F. F. "How to Combine Centrifugal and Absorption Units for Low Operating Costs," *Heating, Piping & Air Conditioning,* (December, 1962), 123-127.

——. "How to Design Centrifugal-Absorption Systems for Low Operating Costs," *Heating, Piping & Air Conditioning,* (January, 1963), 160-166.

——. "How to Estimate Operating Costs for Centrifugal-Absorption Systems," *Heating, Piping & Air Conditioning,* (March, 1963), 126-130.

——. "How to Control Centrifugal-Absorption System,' *Heating, Piping & Air Conditioning,* (July, 1963), 122-127.

Tanzer, E. D. "Comparing Refrigeration Systems," *Chemical Engineering,* (June 10, 1963), 215-220.

Wilson, Maurice J. "Relative Economics of District Vs. Decentralized, Absorption Vs. Combination," *Heating, Piping & Air Conditioning,* (September, 1968), 113-116.

Chillers

Barker, Perley K. "Let's Reassess Screw Compressors," *Power,* (April, 1974), 76-78.

Briley, George C. "Refrigerate with Waste Heat," *ASHRAE Journal,* (December, 1976), 21-25.

Cahill, J. A. "Selecting Centrifugal Compressors for Maximum Efficiency," *Building Systems Engineering,* (August/September, 1974), 13-15.

Centrifugal Compressor System Depression. York Corp., Application Manual SA-3.1(1068), no date.

Cooper, Kenneth W. and Richard A. Erth. "Centrifugal Water Chilling Systems: Focus on Off-design Performance," *Heating, Piping, Air Conditioning,* (January, 1978), 63-67.

Hicks, Tyler. "A Special Report, Refrigeration," *Power*, (October, 1955), 75-94.

Holbay, Nick. "Energy Conservation Techniques for Centrifugal Chillers," *Heating, Piping, Air Conditioning*, (May, 1976), 75-77.

Hughes, Paul L. "How to Apply Motive Power to Open Centrifugal Chillers," *Heating, Piping, Air Conditioning*, (April, 1968), 97-102.

Morabito, Bruno P. "Why Specify Chiller Tube Surface Area?" *Heating, Piping, Air Conditioning*, (September, 1971), 88-92.

Morabito, Bruno. "Refrigeration Machinery for Today's Buildings," *Heating, Piping, Air Conditioning*, (June, 1973), 63-72.

Polimeros, George. "There's Energy for Use in Refrigeration Waste Heat," *Specifying Engineer*, (June, 1979), 96-103; (November, 1979), 122-126; (January, 1980), 74-78 (Series of three papers).

Spencer, Elliot. "New Development in Steam Vacuum Refrigeration," *ASHRAE Journal*, (November, 1961), 59-65.

Staroselsky, Naum and Lawrence Ladin. "Improved Surge Control for Centrifugal Compressors," *Chemical Engineering*, (May 21, 1979), 175-184.

Stillson, Timothy. "Helical Screw Compressor Applications for Energy Conservation," *ASHRAE Transactions*, 1977, Part 1, 185.

Soumerai, Henri. "Large Screw Compressors for Refrigeration," *ASHRAE Journal*, (March, 1967), 38-46.

CHAPTER **4**

Power Cycles

This chapter deals mainly with turbine applications, illustrative plant cycles, refrigerant and steam condensers, heat balance, bottoming organic fluid cycles and solid waste as an energy source.

Steam turbines are the workhorses of central and industrial plants; hence, detailed knowledge of their uses is fundamental for central plant design. The various plant cycles referred to herein are one of a kind, as are practically all central and industrial plants, but they are illustrative of the variety of components and their combinations in central plants. Central plants combine such diverse functions as generation of chilled and hot water, district and process steam, power by steam or gas turbines or internal combustion engines. Other central plant functions encompass solid waste disposal, heat recovery and incineration. Large energy parks in the future will also include sewage disposal and water management.

Refrigerant condensers are examined as components of large chillers and large water and power users. Steam condensers, on the other hand, are an integral part of the steam turbine and play an important role in the cycle heat balance. Most of the illustrations are rudimentary and many cycle components are missing. The intent is to demonstrate the main cycle functions.

4.1 Steam and Combined (Gas-Steam) Cycles

There are an infinite number of methods for meeting steam and electrical loads. Some recognized and basic methods are discussed on the following pages. In the application of these methods, there are combinations that have yet to be devised and some others that have been used successfully in process work but which, because they are of limited interest and outside normal central plant practice, are not discussed here.

For instance, storage batteries not practical for A.C. systems are not covered. Also not discussed is purchasing electrical energy from neighboring systems which, when combined with self-generation, is frowned upon by utility companies and, where allowed, is usually conditioned on heavy demand charges that render it useless. However, federal energy policies are changing rapidly and, in the not too distant future, exchange of electrical energy between utilities and central energy plants, on federally regulated terms, will be a common occurrence.

Condensing turbines, by throwing away all the latent heat of steam, seriously contribute energy waste and, in the smaller capacities, suffer from low engine efficiencies. Internal combustion engine drives are expensive but compensate by not requiring all the costly accoutrements and auxiliaries of the steam cycle. In any case, internal combustion engines, some at full load and almost all at partial loads, exhibit low thermal efficiencies.

One important criterion in industrial power plants, and by inference in central heating and cooling plants, is the balance between process steam load and power demand load, known as the steam power balance or ratio. This ratio is an indication of the build-up of power in comparison to steam and has a direct relationship to unit fuel consumption or heat rate, Btu/kW-hr. An attempt has been made to correlate the steam power ratio to unit fuel consumption as a criterion for equipment selection.[1] In the final analysis, though, a full-fledged study is required that considers capital costs, owning and operating costs, maintenance, fuel availability and a heat balance for all load combinations.

---
[1] Geiringer, Stefan L., "How to Weigh On-site Generation When Designing a New Industrial Plant," *Power*, February, 1970, pp. 33–35.

There are three basic types of combined cycles:

1. *Unfired combined cycle*. Exhaust gases from a gas turbine are fed to a waste heat recovery boiler to generate steam. The smallest approach temperature between gases and superheated steam is of the order of 60 F (133.3C) to 100 F (155.5C). Under such small temperature approach circumstances, steam conditions may vary from 600 to 850 psig (4134-5856 kPa) and 700-900 F (371-482C) for large plants of 30-40 MW and up. The most efficient cycle with the lowest heat rate has an ideal ratio of 70% power produced by the gas turbine to 30% by the steam turbine.

   In cases where all the steam is used to power absorption chillers operating with 12 psig (kPa) steam, the approach temperature can be increased to 250-300 F, resulting in small heat recovery boilers, reducing exhaust gas temperatures to 300-325 F and extracting the most heat practically possible from the gases.

2. *Supplementary fired combined cycle*. Because gas turbine exhaust gases contain a large fraction of excess air, additional fuel can be burned in the duct connecting the gas turbine to the waste heat boiler boosting exhaust gas temperatures to 1200-1400 F (648-760C). This cycle is used primarily in large combined cycle plants, whose steam turbines operate in the range of 900-1250 psig (6201-8625 kPa) and 900-950 F (482-510C). As a result, the percentage of power produced by the steam turbine is increased.

3. *Exhaust fired combined cycle*. The exhaust gases from the gas turbine are used as combustion air for a conventional power boiler. All the excess oxygen is consumed in the boiler firebox. The arrangement is flexible in regards to boiler pressure rating. Steam conditions can vary from a few hundred pounds to a high pressure rating of 2600 psig (17 914 kPa) and 1000 F (537C). This type of boiler ordinarily requires waterwall construction with convection and radiant heat transfer sections for the high pressure boilers. This arrangement is resorted to when a high percentage of steam (70-90%) is used by the steam turbine.

The combined cycle systems are intermediate stages between simple gas turbine and the more complex steam turbine cycles. As more components are added to the simple gas turbine cycle, the arrangement looks more like the steam cycle. For each of the combined cycle types discussed, different boilers and turbines are appropriate.

Specific circumstances must dictate the use of one kind or the other.

Maintenance costs are highest in the case of the unfired cycle and lowest for the exhaust fired cycle. Plant installed costs are higher for the exhaust fired cycle by roughly 35%. Unfired and supplementary fired cycles, on the average, are about the same. The idealized 30-70% of steam to gas turbine participation arrangement is in the high installed cost range. These costs decrease, for the unfired cycle, as the percentage of steam turbine work is increased to 90%.

Quite a few manufacturers package combined cycle plants, which are well-engineered and have low heat rates. A competitive bid ought to establish the advantages of each one.

Some of the main advantages of combined cycles are:

1. High ratio of power output to area of plant space required by equipment.
2. Low heat rates in comparison to other cycles and plants of comparable size. Most combined cycles for industrial applications exhibit heat rates in the 9000-10,000 Btu/kW-hr (2.40-2.51 kW/kW) range, based on the high heating value of fuel. Utility combined cycles, even at low loads, are in the 8200-8600 Btu/kW-hr (2.63-2.93 kW/kW) category. Heat rates are rated at sea level and 59 F (15C) according to the International Standards Organization; they depend on ambient temperature, site elevation, type of fuel; the steam to gas turbine power participation ratio is the key to combined cycle optimization.
3. By equipping either the gas turbine or steam turbine or both with certain devices, it is possible to operate one without the other, avoiding a complete breakdown when one of the units is tripped or is down for repairs. The concept of partial redundancy is reinforced by the fact that most units are manufactured in limited sizes. For larger installations, multiple units are required to accommodate maximum demand, thus providing partial redundancy at partial loads.
4. The use of heat recovery equipment contributes toward decreased thermal discharges to the environment and reduced air pollution because of increased residence time for all combustion products.
5. Pollution is also minimized because heavy oils and coal cannot be used, although the use of pulverized coal is being explored as a possible gas turbine fuel. Heavy oils cause severe internal corrosion and galling.

6. Lead time for delivery of equipment is relatively short.

7. Plant start-up and assumption of full load are relatively quick.

8. Installed costs are low—roughly $200–$250 per installed kW in 1977 dollars. These costs are direct construction costs and do not include engineering, legal, land or interest costs. In addition to gas turbine cycles, some steam turbine combinations are now described. The steam cycle is one of the most developed, and equipment for it is readily available for most basic turbine versions.

4.2 Steam Cycles and Steam Turbine Combinations

4.2.1 *Condensing Turbine (Figs. 4-1 and 4-2)*

This type of plant is rarely used because of its high cost and inefficiency. At least half of the input to the turbine, the latent heat, is lost in the condenser and, secondly, the steam that goes directly from boiler to process does not produce power. On the plus side, this cycle can satisfy a wide variety of process and generation loads in any combination and, therefore, is very flexible.

4.2.2 *Back Pressure Turbine and Load Condenser (Fig. 4-3)*

The back pressure turbine satisfies any electric load demand while the pressure reducing valve (PRV) makes up for any deficiency in process steam when the electric load is not large enough for turbine exhaust flow to satisfy all of the process steam demand. (See Fig. 4-3). The opposite can also happen when electric demand is so large that all exhaust flow cannot be absorbed by the

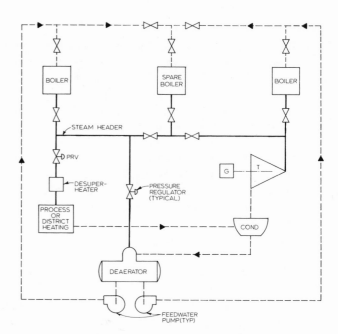

Figure 4-2. Arrangement *b*: Separate boilers for process and generation.

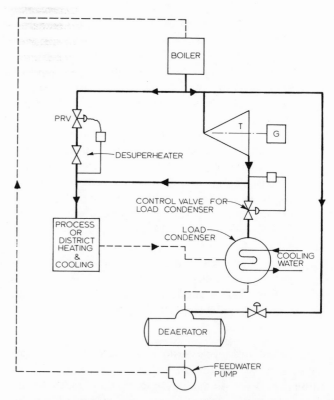

Figure 4-3. Generating and process or district heating and cooling needs are satisfied simultaneously or separately at the expense of efficiency.

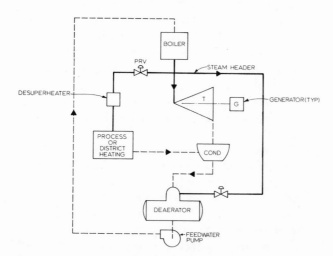

Figure 4-1. Arrangement *a*: Common boiler for process and generation.

process. In such circumstances, the load condenser control valve opens on a high pressure signal and allows the excess steam to bypass to the condenser to be liquefied Since the load condenser handles waste steam, this operation must be kept to a minimum in a well-designed plant. This arrangement is less wasteful than that of a straight condensing turbine. The generated power is small in comparison, since the steam rate of a back pressure turbine is high. Ideally, process steam requirements should match electrical demand. This system also offers flexibility at the expense of efficiency.

4.2.3 *Back Pressure Turbine with Steam Accumulator (Fig. 4-4)*

The accumulator absorbs steam during the demand valleys and releases steam during the peaks, smoothing the demand curve and maintaining steady boiler firing rates through transient conditions. This cycle is of limited use because of the expense and space required for the accumulators. In the past era of field-erected boilers, accumulators had some financial attraction. Currently, package boilers are fairly inexpensive, so an additional boiler may be a reasonable alternative.

It is also possible to use extraction steam to load the accumulator. An accumulator may be a reasonable solution for electric steam boilers for which peak electric demand, if it happens often enough, can become expensive. The accumulator would eliminate these peaks.

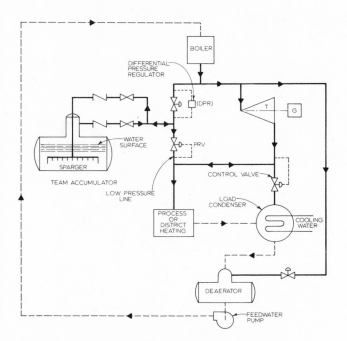

Figure 4-4. Back pressure turbine with steam accumulator.

The accumulator is loaded by injecting steam into the vessel and raising its water content to the saturation temperature and pressure of the steam line. The pressure regulating valve (PRV) admits steam into the low pressure line by maintaining constant downstream pressure. The differential pressure regulator (DPR), on the other hand, passes steam into the accumulator when the PRV closes or during steam draw. The DPR also acts as a PRV, passing steam when the accumulator is not able to supply all the steam demanded during a peak. At that time the DPR setting drops to a predetermined level, so that at times, there is a two-step pressure reduction, *i.e.*, one step at the DPR and a second one at the PRV.

The accumulator arrangement is also used in European practice as a steam booster station for large steam distributing systems.

4.2.4 *Tandem Compound Extraction Turbines (Figs. 4-5 and 4-6)*

These turbine arrangements can control both electrical power generation and process extraction pressure. For large process flows, the arrangement of Fig. 4-6 is preferable because, if electrical power requirements are satisfied by the high pressure machine, the windage losses of the low pressure machine are avoided. Besides, the shafts in Fig. 4-6 can be in tandem or side-by-side with two pinions meshing onto a common gear so that each machine can run at a different and more economical speed. The low pressure turbine can absorb additional steam not required by process, provided there is demand for additional power load. The arrangement of Fig. 4-6 is more expensive than that of Fig. 4-5 because of the two machines. In both arrangements, work is obtained from extraction steam and makes this cycle more efficient than a straight condensing operation.

4.2.5 *Combination Back Pressure Turbine and Condensing Turbine (Figs. 4-7 and 4-8)*

In Arrangement *a*, Fig. 4-7, the condensing turbine is called an exhaust turbine at times because it takes its steam from the process header. In function, it is similar to the low pressure section of a tandem compound extraction turbine. The condensing turbine comes into play when process steam demand is down and power demand is up. Since the two turbines can operate independently of each other, the condensing turbine can be shut down during periods of low electrical demand and high process steam demand. This combination of turbines favors an unbalanced steam power ratio. Thus, when the condensing turbine is shut down, there are no windage losses through the condensing turbine as there are in Fig. 4-6, the tandem compound extraction turbine. Arrangement

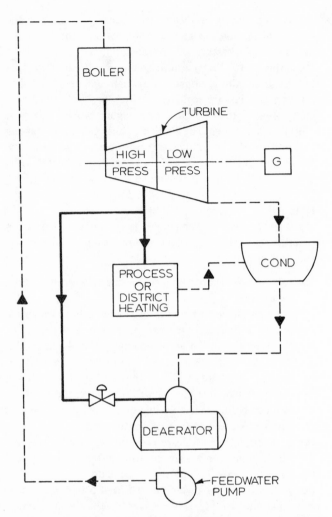

Figure 4-5. Single cylinder with controlled extraction pressure.

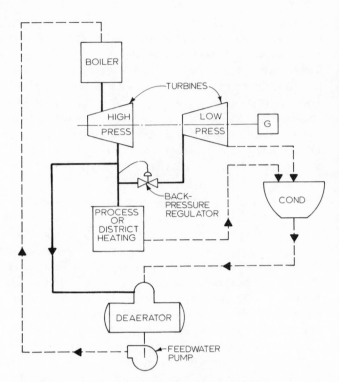

Figure 4-6. Two cyliners with controlled extraction pressure.

4-6, nevertheless, is superior in that the condensing turbine is always running and ready to receive an increase in electrical load, whereas the condensing turbine of Fig. 4-7 must be warmed and turned on. However, when the condensing turbine in Fig. 4-7 is down, some of the accompanying auxiliaries are disengaged, contributing toward a lower energy use level. The PRV is a form of back-up to keep essential services going. The PRV can also supply additional process steam, above the capacity of the back pressure turbine. Arrangement *a* of Fig. 4-7 is advantageous when a high flow operation prevails for most of the year, *i.e.*, at optimum efficiency conditions.

Arrangement *b* (Fig. 4-8) avoids the warm-up problem and is flexible, but includes an expensive high pressure turbine and is saddled with partial flow for most of the year through the condensing turbine, at the expense of efficiency. Annual steam consumption is higher in Fig. 4-8 than in Fig. 4-7.

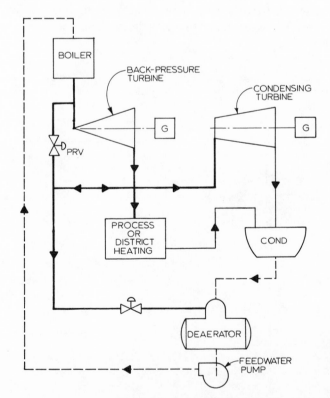

Figure 4-7. Arrangement *a*.

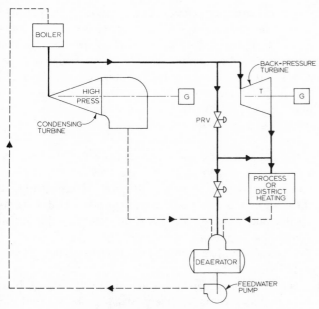

Figure 4-8. Arrangement *b*.

4.2.6 *Mixed Pressure Turbine (Fig. 4-9)*

A mixed pressure turbine may be of the back pressure or condensing type. In addition to high pressure steam, this turbine is capable of receiving low pressure steam, either from a process or plant surplus, at a point in the turbine, corresponding in pressure to that of the low pressure steam.

The pressure controller permits the admission valve in the high pressure steam line to open only after all available low pressure steam is used first, on condition that process steam pressure is maintained. This arrangement contributes to energy conservation.

4.2.7 *Topping Turbine (Fig. 4-10)*

The need for additional power in an existing installation has given birth to the term "topping." In other words, a new back-pressure turbine and high pressure boiler are added or the existing turbine is topped by a new turbine whose exhaust pressure is equal to the existing plant steam pressure. The old boilers can now serve as emergency boilers. Depending on circumstances, additional power may be obtained by internal combustion engines or purchased power.

4.2.8 *Double Extraction, Condensing Turbine (Fig. 4-11)*

This cycle is fully analyzed in Section 4.7.2, "Steam Surface Condensers and Heat Balance and Heat Rate." Any fuel may be used and the cycle is flexible in handling varying electric and process steam loads and has fair efficiency throughout its load range. It is more efficient

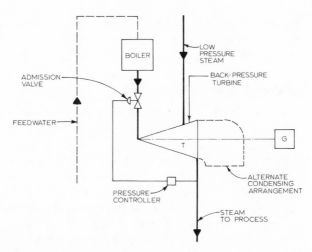

Figure 4-9. Waste low-pressure steam is used first before boiler high pressure is introduced. This arrangement conserves energy and is used when substantial waste steam is available.

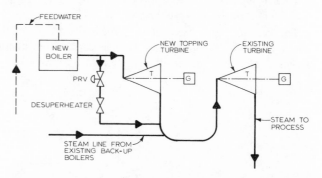

Figure 4-10. A new boiler and new back pressure (topping) turbine satisfy additional power requirements while feeding an existing turbine with back pressure steam.

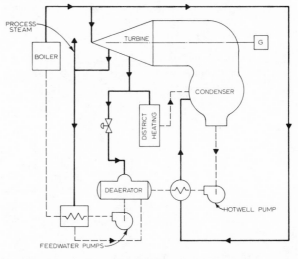

Figure 4-11. Electric generation is combined with district heating.

than the arrangement of Fig. 4-1. This cycle is used where large generating capacities are required, with district heating a convenient by-product.

4.3 Combined and Other Cycles

The term "combined cycle" has assumed the meaning of combining gas turbines with steam turbines. The utility companies use the simple gas turbine, usually 30 to 80 MW, for peaking power service. For some base load applications, the combined cycle is used widely; the use of exhaust gases in waste heat boilers, with or without supplementary firing, generates steam to drive steam turbogenerators. However, steam or gas turbines can be combined with a number of other machines, in what by extension can also be combined cycles. Some of the following examples illustrate these statements.

4.3.1 *Combined Cycle. Case 1: Exhaust Gas Is Equal to Boiler Combustion Air Needs (Fig. 4-12)*

The gas turbine is selected to furnish the exact maximum amount of combustion air needed by the steam boiler. The gas turbine discharges to atmosphere only

when the combustion gases are not needed, an event that is kept to a minimum, or in emergency cases. As a consequence of having the electric load shared between the gas and steam turbines, the sizes of the boiler and the steam turbine are reduced, in comparison to Fig. 4-11, where all the electric load is satisfied by the steam turbine. Steam flow to the turbine, steam piping and component size are reduced, although not the number of components.

The boiler is not a waste heat boiler but a steam boiler adapted to use exhaust gases as combustion air. The turbine gases contain between 300–600% excess air and, at their high temperatures, are exceedingly useful as combustion air. This arrangement, depending on particular conditions, conserves energy in the neighborhood of 5–10% less fuel than is required in the arrangement of Fig. 4-11.

The forced draft fan allows the boiler and steam turbine to operate as an independent steam cycle, without aid from the gas turbine. As a conservation measure, an alternate service may be devised for the forced draft fan, *e.g.*, that of furnishing cooled inlet air to the compressor with

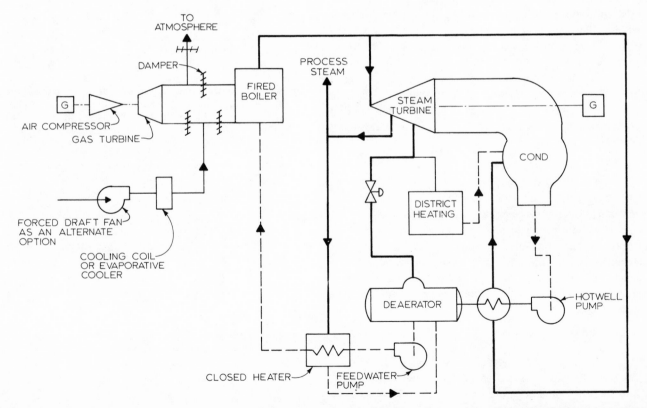

Figure 4-12. The electric load is shared by the gas and steam turbines. The steam turbine provides steam for district heating. Turbine exhaust gases are used as boiler combustion air. The boiler can operate without the gas turbine by using its own forced draft fan.

the aid of a cooling coil or evaporative cooler. Gas turbine efficiency will thus be improved and its output increased by as much as twice the fan input. On a reduction in load, it is best to shut down the gas turbine when the output of the steam turbine is reached. The gas turbine contributes to a fast start from a cold position. Fuel economy is better than that of a condensing turbine, shown in Figs. 4-1 and 4-11.

4.3.2 *Combined Cycle. Case 2: Exhaust Gas Exceeds Boiler Combustion Air Needs (Fig. 4-13)*

This cycle is attractive for fairly constant full loads. The major electrical load is now borne by the gas turbine. The size of the boiler, steam turbine and steam cycle, in general, are further reduced from the sizes of those depicted in Fig. 4-12. These gains are offset by sizable stack losses, as the gases are in excess of boiler needs. The dampers are so arranged as to facilitate discharge to atmosphere. Not all gas turbine gases go out the stack, as some portions are used in the boiler economizer. A back pressure steam turbine is preferable as it reduces capital investment in condensers and auxiliaries. Since make-up is reduced to the steam cycle, it may be cost-saving to investigate the application of a deaerating condenser, instead of both a deaerator and a condenser.

4.4 Gas Turbine with Waste Heat Boiler (Fig. 4-14)

The basic cycle consists of a gas turbine, waste heat boiler and a process or district heating load. This cycle is a full-load cycle. At part loads, the efficiency of the gas turbine goes down dramatically. Two conditions must prevail: first, process steam demand must more or less match the waste heat boiler capacity, based on full gas turbine flow and, secondly, electric demand must be as close to the plate rating of the gas turbine as possible, and constant. Because the heat rate of this arrangement is superior to that of all other arrangements at full load, this cycle merits consideration under all circumstances.

There are various alternate additions to the cycle, as indicated in Fig. 4-14. The basic addition concerns a turbogenerator. The combined cycle in its purest form con-

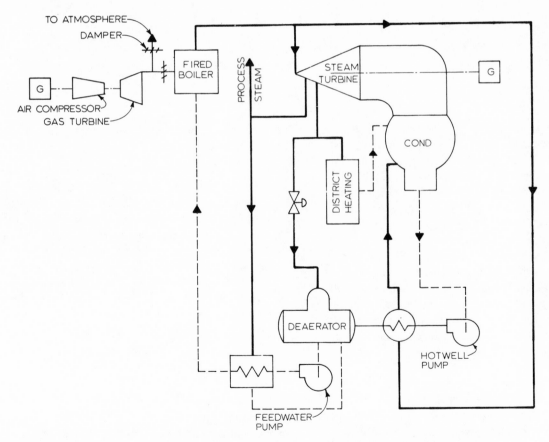

Figure 4-13. In this cycle the gas turbine predominates by assuming a major base electric load at its highest efficiency. The steam turbine provides steam for district heating and serves the variable part of the electric load.

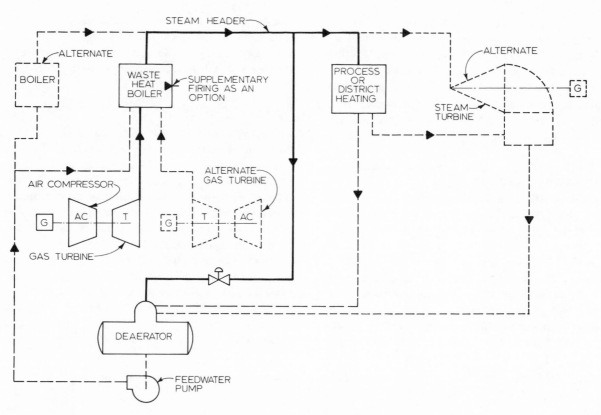

Figure 4-14. The gas turbine combined with a waste heat boiler provides excellent plant efficiency. This cycle can be applied advantageously at constant electric and process loads.

sists of a gas turbine, waste heat boiler and steam turbine with or without steam extraction, with or without waste heat boiler auxiliary firing. This basic cycle without regeneration or supplementary firing is used extensively by the utility companies. The ratio of gas turbine to steam turbine power output is roughly two to one. Another alternative is the addition of a fired boiler to effectuate a high steam power ratio, if it is desired.

A third alternative solution is the installation of two gas turbines in the rough power ratio to the steam turbine of two to one, at increased capital cost, to minimize the disadvantages of a single gas turbine. The premise is that two gas turbines can run closer to rated conditions and can offer greater and more sustained efficiency over a wider range than a single gas turbine, rendering the cycle more useful at other than full load conditions.

In some instances, large diesel engines may take the place of gas turbines. However, despite their dependability, they are used less for larger plants because of their complex auxiliary systems, jacket cooling requirements, large space requirements, high total plant costs and limitation in size.

4.5 Gas or Diesel Engines with Waste Heat Boiler and Steam Turbine (Fig. 4-15)

This cycle is a variation in equipment arrangement of the cycle represented in Fig. 4-14. Depending on the steam-power ratio, either the steam turbine or the internal combustion engines can supply the base load. In most instances, one can expect the steam turbine to handle the base load. This cycle is very flexible but a minimum steam and electrical load is a prerequisite. Fuel economy is far greater than can be expected of the arrangement in Fig. 4-14. It is entirely possible to substitute a chiller for the generator to be driven by the back-pressure turbine, depending on the need for process steam during the summer.

4.6 Heating and Cooling Cycles

The steam and combined cycles we have dealt with up to this point relate only to heating and power generation in well-known combinations that developed historically over many decades. Cooling cycles are not distinguished by the same degree of pattern development and clarity.

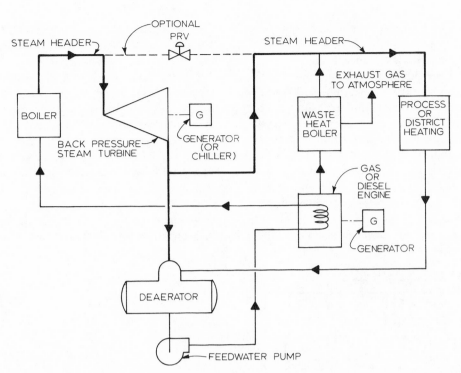

Figure 4-15. Either the steam turbine or diesel engine can handle the base electric load. Flexibility is the main characteristic of this cycle but a minimum electric and steam load is a requirement.

Large central plant cooling cycles developed with a certain degree of experimentation over the past 25 years. The choices are still not clear-cut and require analysis each time. Growth of large refrigeration plants coincided with a number of other developments; abundantly available natural gas, at least for a time, thrust total energy plants, gas turbines and absorption machines to the forefront. Additional factors include the displacement by package boilers of field-erected boilers due to high labor costs, inflation and the gradually darkening fuel picture punctuated by fuel unavailability and spiraling costs.

The development of large refrigeration systems for air conditioning was uneven; for the most part, limited to educational and medical complexes and some shopping centers and office concentrations.

The extreme competitiveness of the package boiler manufacturers brought about a deterioration in product quality, entailing reduced longevity and increased maintenance, to the point where one thinks of them as extremely expendable. On the other hand, refrigeration machine design has progressed immensely, including simplification, due to the fact that only three major manufacturers exist. The larger machines are more dependable and compact than ever, incorporating more useful subcomponents, all geared to serve the central energy plant. The newly developed two-stage absorption machine has

become a direct competitor of the steam turbine-driven centrifugal and, in some cases, the motor-driven centrifugal.

Also, the central plant is being positively affected by improved electronic and solid state instrumentation and computerization. In the end, we will see more economical and better organized central plants.

The current stage is one of continuing development. It is worthwhile to review some of the past and present arrangements in heating, refrigeration, and generation.

4.6.1 *Heating, Refrigeration and Power: Plant No. 1 (Fig. 4-16)*

This cycle can heat, cool and generate power simultaneously or any one or two of these functions separately or in combination. This ultimate flexibility is accompanied by some inefficiencies. The cycle is typical and not singular in any respect. Exhaust gases from the diesels driving the generators pass through the waste heat boilers. The jacket water pump removes heat from the diesel block and carries it to the waste heat boiler which, by ebullient cooling, removes heat from the jacket water and exhaust gases. Jacket water is at a maximum of 250 F (121C) as it goes into the engine block, where it rises approximately 3 F (1.7C). However, for most diesel engines, jacket water temperatures are limited to 150–190 F

(65.6-87.8C). Gases enter the waste heat boiler at approximately 850 F (454C) and are reduced to about 325 F (162.8C) before being discharged to atmosphere.

Figure 4-16 shows three 3000 kW diesel engines, three 25,000 lb/hr (3.15 kg/s) fire tube boilers, three 8500 lb/hr (1.07 kg/s) waste heat boilers and three 850-ton (2992 kW) absorption machines that serve a school-hospital complex. Steam pressure at 15 psig (103 kPa) is adequate for the absorption machines and the 75×10^6 Btu/hr (219,975 kW) capacity medium temperature 170–230F (76.6-110C) hot water system. When steam demand exceeds the capacity of the waste heat boilers, the auxiliary boilers come on line. Inefficiency exists whenever steam demand is lower than the steam capacity

of the waste heat boilers. Then, steam is passed on to the load condenser where the heat of condensation is carried away to the cooling tower, a sheer waste. A well-designed plant should have as little of it as possible.

In general, this arrangement is conducive to high plant efficiency, if all of the steam produced by the waste heat boilers can be absorbed at all times. A disadvantage of this cycle is the low stream pressure of 15 psig (103 kPa), dictated by the absorption machines, which are normally characterized by a relatively high steam rate. This low steam pressure, in turn, dictates a medium temperature water system, which does not have all the benefits of a high temperature water system. However, there is an inherent benefit in low pressure steam in that boiler exhaust

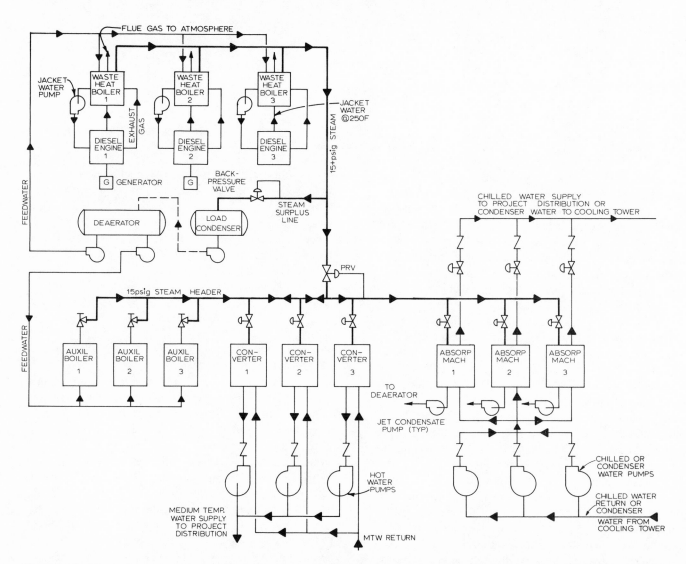

Figure 4-16. Diesel generators with complete heat recovery through waste heat boilers, absorption refrigeration and medium temperature water and auxiliary steam boilers. Mental health complex application.

gases are reduced to the lowest possible temperature, thus promoting maximum heat recovery.

The multiplicity of components in this cycle constitutes both its strength and weakness. Multiple components provide above average flexibility. At the same time, their small size precludes any sizable future expansion.

The basic cycle can be improved by modifications. One would be to use two sets of waste heat boilers; one set to supply low pressure steam to the absorption machines and a second set for high pressure steam for high temperature water. The cost-benefit aspects of such modification must be calculated.

In a second alternate solution, high pressure steam might be used throughout the cycle, *i.e.*, by using high temperature water and steam turbine-driven centrifugals. A bottoming organic fluid cycle might be added to take advantage of the high temperature diesel exhaust gases, thus, reinforcing station electric capability.

4.6.2 *Heating, Refrigeration and Power: Plant No. 2 (Fig. 4-17)*

This is the total energy plant serving the cooperative housing complex, Rochdale Village, in the New York City borough of Queens, with 5840 apartments and 300,000 square feet (27,870 sq m) of shopping area over 170 acres (68.8 ha). It is a highly efficient operation, claiming 60–70% plant efficiency, and is very economical for the apartment tenants. It has no connection to the local electrical utility and, during the total New York City blackouts of 1964 and 1977, it was one of the few places that was bright. Its efficiency is due to 1) the high pressure steam turbines and accompanying low steam rate and 2) the fact that either the absorption machines or the hot water heat exchangers act as steam condensers most of the time. Two diesel generators provide peaking service. The distribution piping serves a dual purpose, heating and

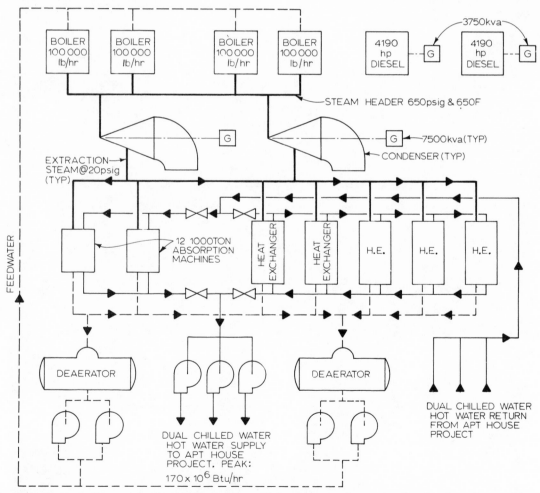

Figure 4-17. High pressure boilers, condensing turbogenerators with low pressure steam extraction for absorption machines and low temperature water heat exchangers. Apartment house complex application.

cooling by simply valving off either the absorption machines or the hot water heat exchangers. One boiler is a complete standby. Steam is fed to the turbines and to domestic hot water generators in each building; and some of it is condensed if it is not needed for refrigeration or heating. The cycle is very simple in concept and operation and is designed to use natural gas or number 6 oil.

4.6.3 *Heating, Refrigeration and Power: Plant No. 3 (Fig. 4-18)*

Fifteen hospitals and a 774 unit housing complex for medical personnel are served by this well-planned central utilities facility, capitalizing on heat recovery and by-product use, such as solid waste and pneumatic air. The capacity of the plant is 51.9 MW, 994,000 lb/hr (125 kg/s) of steam, 20,000 tons (70,400 kW) of refrigeration and 40 tons/day (36.3 mg/day) of refuse disposal. It boasts a 5200 Btu/kW-hr (1.52 kW/kW) heat rate and a low $205 per installed kW.

The heat balance is such that the 15,000-ton (52,800 kW) refrigeration load is primarily served by two 5000-ton, motor driven machines, with one 5000-ton (17,600 kW) steam turbine-driven machine for peaking service and one full standby unit. The 450 psig (3100 kPa) pressure of the boilers is dictated by the corrosive nature of the incinerator exhaust gases, which require that tube surface temperature of the low pressure boilers be kept at 500 F (260C) or above.

Heat recovery is accomplished by six waste heat boilers serving the diesel engines and two serving the incinerators. The diesels are doubly redundant; one for service and the other as full standby. In addition, there is further heat recovery in combustion air preheating and diesel jacket heat recovery in heating all domestic hot water to 140 F (60C). Energy conservation is further pursued by burning unshredded refuse at a controlled minimal rate to keep flyash size and treatment to a minimum. Burning incinerator gases in the low pressure boiler eliminates all odors.

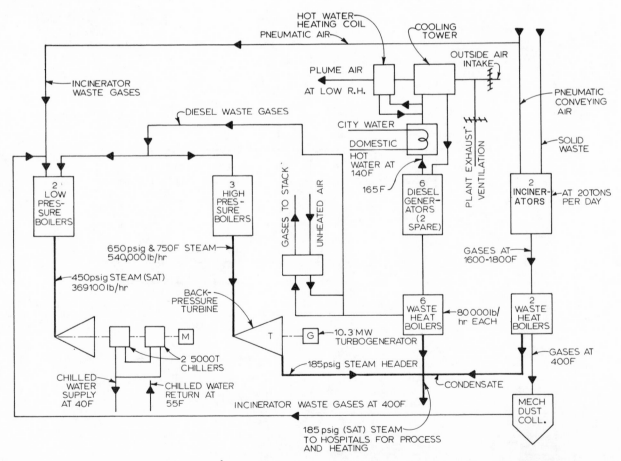

Figure 4-18. Diesels and incinerators with waste heat boilers, high and low pressure boilers with turbine driven chillers and generators, serving the Harvard medical and apartment house complex in the original arrangement.

4.6.4 *Heating, Refrigeration and Power: Plant No. 4 (Fig. 4-19)*

This state-chartered plant has two 360 ton per day (326 mg/day) refuse incinerators with a 215,000 lb/hr (27.1 kg/s) steam capacity and a 14,000 ton district cooling plant serving the district heating and cooling needs of 40 city, state and private buildings with the ability to expand to five such incinerators and 500,000 lb/hr (63 kg/s) steam capacity. It is envisioned that the entire county will deliver solid waste to three compacting centers from whence trailer trucks will deliver the waste to the central plant. This downtown project combines current technology for energy conservation, environmental protection and lower utility charges. The cost of steam and chilled water to plant clients is 25–50% less than it would be if they operated their own facilities.

4.6.5 *Heating and Refrigeration: Plant No. 5 (Fig. 4-20)*

The cascade heater raises the temperature of the return water by mixing the return water, cascading over stainless steel trays, with steam. The water nearly assumes the saturation temperature of the steam. The cascade heater also acts as an expansion drum and deaerator as well and is well-suited for new systems which supply both high pressure steam and high temperature water. The steam turbines, although condensing, achieve an excellent steam rate due to the high pressure steam and low deaerator vacuum. One can expect good equipment performance, simple controls and excellent cycle reliability in using this arrangement. One should notice the independent operation of the high temperature water primary pumps, paucity of equipment and lack of generating equipment, which contributes to the uncomplicated nature of the cycle.

4.6.6 *Heating and Refrigeration: Plant No. 6 (Fig. 4-21)*

The refrigeration load is shared by the centrifugals and the absorption machines. It is assumed that the gas tur-

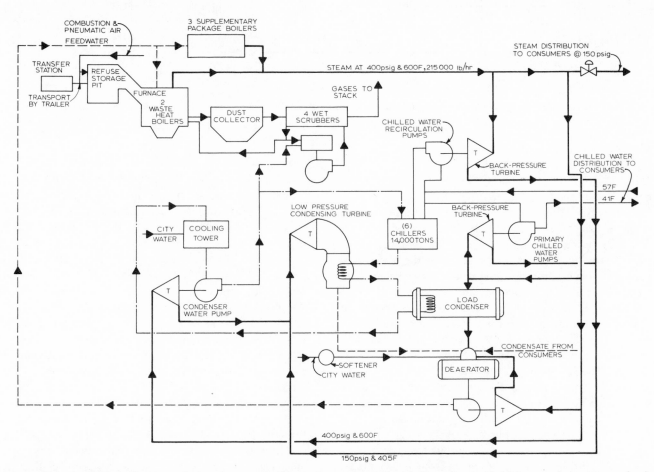

Figure 4-19. At the Nashville, Tenn., incinerator central plant, incinerator waste heat boiler units provide high pressure steam to mechanical drive back pressure steam turbines for condenser and chilled water pumps; exhaust steam from the back pressure turbines is fed into condensing turbines driving the centrifugal chillers.

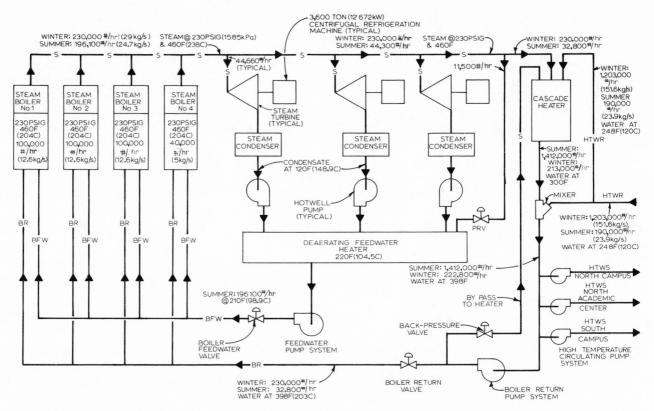

Figure 4-20. High pressure boilers, steam turbine driven centrifugal refrigeration and cascade heaters for high temperature water.

bine-driven centrifugals serve the base load so that turbine exhaust is always fed into the boilers, hence, heat recovery is continuous in this arrangement.

High temperature water is bled into the absorption machine hot water circulating system to provide 280 F (137.7C) water. Thus, water returning to boiler is at 228 F, whether from district heating or the absorption machines. The absorption and centrifugal machines are piped in parallel for condenser water, because of different temperature and flow requirements. Flow is limited to 2 gpm/ton (0.035 l/s kW) for the centrifugals and 3–3.7 gpm/ton (0.053-0.066 l/s kW) for the absorption machines. The heat rate of the gas turbines is 10,000 Btu/hp-hr (3.93 kW/kW) at 100% load and 29,000 at 30% load in almost a straight line. It is apparent that this cycle is terribly inefficient at any load point other than 100% for the gas turbines. Using the absorption machines for a base load is also inefficient because of their high steam requirements. This cycle is advantageous for at least 50% guaranteed refrigeration load. One of its main advantages is the complete absence of electrical drives except for pumps and minor auxiliaries.

4.6.7 *Heating and Referigeration: Plant No. 7 (Fig. 4-22)*

This cycle consists of a combination of absorption and centrifugal refrigeration machines. The centrifugals act as back pressure turbines feeding the absorption machines, whose condensate is pumped to the deaerator. This arrangement results in a low steam rate of 10.68 lb/ton-hr $(0.38 \times 10^{-3}$ kg/s) at 100% load, 11.13 at 50%, 13.82 at 25% and then a steep rise to 23,000 $(0.82 \times 10^{-3}$ kg/s) at 10%. This combination works in a straight line relationship down to about 40%; *i.e.*, excellent economy throughout most of the load range. The machines may be piped, either for chilled water or condenser water, in parallel or series, either machine doing low or high duty. An analysis will indicate the most advantageous arrangement.

Figures 4-23, 4-24 and 4-25 demonstrate the performance of the piggyback cycle of Plant No. 7.

There is a minimum load point at which the last centrifugal drops out. Just before this point is reached, one centrifugal machine works with a number of absorption machines. This point of minimum load implies that centrifugal machine steam consumption is minimum and that a balance can still be maintained between centrifugal and absorption machines. Minimum steam consumption for the centrifugal machine occurs at 11% load, *i.e.*, 265 tons (933 kW) at 75 lb/ton-hr (2.69×10^{-3} kg/s) or 19,800 lb/hr (2.49×10^{-3} kg/s). In turn, maximum consumption for a 975-ton (3432 kW) absorption machine is 18,966 lb/hr (2.39×10^{-3} kg/s). Therefore, no less than two absorption machines can work in conjunction with one centrifugal machine, a well-established rule of thumb for the piggyback arrangement. Figure 4-24 shows that the minimum load point is reached at 25% of plant load. Any load below 12.5% must be handled by the absorption machines exclusively. Figure 4-24 indicates the combina-

tion of the various machines at partial loads to obtain steam balance and a low combined steam rate.

A load-sharing relationship exists between the centrifugal and absorption machines, illustrated in Fig. 4-25. The centrifugal must satisfy all the needs of the absorption machine. Each 2412-ton (8490 kW) centrifugal machine tops three 975-ton (3432 kW) absorption machines. Due to the varying efficiency of the steam turbine and the varying ratio of brake horsepower per ton at partial loads, the percentage of load shared by the two machines is not necessarily a linear relationship. In actuality, this relationship is linear between 100% and 25% load

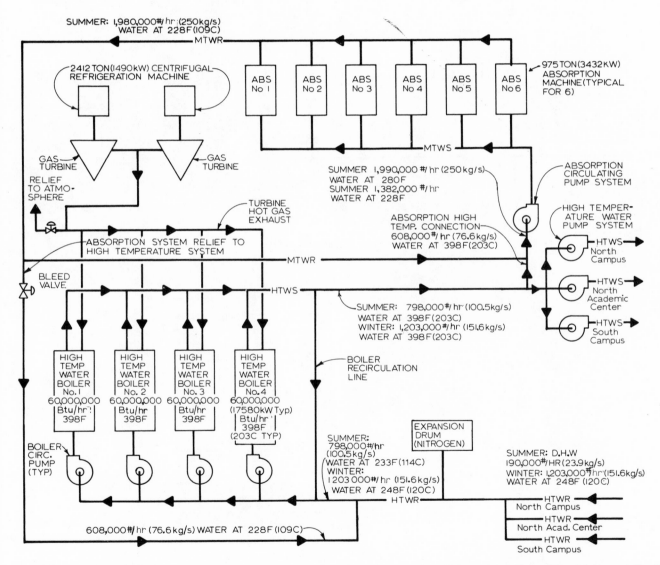

Figure 4-21. High temperature water boilers with individual circulating pumps and primary HTW pumps. Gas turbines driving centrifugals and waste heat boilers with supplementary firing serving district heating needs and the absorption refrigeration machines.

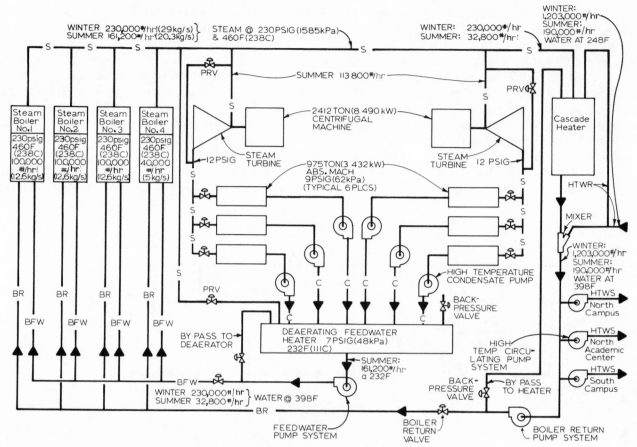

Figure 4-22. Piggyback arrangement of centrifugal and absorption refrigeration machines. A cascade heater provides high temperature water for district heating.

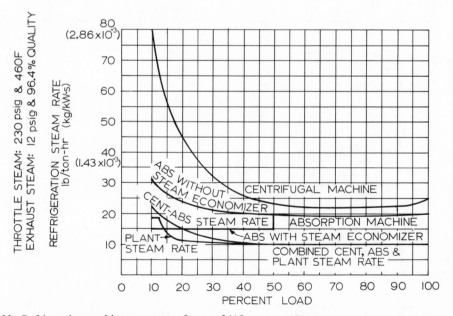

Figure 4-23. Refrigeration machine steam rates for two 2412-ton centrifugal and six 975-ton absorption machines.

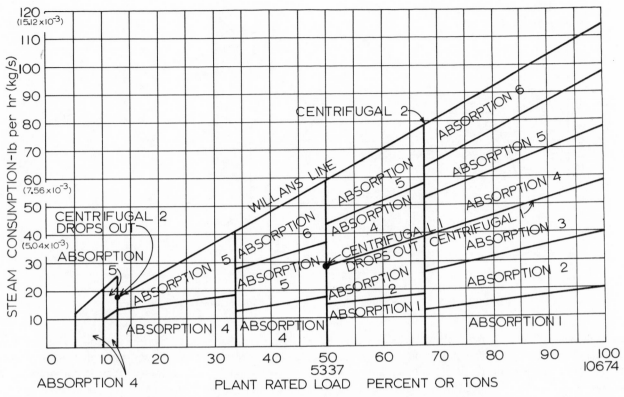

Figure 4-24. Steam consumption and refrigeration machine load participation.

and logarithmic between 25% and 12% of plant load, as indicated in Fig. 4-25.

4.7 Condenser Water Systems

Condenser water in central heating and refrigeration plants, unlike power plants, flows not only to steam condensers but to refrigerant condensers. Items of condenser design to be considered in all cases are tube diameter, tube material, water velocity, number of water passes and fouling factor, as these design features affect cost.

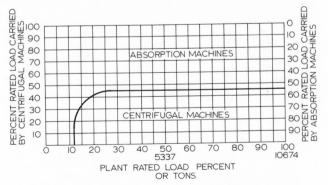

Figure 4-25. Centrifugal-absorption machine load sharing.

Water flow depends upon many factors and variables. A recent trend to reduce flow rate to 2 gpm/ton of refrigeration should be regarded with a great deal of caution. Although it reduces piping and pumping costs, it affects the refrigerant condenser and steam turbine performance negatively. Compromising all the diverse elements in the selection of final condenser water flows is a task that requires considerable thought, time and effort. Condensing conditions, whether steam or refrigerant, affect operating costs and energy consumption and, therefore, merit the utmost attention at the initiation of each project. Condenser water flows must be subject to an analysis. (See Section 4.9, "Combination of Refrigerant and Steam Condensers" and Section 4.10 "Optimization of Condenser and Chilled Water Flow.") The mathematical solution proposed in Section 4.7.1, following, is adequate for preliminary approximations only.

4.7.1 *Refrigerant Condensers*

An optimum condition exists when pumping and condensing costs are minimal. This condition is defined and illustrated mathematically taking into consideration only the costs of operating the condenser pump and the compressor. The advantage of this elementary approach is that a quick first approximation of con-

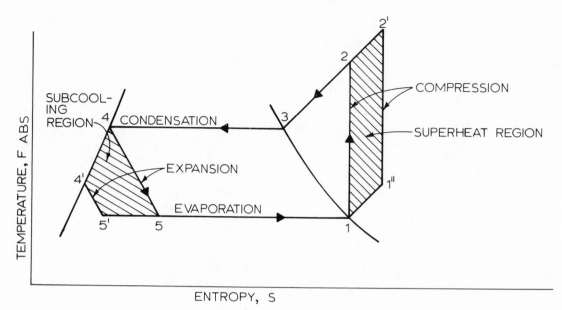

Figure 4-26. Temperature-enthropy diagram of basic refrigeration cycle.

denser flow can be made. A more complete examination should include chiller and chilled water pump operating costs and fixed charges for the entire chiller and condenser system. This latter approach is further explored in Section 4.10.

The mathematical formulas serve two purposes: first, to define the operating costs, if equipment has been selected and, second, to define some equipment parameters once certain thermodynamic and hydraulic criteria are set. Figures 4-26 and 4-27, which indicate the temperature-entropy and pressure-enthalpy relationships of the basic refrigeration cycle, are helpful in following some formulas.

Although there is an optimum condenser flow for a given selection of condenser and compressor, it is advantageous to select the most efficient compressor. An economic balance must be struck between condenser surface and compressor motor horsepower. A higher condensing pressure and temperature requires more horsepower and less condenser surface, and *vice versa*. Lower condensing pressures and temperatures are preferable, since increased condenser tube surface is a first cost item but lower motor horsepower offers continuing savings in energy and operating costs for the life of the compressor.

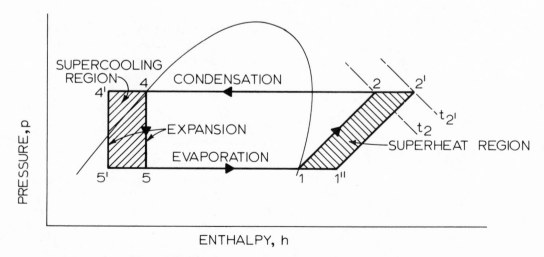

Figure 4-27. Pressure-enthalpy diagram of basic refrigeration cycle.

The total cost of pumping and compressing is expressed by the following equation:

$$C = A \left[(mt_d)/e_c + 8.33GL \, (\text{sp. gr.})/33{,}000 e_p \right]$$

$$= A \left[(mt_d)/e_c + GL(\text{sp. gr.})/3960 e_p \right] \qquad (4.1)$$

Where

A = Power cost, $/hp-hr

C = Total cost, compressing and pumping, $/ton-hr

e_c = Efficiency of compressor with its driver, dimensionless

e_p = Pump efficiency, 0.70-0.75

G = Condenser water flow, gal/ton-min

$200/(h_1 - h_4)$ = Weight of refrigerant, lb/ton-min with h_1 and h_4 enthalpies as defined in Fig. 4-27

L = Pump head, ft

m = hp/ton-deg F rise in condensing temperature. See Equation 4.2, below

t_d = Rise in condensing refrigerant temperature deg. F or rise in leaving condenser water temperature, deg. F

c_p = Specific heat, Btu/lb-F

h_1 = Heat removed from refrigerant, Btu/ton-min.

The term m or hp/ton of refrigeration can be calculated as follows[2]:

$$\text{hp ton of refrigeration} = \left(\frac{200}{h_1 - h_4} \right) \left(\frac{h_2 - h_1}{e_s \, (33{,}000/778.3)} \right) \qquad (4.2)$$

Where

$$\left(33{,}000 \, \frac{\text{lb-ft}}{\text{hp-min}} \right) \div \left(778.3 \, \frac{\text{lb-ft}}{\text{Btu}} \right) = \text{Heat equivalent of hp-min}$$

and

$1/e_s$ = Compensation for loss of efficiency during suction stroke.

e_s = Thermal volumetric efficiency, a function of the ratio of condensing to evaporator pressure in the case of hydrocarbon refrigerants as follows:

ratio	2	3	4	5	6	7
e_s	0.85	0.80	0.75	0.70	0.64	0.58

[2]H.J. Macintire and F.W. Hutchinson, *Refrigeration Engineering*, 2nd ed. (New York: John Wiley and Sons, Inc., 1950), p. 55.

The heat removed from the refrigerant is removed by the condenser water. Therefore

$$h_1 L/33{,}000 \, t_d \, e_p = 8.33GL/33{,}000 \, e_p$$

Equation 4.1 can be rewritten:[3]

$$C = A \left[\, mt_d/e_c + (h_1 L/33{,}000 \, t_d \, e_p) \right] \qquad (4.3)$$

Differentiating and setting the derivative equal to zero, we find that the conditions for minimum power cost are:

$$\frac{dC}{dt} = A \left(\frac{m}{e_c} - \frac{h_1 L}{33{,}000 \, t_d^{\,2} - e_p} \right) = 0 \qquad (4.4)$$

Hence

$$m = \frac{h_1 L e_c}{33{,}000 \, t_d^{\,2} \, e_p}$$

$$t_d = \frac{h_1 L e_c}{33{,}000 \, e_p m} \qquad (4.5)$$

4.7.2 Steam Surface Condensers, Heat Balance, and Heat Rate

Matching the performance of a steam surface condenser to that of a steam turbine is extremely important in achieving an efficient power plant. The heat rate, expressed in Btu/kW-hr, is a measure of the efficiency of a generating plant. The smaller the heat rate, the more efficient the power plant.

Section 4.8 "Heat Balance" works out a procedure to obtain the heat balance for the specific power plant illustrated in Fig. 4-28. However, the method is applicable to other load conditions and equipment arrangements. Note that the following is one of the many procedures that may yield equally valid results. Most heat balance procedures are of a proprietary nature. That is why little information on heat rate calculations appears in literature outside of highly specialized manufacturers' bulletins. Each firm has its favorite method, and heat rate calculations are almost always performed on computers, which are ideal for this type of operation.

The designer must identify all possible steam and electric load combinations and operating conditions. The size of generator load, process steam demand, space heating requirements, cooling water to steam condenser and condenser back pressure are items which can affect heat balance and heat rate. Each prevalent electric and steam load combination requires a new heat balance.

[3]*Ibid.*, pp. 491–492.

The number of heat balances is a matter of judgment on the part of the engineer in order to define plant performance during a desired period. The various plots of Section 4.8 were constructed by performing four complete heat balances and heat rate calculations, *i.e.*, 7375 kW net generator output (8000 kW actual output) illustrated in Section 4.8, and 6011 kW, 4858 kW and 4641 kW, which are only recorded.

However, in lieu of this lengthy method, a shortcut is presented in Table 4-4 which allows performance at other back pressures to be predicted by plotting exhaust pressure versus change in load with exhaust flow as a parameter.[4,5]

4.8 Heat Balance

The starting point for a steam power plant design is the preparation of a heat balance, wherein are entered flow, pressure, temperature, and enthalpy of the working fluid at every point in the cycle as well as heat input into the cycle, work done and, by calculation, the ratio of the heat input over the work done, *i.e.*, the heat rate.

To define a cycle, one must consult the turbine manufacturer for limits to extraction pressures at the various turbine interstages. These pressure data are usually supplied by the turbine manufacturer as curves, with static pressures at bucket discharge (or *stage pressure* as it is usually termed) plotted against flow to the following stage.

Stage pressure is roughly proportional to mass flow to the following stage and is expressed as

$$P = \frac{K_x \text{ (Flow to the following stage)}}{10^6} \qquad (4.6)$$

$$K = \text{Constant}$$

The manufacturer furnishes either the coefficient or the graph for the extraction stage, giving the relationship between stage pressure and flow.

Typical pressure drop values for extraction are 3% from turbine stage to turbine flange and 3% to 5% from turbine flange to heater. In our calculations, a total of 10% is

used. For other than top load, use the following empirical relationship to determine pressure drops in the extraction line:

$$\frac{\Delta P_1}{\Delta P_2} = \left(\frac{Q_1}{Q_2}\right)^{1.2} \qquad (4.7)$$

Where subscript 1 refers to design conditions or peak load and subscript 2 refers to any other condition,

ΔP = pressure drop in extraction line, and
Q = extraction flow, lb/hr (kg/s)

Closed feedwater heaters use a heating surface so that the two media do not mix. Condensate flows through the tubes and the extraction steam enters the shell side of the heater.

In open or direct contact heaters, extraction steam mixes intimately with feedwater, which is discharged from the heater at saturation conditions. Deaerators are direct contact heaters. Besides deaerating feedwater, they act as a reservoir to ensure adequate feedwater supply to the boiler feedwater pumps under all conditions.

Closed feedwater heaters are subject to some design limitations, if an economic selection is to be made. The relationship between shell temperature and leaving feedwater temperature is usually limited to 5F (2.8C) and is called feedwater terminal difference or FTD. The following empirical relationship is useful in adjusting closed heater performance to other than design conditions:

$$\frac{FTD_1}{FTD_2} = \left(\frac{Q_1}{Q_2}\right)^2 \qquad (4.8)$$

Where

Q = feedwater flow (tube side), lb/hr (l/s)

Another important parameter is the difference in temperature between entering feedwater and drain liquid leaving the heater, commonly referred to as drain cooler difference of approach or DOA, which varies between 5 and 15F (1.8–8.3C) and which, in this example, is taken as 10F (5.6C). By definition,

$$T_{\text{drain}} = T_{\text{entering feedwater}} + \text{DOA}$$

In the following example it is assumed that DOA varies directly with drain flow.

$$\frac{DOA_1}{DOA_2} = \frac{DF_1}{DF_2} \qquad (4.9)$$

[4]G.B. Warren and P.H. Knowlton, "Relative 'Engine Efficiencies' Realizable from Large Modern Steam-Turbine-Generator Units," *Transactions of the American Society of Mechanical Engineers* (ASME), Vol. 63, 1941, 125.

[5]J. Kenneth Salisbury, *Steam Turbines and Their Cycles* (New York: John Wiley & Sons, Inc., 1950), p. 566.

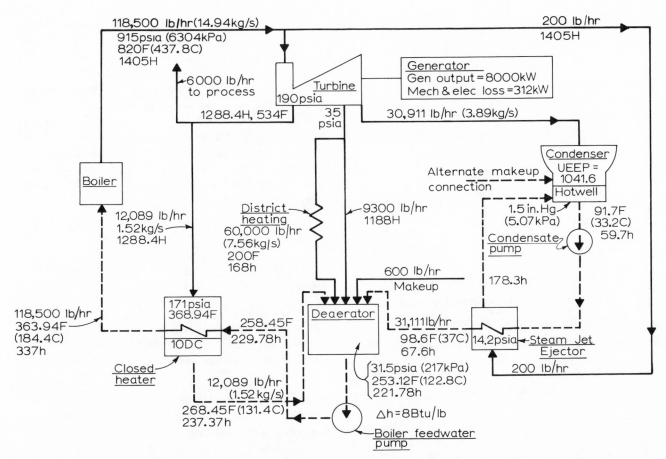

Figure 4-28. Schematic diagram of a hypothetical industrial plant, incorporating district steam heating and process steam. The deaerating heater and closed heater arrangement is typical for this application. The heat balance as shown refers to the 8000 kW generator output only. Section 4.8 deals with heat balance procedure.

Where

$$DF = \text{Drain flow, lb/hr.}$$

Referring to Fig. 4-28, an energy and a mass balance can be set up at the closed and open (deaerator) feedwater heaters, respectively.

The closed heater balance of Fig. 4-28 is indicated in Fig. 4-29.

Applying an energy balance to the closed heater, we obtain

$$W_{E_1} H_{E_1} + W_{FW_2} h_{FW_2} = W_{FW_3} h_{FW_3} + W_d h_d$$

But since

$$W_{FW_2} = W_{FW_3} \text{ and } W_{E_1} = W_d$$

$$W_{E_1} = \frac{W_{FW_2}(h_{FW_3} - h_{FW_2})}{H_{E_1} - h_d} = \frac{118{,}500\,(337 - 229.78)}{1288.4 - 237.37}$$

$$= 12{,}089 \text{ lb/hr (1.52 kg/s)}$$

The heat balance around the deaerator of Fig. 4-28 is shown in Fig. 4-30 and is resolved by setting up two simultaneous equations for energy and mass balance.

$$W_{FW_1} h_{FW_1} + W_{MU} h_{MU} + W_{E_2} h_{E_2} + W_H H_H$$
$$+ W_d h_d = W_{FW_2} h_{FW_2}$$

$$W_{FW_2} = W_{FW_1} + W_{MU} + W_{E_2} + W_H + W_d$$

$$= 118{,}500 \text{ lb/hr (14.94 kg/s)}$$

Solving the two simultaneous equations for W_{E_2} and W_{FW_1}, we find that the second extraction flow and condensate flow are

$$W_{FW_1} = 31{,}111 \text{ lb/hr (3.92 kg/s)}$$
$$W_{E_2} = 9300 \text{ lb/hr (1.17 kg/s)}$$

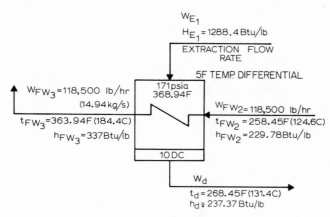

Figure 4-29. Balance at closed heater.

Where

W_{E_2} = Second extraction steam flow, lb/hr (kg/s)

W_{FW_1} = Condensate flow from Rotwell, lb/hr (kg/s)

W_{MU} = Make-up water flow, lb/hr (kg/s)

W_H = Condensate flow from district heating system, lb/hr (kg/s)

W_d = Closed heater condensate flow, lb/hr (kg/s)

H_{E_2} = Second extraction steam enthalpy, Btu/lb (J/kg)

h_{FW_1} = Enthalpy of condensate from Rotwell, Btu/lb (kg/s)

h_{MU} = Make-up water enthalpy, Btu/lb (J/kg)

h_H = Enthalpy of district heating condensate Btu/lb (J/kg)

h_d = Enthalpy of closed heater condensate, Btu/lb (J/kg)

Control of steam leakage is one of the major problems in turbine design because of the sizable effect which leakage has on turbine performance. Interstage leakage is usually accounted for in the turbine internal efficiency. However, external leakage, such as valve steam leakage, shaft-end packing leakage, shaft-packing leakage and steam-seal leakage must be considered in heat balance calculations. (See Fig. 4-31.)

The following expression may be used to calculate valve and shaft-end steam leakages, which are ordinarily supplied by the turbine manufacturer along with leakage diagrams and arrangement.

$$F = C \sqrt{\frac{P}{v}} \qquad (4.10)$$

Where:

F = Leakage flow, lb/hr
P = Initial pressure in packing, psia
v = Initial specific volume, cubic feet per pound
C = Packing leakage constant, obtained from manufacturer

In this example, zero leakage was assumed.

Figure 4-32 illustrates initial and final points of steam expansion in the turbine.

Turbine flow distribution of Fig. 4-28 is indicated in Table 4-1. The power produced by the turbine generator must now be calculated by multiplying the flow of each stage by the net unit energy output of that stage divided by the conversion factor 3412.75 Btu/kW-hr abbreviated to 3412 or 3413 in various textbooks. The term kilowatt

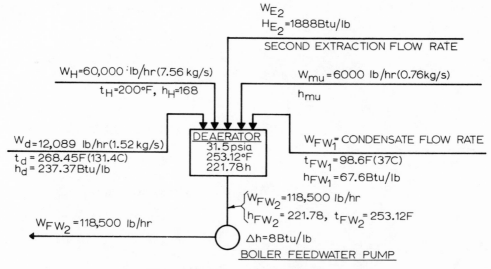

Figure 4-30. Balance at the deaerator.

is based on the international kW (1.0003 abs kW). The Btu is 1/3412.75 kW-hr or 778.26 lb-ft.

$$\text{Output of first stage, bowl to 1st extraction} = \frac{(118,300 \text{ lb/hr})(1405\text{-}1288.4 \text{ Btu/lb})}{3412 \text{ Btu/kw-hr}} = 4043 \text{ kW}$$

$$\text{1st extraction to 2nd extraction} = \frac{100,211\ (1288.4\text{-}1188)}{3412} = 2949 \text{ kW}$$

$$\text{2nd extraction to expansion line end point} = \frac{30,911\ (1188\text{-}1041.6)}{3412} = 1326 \text{ kW}$$

Total	8318 kW
Mechanical and generator losses (from manufacturer)	318 kW
Generator output	8000 kW

The step-by-step procedure of Table 4-2 is a systematic approach to determine turbine performance, applied to Fig. 4-28.

Figure 4-32 indicates steam expansion through the turbine of Fig. 4-28 and Table 4-2. In establishing the initial point of the expansion in Fig. 4-32, no allowance was made for pressure drops through the stop valves, which may be negligible or considerable, depending on turbine

Table 4-1
Turbine Flow Distribution for Fig. 4-28

Flow designation	Flow lb/hr (kg/s)
Throttle flow	118,500 (14.94)
Steam jet ejector	− 200
First valve-steam leakage	0
Second valve-steam leakage	0
Flow to governing stage	118,300
High pressure shaft-packing leakage	0
Flow to second stage	118,300
First extraction flow (calculated)	− 12,089 (1.52)
	@ 190 (1.52)
Process steam	− 6,000 psia
Flow to stage following first extraction	100,211 (12.6)
Second extraction flow (calculated)	− 9,000 (1.17) @ 35
District heating steam	− 60,000 (7.56) psia
Condenser flow	30,911
Steam from jet ejector	+ 200 (3.92)
	31.111

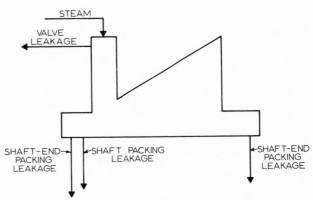

Figure 4-31. Diagram of turbine steam leakages.

size and percent load. By convention, the stop valve drop is not charged against the turbine, since the drop is isenthalpic, but it makes less energy available for work.

Working with Knowlton's vacuum correction curves of Fig. 4-33 (see footnote 4, this chapter), we can generate Table 4-3. It should be noted that the curves of Fig. 4-33 are based on 1 in. Hg abs. exhaust pressure. Figure 4-33 is based on averages culled from a wide range of machines. Table 4-3 is based on an "average" exhaust flow of 5000 lb per hour per square foot (6.85 kg/m^2) of annulus area.

Table 4-4 includes items 33, 38, 59, and 60 of Table 4-2 for the same kW loads but other steam flow conditions than those in Table 4-2. Item 60 is valid for 1½ in. Hg abs (5.07 kPa). All other heat rates are adjusted on the basis of Table 4-3. The four load categories of Table 4-4 are kw and steam flow combinations other than that shown in Table 4-2. In all four combinations the district heating steam flow was removed to facilitate heat rate calculation. However, these combinations are arrived at in the same manner as the combinations of Table 4-2 but are not detailed here.

Figure 4-34 is drawn from data of Table 4-4 and is an auxiliary graph. A plant can be instrumented to give an automatic readout on heat rate and the variable of Figs. 4-35 and 4-36, which are convenient for checking plant performance. Please note that in Table 4-2, heating is included in heat rate calculations. Heating steam is the major part of total steam flow and is extracted without contributing to electrical generation in the third turbine stage. Because it is included in calculating heat rate, the heat rate of Table 4-2 is very high. True plant heat rate, on the other hand, must be calculated without considering heating steam, as in Table 4-4 and, hence, are significantly lower than those shown in Table 4-2.

In constructing Fig. 4-34 the following items are used from Table 4-5: Item 15 "condenser steam flow" and Item

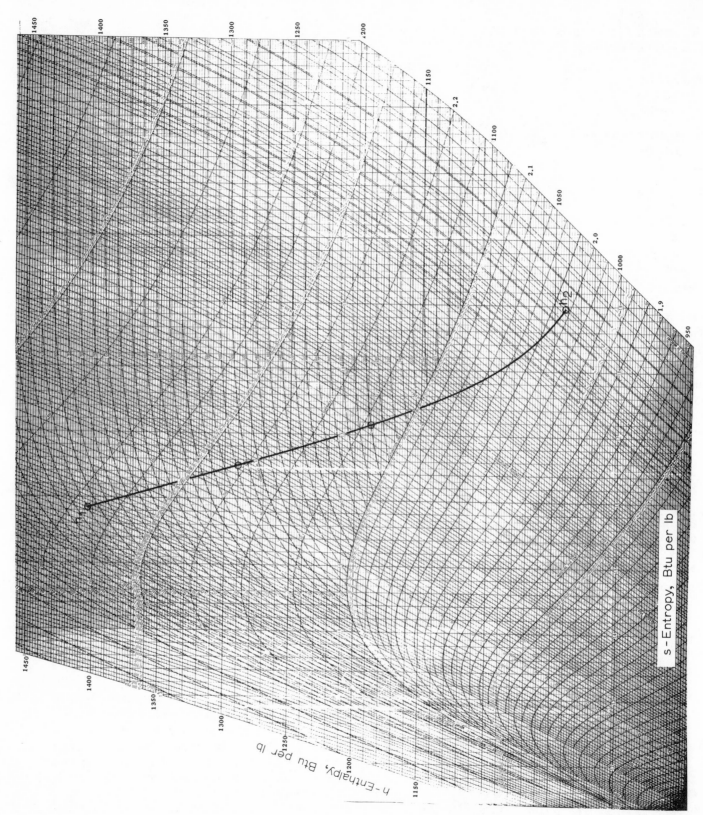

Figure 4-32. Expansion line shown on Mollier diagram.

Table 4-2

Calculation Procedure for Determining Heat Balance and Heat Rate for Only One Combination of Heating, Process and Electrical Loads for the Cycle of Fig. 4.28

Item	Source	Description	
1	Design condition	Approximate load, kW	8000
2	Assumed	Approximate steam flow, lb/hr	120,000
3	Design condition	Closed heater terminal TD, deg F	5
4	Design condition	Closed heater DOA, deg F	10
5	Design condition	1st extraction shell pressure, psia	190
6	Item 5 x 0.90	Closed heater pressure, psia	171
7	Steam tables	Closed heater saturated pressure, psia	368.94
8	Steam tables	Closed htr. condensate enthalpy, Btu/lb	341.83
9	Items 7–3	Feedwater temp. lvg. closed htr., F	363.94
10	Design condition	2nd extraction shell pressure, psia	35
11	Item 10 x 0.90	Deaerator pressure, psia	31.5
12	Steam tables	Deaerator saturation temperature, F	253.12
13	Design condition	Deaerator design temperature, deg F	250
14	Steam tables	Deaerator liqu. enth., Btu/lb @ 31.5 psia	221.78
15	Same as Item 12	Feedwater temp. lvg. deaer., deg F	253.12
16	Turbine curves by manufacturer	1st extraction temperature, deg F	532
17	Steam tables at Items 16 and 5	1st extraction enthalpy, Btu/lb	1288.4
18	Turbine curves by manufacturer	2nd extraction temperature, deg F	327
19	Steam tables at Items 18 and 10	2nd extraction enthalpy, Btu/lb	1188
20	Manufacturer	Last turbine stage efficiency	0.65
21	Manufacturer	Generator efficiency	0.95
22	Steam tables	Enth. @ initial state, 915 psia & 820 F (437.8 C)	1405
23	Steam tables, 190 psia @ Item 15	Enth. 1st extraction, Btu/lb	1288.4
24	Steam tables, 35 psia @ Item 18	Enth. 2nd extraction, Btu/lb	1188
25	Items 53 + 45	Used energy endpoint, Btu/lb	1041.6
26	Items 22–23	Used energy, 1st stage, Btu/lb	116.6
27	Items 23–24	Used energy, 2nd stage, Btu/lb	100.4
28	Items 24–25	Used energy, 3rd stage, Btu/lb	146.4
29	Steam tables	Hotwell temp. @ 1 1/2 in. Hg, deg F	91.72
30	Steam tables	Hotwell enth. @ 1 1/2 in. Hg, deg F	59.71
31	Calculations	Heat added by jets, deg F	6.9
32	Items 29 + 31	Temp. of water entering deaer., deg F	96.72
33	Simultaneous equation	Total steam flow, lb/hr (1st check)	118,500
34	Simultaneous equation	1st extraction steam flow, lb/hr	18,059
35	Simultaneous equation	Steam flow to deaerator, lb/hr	9300
36	Item 35 + district heating steam	2nd extraction steam flow, lb/hr	69,300
37	Manufacturer	Steam flow to jets, lb/hr	200
38	Items 33–34–35–36–37	Steam flow to condenser, lb/hr	30,911
39	Item 34–6000 lb/hr to process	Steam flow to closed heater, lb/hr	12,089
40	Items (33–37) x (22–23) ÷ 3412	kW produced by 1st stage	4043
41	Items (33–37–34) x (23–24) ÷ 3412	kW produced by 2nd stage	2949
42	Items (33–37–34–36) x (24–25) ÷ 3412	kW produced by 3rd stage	1326
43	Items 40 + 41 + 42	Total internal power	8312
44	Manufacturer	Internal efficiency	0.705
45	Manufacturer or literature	Exhaust loss, Btu/lb	7.1

Table 4-2 (Cont.)

**Calculation Procedure for Determining Heat Balance and Heat Rate for Only One Combination of
Heating, Process and Electrical Loads for the Cycle of Fig. 4.28**

Item	Source	Description	
46	Calculation	Boiler F.W. pump disch. pressure, psig	1300
47	Pump manufacturer	Boiler F.W. pump efficiency	0.50
48	Calculated, Items 61 ÷ 47	Enthalpy increase of F.W., Btu/lb	8
49	Pump manufacturer	Efficiency of F.W. pump motor	0.92
50	Calculated, Items 48 x 33 ÷ 49 x 3412	Power by boiler F. W. pump motor, kW	302
51	Manufacturer or literature	Mechanical and generator losses, kW	312
52	Items 43−51	Generator output, kW	8000
53	Items 22−[(22−62) x (44)]	Endpoint of actual expansion from Item 22 to exhaust press of 1 1/2 in. Hg	1034.5
54	Items 7−3 (no other loss)	F.W. temp. entering boiler, deg F	363.94
55	Steam tables	F.W. enthalpy, Btu/lb	337
56	Items 22−25	Heat added in boiler, Btu/lb	1068
57	Items (33 x 56) ÷ 52	Turbine heat rate, Btu/kW-hr	15,820
58	Calculated	Lighting & aux. power, kW (motors, etc.)	323
59	Items 52−50−58	Net generator output, kW	7375
60	Items (33 x 56) ÷ 59	Net station heat rate, Btu/kW-hr	17,160
61	Manufacturer or literature	Isentropic work of liquid pump compression, Btu/lb	4
62	Steam tables or calculation	Isentropic expansion from Item 22 to exhaust press of 1 1/2 in. Hg, Btu/lb	879

Table 4-3
Vacuum Corrections

Operating vacuum, in. Hg abs	0.5	1	1.5	2
Change in load, kW/1000 lb/hr exhaust flow	+2	0	−3	−6.2
Relative gain per 1000 lb exhaust flow	+5.2	+3	0	−3.2

26 "saturated temperature of condenser steam" are plotted for various condenser water temperatures, Item 12. On the right ordinate, Item 27 "condenser exhaust pressure" is plotted against the same abscissa. This serves as a link for plotting load lines based on data of Tables 4-4 and 4-5 for various steam flows. Figure 4-34 does not give a complete evaluation of turbine and condenser working as a unit. The important parameter of heat rate, which includes losses, must be introduced to obtain an integrated performance. This is accomplished in Figs. 4-35 through 4-37.

Table 4-5 is an example limited to a circulating water temperature of 40 F (4.44 C), assuming that river water in wintertime can be at this low temperature. The same tedious procedure must be repeated for 50, 60, 70 and 80 F cooling water.

Heat rate is a distinct indication of plant efficiency, as it includes losses. See Fig. 4-37. Having foreknowledge of heat rate during the design stage not only helps to estimate fuel costs but also to decide if additional equipment to improve heat rate is needed. Plant designers consider heat rate the basis for formulating design parameters and, along with load duration curves, a foundation for fuel cost estimates.

In air conditioning practice, tube water velocites run as high as 10 ft per sec (3.04 m/s), which is higher than the ideal velocity for large steam surface condensers. One-pass condensers, at about 4 in. Hg abs (13.5 kPa), incur roughly a 15 ft (44.7 kPa) pressure drop, while two-pass condensers perform at lower vacua but in the 20–30 ft (59.6–89.4 kPa) pressure drop range. Velocities near 8 fps (2.43 m/s) are desirable, if economically justified, to keep wear and tear down, especially with brackish waters. The shell, since it is under vacuum, is not ordinarily considered a pressure vessel and is, therefore, not subject to ASME certification, whereas the tubes are. Condensers

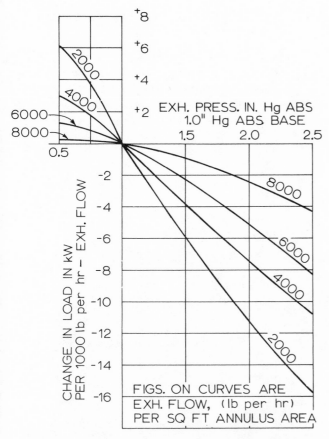

Courtesy General Electric Co.

Figure 4-33. Vacuum correction curves.[6]

are designed according to the standards of the Heat Exchange Institute.

4.9 Combination of Refrigerant and Steam Condensers

This combination is usually encountered in plants utilizing steam turbine-driven centrifugal refrigeration compressors. It can also occur when using a combination of centrifugal refrigeration machines and turbogenerators.

The arrangement consists of passing condenser water through the condenser of the refrigeration machine first and then, in series, through the steam surface condenser. The result might be as shown in Table 4-6. Table 4-6 and Fig. 4-38 pertain to the performance of a given turbine at a specific steam rate. Performance of any other turbine may be plotted easily enough by substituting its steam rate. Table 4-6 represents condenser water flow limits from a thermodynamic point of view without considering an economically optimum rate. In order to achieve this optimization, Equations 4 1 and 4 3 must be amend-

[6] Warren and Knowlton, General Electric Publication GET-1842, *op. cit.,* p. 9.

ed to include the cost of pumping through the steam condenser (see Section 4.10).

The cooling tower temperature approach and the impact of cooling tower selection must be adequately evaluated and weighed against decreased pumping cost, fewer pumps and smaller pipes.

Steam condenser performance as a function of cooling water temperature, as plotted in Fig. 4-39, can be obtained from the condenser manufacturer for any given design. Since cooling water temperature affects equipment costs as well as performance, condenser parameters must be chosen early in the course of a project.

4.10 Optimization of Condenser and Chilled Water Flows

Optimized condenser and chilled water flows are interdependent since each affects chiller horsepower, energy consumption and physical size. In an optimization study, one must consider the cost of the pumps, chiller and cooling tower on the one hand and the costs of the chilled and

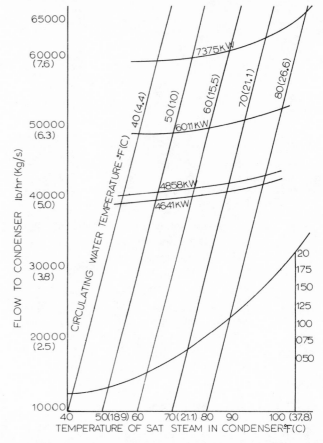

Figure 4-34. A plot of condenser steam flow vs. saturated steam temperature in condenser, as an auxiliary graph for constructing Fig. 4-36. It is based on data from Tables 4-4 and 4-5 for various steam flow conditions.

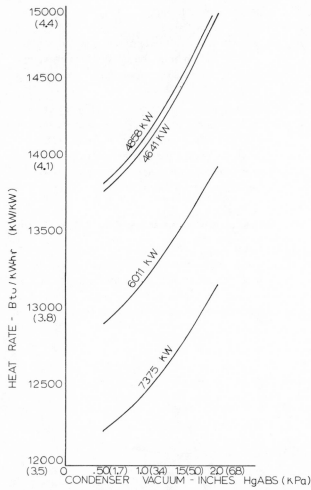

Figure 4-35. A plot of heat rate vs. condenser vacuum, as an auxiliary graph for constructing Figure 4-37. Based on data from Table 4-4 for various steam flow conditions.

ACH$_{fc}$ = Annual fixed charges for chiller, chilled water pump and piping

ACC = ACC$_{op}$ + ACC$_{fc}$

ACC$_{op}$ = Annual operating cost for condenser water pump and cooling tower

ACC$_{fc}$ = Annual fixed charges for cooling tower, condenser water pump and piping

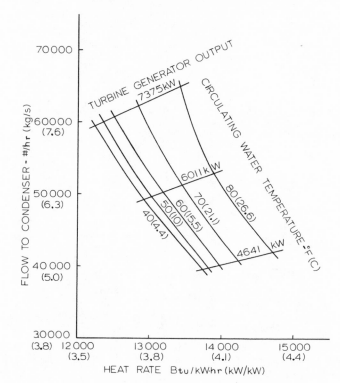

Figure 4-36. A cross plot of Figures 4-34 and 4-35.

condenser water piping systems on the other. Energy costs for operating all equipment must be included. Items such as labor, taxes, insurance, electrical installation, etc., must be considered, if they can be attributed to operating or fixed costs. The optimum chilled and condenser water flows will occur when the combined owning and operating costs of a selected chilled water system and condenser water system are the lowest of all other chilled and condenser water flow combinations.

The optimum point will be obtained when the following expression is true:

ACH + ACC = Minimum cost

Where:

ACH = ACH$_{op}$ + ACH$_{fc}$

ACH$_{op}$ = Annual operating cost for chilled water pump and chiller

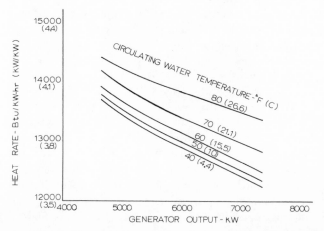

Figure 4-37. Heat rate, plotted from data of Figure 4-36, provides a quick check on plant performance.

Table 4-4
Cycle Conditions at Various Exhaust Pressures and Steam Flows Other Than Those of Table 4-2

Exchange pressure in. Hg abs. (kPa)	Item 59 Load kW	Item 60 Heat rate Btu/kW-hr (kW/kW)	Item 38 Condenser steam flow lb/hr	From Table 4-3 Rel. gain kW/1000 lb exh	% gain	Item 33 Steam flow lb/hr (kg/s)
½ (1.69)	4641	15156 (4.44)	39771 (5.01)	−5.2	−4.67	59848 (7.54)
	4858	15152	40777	−5.2	−4.57	62831
	6011	14164	53870	−5.2	−4.46	72686
	7375[9]	13381 (3.92)[10]	59523 (7.49)[10]	−5.2	−4.38	84438 (10.63)[10]
1 (3.38)	4641	14869	40597	−3.0	−2.69	61091
	4858	14871	41606	−3.0	−2.63	64108
	6011	13908	50244	−3.0	−2.57	74124
	7375[9]	13244[10]	60675[10]	−3.0	−2.53	86066[10]
1½ (5.07)	4641	14480	41720	0	0	62780
	4858	14490	42730	0	0	65840
	6011	13560	51570	0	0	76080
	7375[9]	12820[9]	62250[9]	0	0	88300[9]
2 (6.76)	4641	14898	42925	+3.2	+2.87	64594
	4858	14826	43930	+3.2	−2.81	67690
	6011	13931	52983	+3.2	+2.74	78164
	7375[9]	13116[10]	63930[8,10]	+3.2	+2.70[7]	90684[10]

The formulas can become very unwieldy, if one is to include all the design parameters and variations of the piping and pumping system. It is preferable to use direct calculations, which will be more illuminating than any strict interpretation of the formula. This approach will permit the use of good engineering judgment.

Table 4-7 represents all possible combinations of chilled water and condenser water flows, in increments of 0.2 gpm. If a computer is available, the number of combinations may be increased for greater accuracy. For manual operations, it is advisable to reduce the number of combinations, since the 56 operations of Table 4-7 will be time-consuming.

In Fig. 4-40, fixed and operating cost values are plotted for the various combinations of Table 4-7 and a separate total cost curve is shown. The minimum cost point on the total cost curve represents a combination of chilled

and condenser water flows that result in the most economical chilled and condenser water system design; this is the point of optimization. The smooth shapes of the curves in Fig. 4-40 are idealized; in actuality, they will be irregular because most equipment and pipe sizes cover ranges of performance and, therefore, physical size and cost take discreet jumps. It is entirely possible that a number of combinations may have the same material costs, although operating costs may differ somewhat. Optimization studies of this type can be extremely time-consuming without the benefit of full manufacturer cooperation, a capable estimating department and a computer.

In outlining the course of a study, one ought to consider various options in chiller selection, such as the amount of heat transfer surface or motor horsepower per unit of work. The procedure for selecting and evaluating such options must tend to level any differences in cost by assigning a monetary value to additional system features or lack thereof and adding or subtracting accordingly to the basic cost.

Other ground rules must resolve the eternal dilemma of constant versus variable water flow. Common practice dictates constant flow through the condenser, regardless of season or load, for fear of exceeding maximum head

---

[7] $62250 \times \frac{3.2}{7375} = 2.70$

[8] $62250 + \frac{2.70 \times 62250}{100} = 63930$

[9] From untabulated heat balance calculations similar to those of Table 4-2.

[10] It is assumed that total steam flow and heat rate decrease by the same percentage as condenser flow for ½, 1 and 2 in. Hg, respectively.

Table 4-5
Performance of Steam Surface Condenser, Fig. 4-28

Item	Source	Description					
1	selected	P, number of passes	1				
2	selected	L, tube length, ft	18				
3	selected	dia., in., weight, material	3/4″, 18 gauge, muntz metal				
4	see note 11	tube constant, k	0.188				
5	see note 12	heat transfer to water Btu/lb of steam (kJ/kg)	950 (223)				
6	$\dfrac{\text{Btu/hr}}{500\,(t_d)\,H_2O}$	circ. water flow, gpm (l/s)	12340 (729)				
7	selected	tube vel. of H_2O, fps (m/s)	7 (2.13)				
8	$S = \dfrac{Lk(\text{gpm})}{V}$	condenser surface, ft² (m²)	5980 (556)				
9	Fig. SF-2, Note 13	base heat transfer Btu/ft² hr F (W/mC)	715 (4061)				
10	Fig. SF-1, Note 13	material correction factor	0.96				
11	selected	cleanliness factor	0.85				
12	selected (river)	t_i, circ. H_2O inlet temp. F (C)	40 (4.4)				
13	Fig. SF-2, Note 13	circ. water temperature correction factor	0.75				
14	Items 9 × 10 × 11 × 13	U, heat transfer rate Btu/ft² hr F	715				
15	Design condition	condensed steam, lb/hr (kg/s)	30000 (3.8)	40000 (5.0)	50000 (6.3)	60000 (7.6)	65000 (8.2)
16	Items 5 × 15/10⁶	heat rejected to condenser/10⁶	28.5	38.0	47.5	57.0	61.75
17	Items 16/(8 × 14)	t_m, LMTD, F	11.08	14.77	18.47	22.16	24.01
18	Items (16 × 4 × 2)/(8 × 7 × 500)	temp. rise circ. H_2O, F	4.60	6.14	7.67	9.21	9.98
19	Items 12 + 18	t_d, disch. temp. circ. water, F	44.60	46.14	47.67	49.21	49.98
20	Items 18/17	$\log_e[(t_x-t_i)/(t_x-t_d)]$	0.415	0.415	0.415	0.415	0.415
21	Item 20/2.303	$\log_{10}[(t_x-t_i)/(t_x-t_d)]$	0.180	0.180	0.180	0.180	0.180
22	antilog Item 21	$(t_x-t_i)/(t_x-t_d)$	1.51	1.51	1.51	1.51	1.51
23	Items 22 × 19		67.41	69.98	72.13	74.65	75.64
24	Items 23 – 12		27.47	29.98	32.13	34.16	35.64
25	Item 22 – (1.0)		0.51	0.51	0.51	0.51	0.51
26	Items 24/25	t_x, sat. temp. of condenser steam.	53.54	57.83	62.63	66.92	69.47
27	at Item 26	pressure in condenser, in. Hg abs (kPa)	0.413 (1.39)	0.483 (1.62)	0.572 (1.92)	0.665 (2.24)	0.726 (2.44)

[11] E.V. Pollard, "Figure Combined Turbine-Condenser Curves," *Power*, February, 1957, pp. 80–85 and 202.

[12] William Ellingen, "Approximate Methods for Selection, Sizing and Pricing of Steam Surface Condensers," *ASME paper 54-A-127*, p. 2.

[13] *Standards for Steam Surface Condensers*, 6th Ed. (New York: Heat Exchange Institute, 1970).

Table 4-6
Typical Steam and Refrigerant Condenser Performance

Condensing Pressure Hg Abs	Condensing Temperature F	Heat of Condensation Btu/lb	Steam Rate @ Full Load lb/ton-hr	Total Heat of Condensation Btu/ton-hr (Steam Condenser)	Approximate Heat of Rejection Btu/ton-hr (Refrigerant Condenser)	Total Heat Rejection Btu/ton-hr	Condenser Performance						Cond. Water Total Temp. Diff. F	Condenser Water Flow (in series) gpm
							Refrigerant Condenser			Steam Condenser				
							Ent. Refr. Cond. F	Lvg. Refr. Cond. F	Temp. Diff. F	Ent. Steam Cond. F	Lvg. Steam Cond. F	Temp. Diff. F		
3	115	1028	12.0	12340	15,000	27,340	85	97.62	12.62	97.62	108	10.38	23	2.38
3½	120	1026	12.2	12450	15,000	27,540	85	100.25	15.25	100.25	113	12.75	28	1.97
4	125.5	1023	12.4	12700	15,000	27,700	85	103.18	18.18	103.18	118.5	15.32	33.5	1.65
4½	130	1021	12.6	12870	15,000	27,870	85	105.45	20.45	105.45	123	17.55	38	1.47
5	134	1017	12.8	12950	15,000	27,950	85	107.55	22.55	107.55	127	19.45	42	1.33
5½	137.4	1015	13.0	13200	15,000	28,200	85	109.1	24.10	109.10	130.4	21.3	45.4	1.25
6	140.8	1013	13.2	13360	15,000	28,360	85	110.85	25.85	110.85	133.8	22.95	48.8	1.16

Table 4-7

Chilled Water			Condenser Water, gmp/ton (l/s-kW x10^2)						
Chilled water, gpm/ton (l/s-kW x10^2)	Chilled water temperature, differential, F (C)	Annual cost, $	(3.22) 1.8	(3.58) 2.0	(3.94) 2.2	(4.31) 2.4	(4.66) 2.6	(5.02) 2.8	(5.37) 3.0
						Annual cost, $			
2.4 (4.31)	10.00 (5.6)	ACH_1	ACC_9	ACC_{17}	ACC_{25}	ACC_{33}	ACC_{41}	ACC_{49}	ACC_{57}
2.2 (3.94)	10.90 (6.1)	ACH_2	ACC_{10}	ACC_{18}	ACC_{26}	ACC_{34}	ACC_{42}	ACC_{50}	ACC_{58}
2.0 (3.58)	12.00 (6.7)	ACH_3	ACC_{11}	ACC_{19}	ACC_{27}	ACC_{35}	ACC_{43}	ACC_{51}	ACC_{59}
1.8 (3.22)	13.33 (7.4)	ACH_4	ACC_{12}	ACC_{20}	ACC_{28}	ACC_{36}	ACC_{44}	ACC_{52}	ACC_{60}
1.6 (2.86)	15.00 (8.3)	ACH_5	ACC_{13}	ACC_{21}	ACC_{29}	ACC_{37}	ACC_{45}	ACC_{53}	ACC_{61}
1.4 (2.51)	17.14 (9.7)	ACH_6	ACC_{14}	ACC_{22}	ACC_{30}	ACC_{38}	ACC_{46}	ACC_{54}	ACC_{62}
1.2 (2.15)	20.00 (11.1)	ACH_7	ACC_{15}	ACC_{23}	ACC_{31}	ACC_{39}	ACC_{47}	ACC_{55}	ACC_{62}
1.0 (1.79)	24.00 (13.3)	ACH_8	ACC_{16}	ACC_{24}	ACC_{32}	ACC_{40}	ACC_{48}	ACC_{56}	ACC_{64}

Table 4-8
Compressor Energy Consumption

Percent of full load	Percent of operating time	Full load kW	Percent full load power	Part load kW	Number of hours in operation	kW-hr consumed
100.00	6.11	3069	94	2886	347	1.001,440
75.11	59.83		64	1965	3367	6,616,155
56.25	7.41		46	1412	417	588,906
34.03	17.77		29	890	1002	890,000
17.36	8.88		19	583	500	291,500
	100.00				5631	9,387,901

pressure in case of inadequate flow. Variable condenser water flow amounts to a control problem and is thus shunned.   On the other hand, a benefit of constant condenser water flow is reduced horsepower per ton whenever water temperature drops.   As for the evaporator, variable chilled water flow is practiced to a limited extent because it contributes to energy conservation and keeps pumping costs low, provided the hp/ton requirement can be kept within specified limits.   Energy for pumping at variable flow rates through the evaporator can be calculated in the manner of Table 4-8.

In the following example, flow of chilled water is assumed to be constant, and condenser water is taken at a constant 2 and 3 gpm per ton (0.0358 l/s kW) (0.9537 l/s kW), to shorten the calculations while presenting the technique for cost optimization.

Problem: A 4500-ton (15 840 kW) chiller can use condenser water between 65 F (18.3 C) and 85 F (24.5 C) at a

condenser water flow between 2 and 3 gpm. What is the optimum flow rate?

Solution: Select two random points at 2 and 3 gpm per ton and test.

Case I, 2 gpm of condenser water per ton of refrigeration. Required condenser water = 4500 tons × 2 gpm/ton = 9000 gpm

Pump motor input (kW)
$$= \frac{gpm \times head\ (ft)}{Efficiency \times 3960} \times \frac{0.746\ kW/hp}{motor\ eff.}$$

$$= \frac{9000\ gpm \times 60\ ft \times 0.746}{0.70 \times 3960 \times 0.90} = 162$$

Pump annual energy consumption = 162 kW × 5631 = 912,222 kW-hr

The next task is to find the annual energy consumption for the compressor.   Assume that Table 4-8 indicates

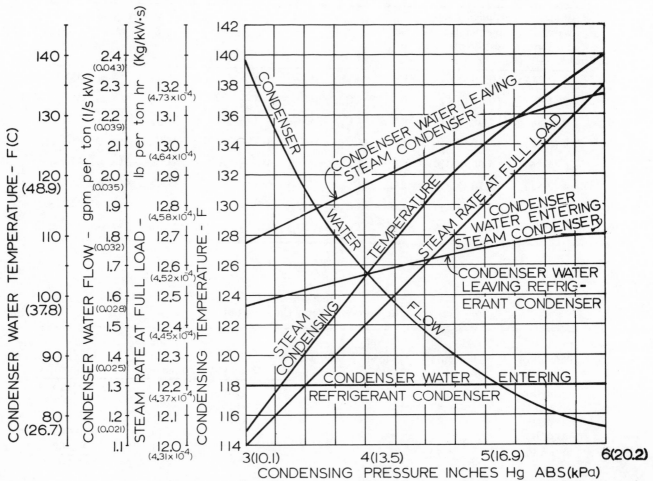

Figure 4-38. Plot of Table 4-6 data, typical condenser performance.

actual compressor performance and annual consumption for this example

Total annual
energy
consumption = compressor input + pump input
= 9,387,901 + 912,222
= 10,300,122 kW-hr

Case II, 3 gpm of condenser water per ton of refrigeration. Required condenser water = 4500 tons × 3 gpm/ton = 13,500 gpm (852 l/s)

From a computation similar to that of Table 4-8, the total compressor input equals 8,642,135 kW-hr.

$$\text{Pump motor input} = \frac{13,500 \text{ gpm} \times 60 \text{ ft} \times 0.746}{0.70 \times 3960 \times 0.90} = 243 \text{ kW}$$

Pump annual
energy
consumption = 243 kW × 5631 hrs = 1,368,333 kW-hr

Total annual
energy
consumption = 8,642,135 + 1,368,333
= 10,010,468 kW-hr

Estimating the payback period can be accomplished by the following calculations:

Additional
kW-hr for 2
gpm per ton   = energy consumption, Case I—energy
(0.0358 l/s-kW)   consumption, Case II
= 10,300,123 − 10,010,468
= 289,655 kW-hr

Cost of additional power @ $0.05/kW-hr
= $0.05 × 289,655
= $14,483

Assuming that the cost of pumps and chillers for Cases I and II are identical, it is necessary to find any differential installation costs so that they can be compared to electrical operating savings. Dividing an increased outlay by the incurred savings gives the payback period. In this case, the sum of $14,483 represents higher operating costs for Case I or savings for Case II.

Cost of piping in Case I (2 gpm) = $110,000 (estimated)
Cost of piping in Case II (3 gpm) = $140,000 (estimated)
Full load kW difference of Case I–Case II = 3069 − 2826 = 243 kW
Estimated additional electrical work for 243 kW = $7,000
Net savings in favor of Case I = Installation cost of

Case II–Case I = $140,000 − ($110,000 + $7,000)
= $23,000

$$\text{Payback period} = \frac{23,000}{14,483} = 1.59 \text{ years}$$

Visual inspection of the figures gives no indication if zero payback period is at more or less than 3 gpm flow rate. Since the payback period is so short, it is appropriate to select 3 gpm per ton (0.0537 l/s-kW) as the design flow rate. However, it is entirely up to the engineer whether in his judgment he ought to seek another approximation in order to reduce the payback period to zero and obtain a more accurate flow rate.

4.11 Organic Fluid Rankine Cycle Systems

The organic Rankine cycle is a power-producing cycle similar to the steam Rankine cycle used in steam power plants. The organic cycle is fully described in the following pages and is illustrated in Figs. 4-41 and 4-42. These systems are ideal for capturing waste heat from effluent fluid or gas streams which are normally released to the environment or to the atmosphere. Recycling energy increases the overall efficiency of a plant. Efficiency depends on the amount of waste heat recaptured. Bottoming cycles, whether using steam or an organic working fluid, can play an enormous role in national energy conservation programs. Immense amounts of waste heat are released to atmosphere. In the steel industry, off-gases from blast furnaces contain 550,000 to 800,000 Btu per ton of steel produced. Twenty to thirty percent of this amount is recaptured in the form of steam. The rest of the gas (carbon monoxide) is burned, mixed with water spray and discharged to atmosphere.[14] In the chemical refinery and paper industries, 80% of the thermal releases in the 300 F to 1000 F (149-538 C) range are reflected to atmosphere.[15] The practice of dumping heat sources above 300 F continues unabated.

Organic cycles are a fairly recent development and experience with them has been mostly experimental. Hesitancy in using these systems was based not only on the high cost of prototype units but also on the lack of operating experience and questionable reliability. However,

[14]*Potential for Energy Conservation in the Steel Industry.* Batelle Columbus Laboratories. Prepared for the Federal Energy Administration. Document PB-244-097. Springfield, Va.: National Technical Information Service, 1975, pp. V-45.

[15]S.E. Nydick, J.P. Davis *et al.* A Study of Inplant Electric Power Generation in the Chemical, Petroleum Refining and Paper and Pulp Industries. Report No. TE5429-97-96. Prepared for the Federal Energy Administration. Waltham, Mass.: Thermo Electron Corp., 1976?, p. 3-10.

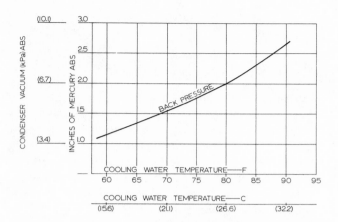

Courtesy Power Engineering magazine.

Figure 3-39. Typical performance curve for large steam surface condenser.

the picture is beginning to change due to the rising cost of fuel and the promise that solar energy can mesh well with the organic Rankine cycle. Even the low water temperatures currently generated by solar systems are adequate to power the organic Rankine cycle, either to generate electricity or to operate pumps and other equipment directly. The development of solar collectors able to withstand high temperatures can promote power cycles employing either water or thermal liquids. One manufacturer is moving into mass production of generator units of up to 10,000 kW, using fluorocarbons. Lessons drawn from operating experience, if incorporated into units designed for mass production, will do a great deal to lower costs. Even today, large units are competitive with gas turbines, provided a free source of energy is available.

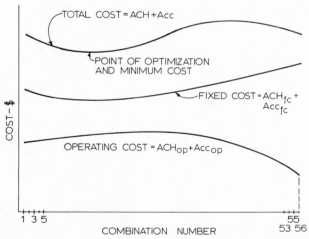

Figure 4-40. Hypothetical cost curves for combinations of chilled and condenser water flows from Table 4-7.

From a cost-effectiveness point of view, the organic cycle can prove advantageous, since the cost of fuel is zero.

Many organic cycles, but not all, are not saddled with expensive, high pressure auxiliaries. However, some organic fluid cycles can have high efficiencies at high temperatures and low power levels, a feature that may compensate for high equipment costs. Steam Rankine cycles are totally inadequate in the low temperature range due to the high latent heat of water and the limited steam expansion one can expect in this range. Latent heat of water is at least nine times larger than that of Freon 12 or 22. To produce steam, the effluent stream must be approximately at 500 F (260 C) or above. However, the organic cycles can operate with one effluent stream being as low as 150–200 F (65–93 C).

Here are some of the advantages that are claimed for the organic fluid turbines (called expanders in organic cycles):

1. Gases of high molecular weight tend not to condense in the turbine, minimizing turbine erosion.
2. Due to the high molecular weight of the organic fluids, gas velocities in the nozzles and blade passages are a fraction of steam velocities under similar circumstances, resulting in decreased wear and tear on the expander.
3. Organic fluids produce high pressure drops across the expander that tend to minimize its physical size.
4. Despite high pressure differences, the heavy molecular structure of the turbine gases contributes toward fewer expansion stages and, therefore, cheaper turbines.

Expanders are associated with gases at pressures above atmospheric but without combustion taking place. Turbo-expanders were first used by the gas processing industries to drive compressors and, in cryogenic applications, served in lieu of expansion valves. Now they are finding another important application in organic cycles. The expander is so called because the high pressure gas in going through it loses its pressure and temperature; *i.e.*, it expands in doing work. A turbine is an expander. A screw or piston engine can also be an expander. Expanders are fitted and controlled differently than generating steam turbines or internal combustion engines.

The Rankine cycle, which converts thermal energy into mechanical work, is employed by both steam and organic fluid power plants. Figures 4-41 and 4-42 illustrate the cycle, which works as follows:

The fluid leaves the boiler at state 3 in gaseous form and is drawn to the turbine or engine, where it expands poly-

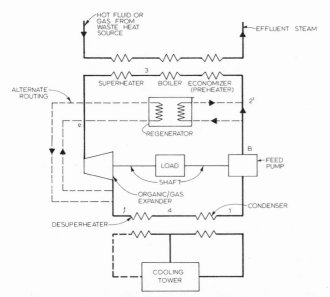

Figure 4-41. Diagram of organic fluid Rankine cycle. The economizer, superheater, desuperheater and regenerator are optional additions to the basic cycle.

tropically to point 4. This expansion produces the actual work of the cycle in turning a generator, a pump or other equipment. The expanded gas begins to condense at point 4. During process 3 to 4, latent heat is removed from the gas, or from f to 4, if there is a desuperheater. Desuperheating before condensing is often required due to fluid properties, especially for dry fluids, which is explained below under "Fluids and their Properties." The latent heat rejected from the cycle (heat of rejection) is transferred to the cooling tower or other heat sink and rejected to the environment at near ambient conditions.

Although not indicated on the temperature-entropy diagram, Fig. 4-42, the liquid is pressurized to point B in the feed pump and is heated in the boiler until it reaches saturation temperature. Heat given up by the effluent waste stream is transferred to the organic fluid in the

boiler until the liquid evaporates at point 3. The superheater heats the gas to point e. The desuperheater work is shown by line f-4. Line B-2' represents the regeneration process during which heat is extracted from the turbine discharge gas before it is condensed. This heat is transferred to the feed liquid, improving cycle efficiency. A superheater, desuperheater or regenerator may be used in the cycle to improve cycle efficiency or equipment dependability or safety. In the case of an economizer, economic justification is not so important because the economizer, most often is required anyway, to raise the condensed fluid temperature to the boiling point. Most of these cycle refinements are desirable energy conservers but should be subjected to cost-benefit analyses for their functional necessity and payback.

Fluids and their properties. There are three basic types of fluids. In Fig. 4-42, it is seen that each imposes its characteristic configuration on the saturation curve or vapor dome in a temperature-entropy diagram. Wet fluids include steam and metal vapors. They produce droplets inside the turbine during expansion, contributing to nozzle and blade erosion. Wet fluid cycles are usually provided with superheaters to prevent wet conditions during expansion.

Dry fluids, on the contrary, become superheated when expanding through the expander. Fluorocarbons F-12, F-114, F-113 and F-215 belong to this category. Dry fluid cycles normally use a regenerator for high cycle efficiency since the mean heat addition temperature is low and the mean heat rejection temperature is high. In addition, the dry types need a desuperheater before the condensation process begins.

Isentropic fluids do not become wet, acquire only slight superheat, if at all, during the expansion process and ordinarily do not need auxiliary heat exchangers. There are very few such fluids; two of them are Fluorinol-85 and thiophene.

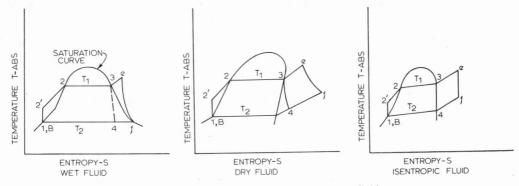

Figure 4-42. Vapor dome configuration of organic fluids.

Fluid selection is dependent not only on the type of operation but on the fluid properties such as thermal stability and decomposition temperature, and on cost, which are of primary importance. Toxicity, fire resistance, compatibility, melting point, transport properties and boiling point are less important properties.

The melting point of the fluid must be below the temperature of the surrounding atmosphere. Most organic fluids have below zero melting points. Benzene, an exception, has a melting point of 41 F.

Boiling point is a moot parameter. The specific volume of the fluid, as it undergoes expansion through the turbine, affects turbine size and speed for peak efficiency. Critical pressure and temperature can be viewed in at least two ways. On the one hand, these parameters can be selected above the heat addition temperature of the cycle, avoiding superheat and superheaters. In this fashion, cycle efficiency approaches Carnot efficiency as much as possible by having heat added isothermally. Another approach would be to operate near critical pressure and temperature and not add heat to the cycle isothermally. By selecting a fluid with low latent heat, most of the heat can be added by preheating the fluid, if the specific heat of the liquid is low. This tends to reduce the thermodynamic irreversibilities of the heat transfer process.

High molecular weight is the key to expander design and is valued because it reduces enthalpy drop per stage, making possible a low-speed, single-stage or, at worst, a two-stage impulse or reaction, high-efficiency turbine. However, at higher power levels, molecular weight should not be too high, lest the turbine have too large a rotor, rotating at too low a speed.

Organic fluids are superior to steam in recovering low temperature heat both in terms of the thermodynamic potential of the cycle and the characteristics of the expander. Some degree of regeneration is advisable to improve cycle efficiency and conserve energy. Unfortunately, most of the organic fluids that have high decomposition temperatures, let us say between 600 F and 800 F, are flammable or are of the wet type or have low critical pressures. There are some liquids that are non-flammable in the high temperature ranges, but they are expensive. Some organic liquids need large regenerators and some others have a number of undersirable characteristics, which limit heat transfer at a high thermal level, as the fluorocarbons do.

Essentially, there is no ideal organic fluid and selections must be a compromise. Hundreds of organic liquids have been explored and tested in the last two decades.[16] The fluorocarbons are favored by at least one manufactur-

er and according to its design philosophy, it offers current development possibilities in the low temperature range. This approach is not shared by another manufacturer , who uses Fluorinol-85 primarily, and feels that use of fluorocarbons must wait until fuel costs rise to the point of making them cost-beneficial. The fluorocarbons have acceptable fire safety properties and a low decomposition temperature of 400 F that makes these gases unsuitable for heat sources above 500 F because of their low cycle efficiency when compared to other fluids at these temperatures. However, the fluorocarbons have workable molecular weights and critical pressures and are readily available.

Liquids that are limited to 400 F are not cost-effective for many upcoming high temperature solar and industrial applications. The manufacturer using Fluorinol-85 has found this liquid to have highly desirable characteristics for working fluid temperatures in the range of 300–650 F, permitting high efficiency expansion, single-stage turbines operating in the superheated region and suitable for power cycles.

In the temperature range of 200-700 F, heat recovery is profitable using the organic Rankine cycle; above that range, steam generation through waste heat boilers for process purposes is preferable. The area between 400 and 700 F is for fluids other than fluorocarbons. The point at which steam becomes competitive is a point of contention. Most diesel engines and gas turbines of the industrial variety have exhaust gases in the 800-850 F range. This fact has encouraged steam combined cycles and this temperature range has become the dividing line. Others contend that 1000 F is the point at which steam Rankine cycles become cost-effective. When organic systems become truly available on a commercial basis for all temperature ranges, a cost-effectiveness study should be performed for any proposed project to establish the economic aspects of all pertinent cycles.

For fluorocarbon cycles in the 150-212 F effluent temperature range, it is estimated that 7% to 15% of the waste heat is recoverable (as power). Using gas steams near 700 F with auxiliary devices such as regenerators, cycle efficiencies jump to 35% and higher; effluent steams near 1000 F boost efficiency to 50%. But such applications are costly and a rather specialized nature. Their development for commercial and industrial applications may not

---

[16]Toluene and trifluoroethanol cycles for power generation are explored by Duane E. Randall, "Conceptual Design Study for the Application of a Solar Total Energy System at the North Lake Campus, Dallas County Community College District," NTIS and Sandia Laboratories Document SAND 76-0521. (Albuquerque, N.M.: Sandia Laboratories, 1976).

be that far in the future, since the power conversion efficiency may be high enough to justify equipment costs at current fuel prices. Economic justification at low cycle efficiencies should be fully explored in each particular case.

In general, application of organic fluids for power is still under development and an area of potential growth. Some organic cycles have already been developed for solar and low level heat recovery applications as well as for high temperature heat recovery and power packages. A demonstration plant organized by the United States Energy Research and Development Administration in conjunction with the New England Electric System and a Massachusetts manufacturer was scheduled to go into operation in 1979. The plant employs a 450 kW organic Rankine turbine bottoming plant, which will be operated from the exhaust gases of one of the 2500 kW diesel-driven powerplants, located at the Utility's Lynn Station. Figure 4-43 represents a commercially developed model.

4.12 Power from Waste

Section 4.12 is an overview of an extensive technological subject and can serve as an introduction to it for the interested reader. Industrial wastes are treated in an entirely synoptic manner. Notwithstanding economics, waste disposal methods not pertaining to the development of power are assumed to be irrelevant to the interests of power engineers. Sewage treatment and disposal are not examined here except for methane production (see Section 4.12.8).

4.12.1 *Energy Recovery from Municipal and Industrial Wastes*

Urban centers, producers of abundant waste, can also be the best markets for waste. In the cities the amount of refuse is dependably available and ever more plentiful. At the same time, landfill sites are becoming scarcer and sludge disposal is a growing problem. However, for waste products to become serious contenders as energy sources for central plants in urban and industrial applications, some of the following conditions must exist:

1. The waste disposal plant must be near or contiguous to the central plant and convenient to purchasers of recovered "recycled" materials. The central plant and the surrounding community must absorb substantially all of the waste as an

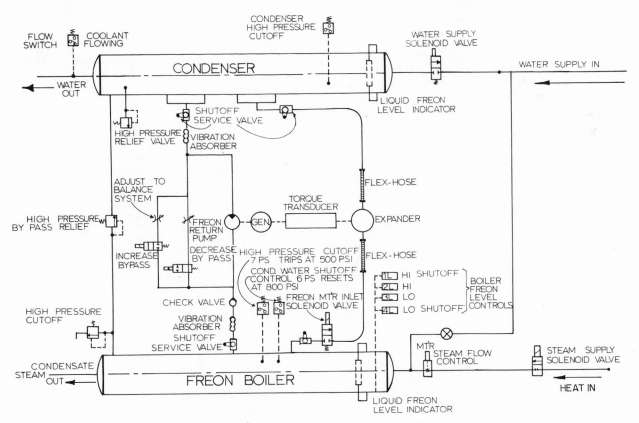

Figure 4-43. Fluorocarbon engine Rankine cycle schematic diagram.

industrial product at competitive market prices.

2. Long term contracts for disposal of recovered materials are mandatory for a financially healthy plant. The contracts should be preceded by well-established estimates of costs, quantities and markets. Similar contracts should be consumated with the local authorities specifying refuse deliveries and accounting for holidays and seasonal irregularities. Many municipalities finance waste disposal plants by creating independent authorities, empowered to float bond issues.

3. Urban authorities may encourage establishment of industries using the byproducts of the municipal stream through financial incentives.

4. Municipalities must not shy away from becoming marketers and consumers of these products.

Outside of social, economic and political aspects of waste disposal, which are not being examined here, there are other factors that promote recycling and recovery of energy from wastes. Here are some important ones:

1. As the problem of waste disposal assumes larger proportions, increasingly beyond the capabilities of municipal authority, state and federal legislation regulating disposal is bound to appear. An example of this trend is the Resource Conservation and Recovery Act of 1976.

2. Technological improvements undoubtedly will be able to handle plastics, chemical discharges, pollutants and other deleterious wastes and render them harmless. Technology will make recovery of materials and energy feasible and profitable.

3. Landfill and other customary disposal operations will become more and more costly, tilting the balance towards material and energy recovery as a profitable alternative. Oil and other fuel costs also promise to rise, augmenting the tendency to tilt the balance faster and further.

4. Research and development is creating new products not only out of waste but also by and out of the incineration process. It can not be foreseen what effects the new products will have either on waste disposal or on the development of power from the incineration process.

4.12.2 *Industrial Wastes*

Industrial Gases. The most important gases, which can be viewed as a source of considerable waste energy, are refinery, coke-oven, regenerator and blast furnace gases. Refinery gas, a by-product of gasoline refining, has a heating value higher than that of natural gas. For this

reason, it is usually blended with other gases for commercial exploitation.

Coke-oven gas is generated during the high temperature carbonization of bituminous coal into coke. Its calorific content ranges between 460-650 Btu/ft^3 (17-24 MJ/m^3). It has to be cleaned of particulate matter before it is burned. Most gas burners can be used to burn coke-oven gas with minor modifications.

Regenerator gas, a by-product of refinery fluid-catalytic-cracking units, has a high inert gas and solids content and is produced at a high temperature of 700–1000F (371-538C). Its calorific value is very low, only 40 Btu/ft^3 (1.5 MJ/m^3) so that supplementary oil firing is necessary. Boilers burning regenerator gas, which is mainly carbon monoxide, are used for base load applications and a minimum steaming rate must be ensured for complete combustion of carbon monoxide.

Blast furnace gas is laden with dust and carbon monoxide (as much as 30% by volume). It must be cleaned of dust and other particulates in order not to clog heat transfer surfaces, hot gas pipes and burners. Depending on the process sequence, blast furnace gas varies in heat content and in volumetric flow rate. Hence, burners must be designated to handle supplementary fuels. Boilers of this type are difficult to design and operate; maintaining steaming rate and superheat are also problems.

Liquids. Liquid industrial wastes include spent solvents and sulfite liquors, tars, waste oils, resins, polymers, chlorinated hydrocarbons, phenols, cresols, greases, combustible chemicals, fats, etc. The properties of these materials vary considerably. If a liquid is homogeneous, the burner can be modified to handle it. Blended wastes of unpredictable homogeneity can be burned in a refractory furnace (Loddby furnace) and the resultant heat can be captured in a downstream waste heat boiler. The refractory furnaces have capacities from 14,600 kW to 58,600 kW (50 to 200 million Btu/hr) and have recirculation patterns that evaporate liquid wastes before they reach the refractory walls.

Solids. Some industries, such as sugar, plywood, paper and furniture, use their wastes (bark and wood wastes, sawdust, wood shavings, bagasse, etc) to generate steam for their process needs.

Table 4-9 gives average calorific and density values that can be used for preliminary estimating for certain industrial wastes. Actual heating and density values must be checked separately for each project, as they depend on the composition of the wastes.

The properities of bark and wood wastes, for example, depend not only on geographic location but also on mill practice, soil composition, type of wood and method of

Table 4-9
Heating Values and Densities of Industrial Wastes

Type of Waste	Heating Value		Density	
	Btu/lb	MJ/kg	lbs/cu.yd.	kg/m^3
Gases[17]				
Blast-furnace	1139	2.7	-	-
Carbon monoxide	575	1.3	-	-
Coke-oven	19700	45.8	-	-
Refinery	21800	50.7		
Liquids[17]				
Black liquor	3700- 4200	2.3- 4.8	-	-
Dirty solvents	10000-16000	23.1-37.0	-	-
Industrial sludge	3700- 4200	2.3- 9.8	-	-
Oily waste and residue	18000	41.9	-	-
Paints and resins	6000-10000	13.9-23.1	-	-
Spent lubricants	10000-14000	23.1-32.5	-	-
Sulfite liquor	4200	9.8	-	-
Solids[17,18]				
Bagasse	3600- 6500	8.4-15.1	-	-
Bark	4500- 5200	10.5-15.1	-	-
Bitumen waste	16570	38.5	1500	890
Brown paper	7250	16.8	135	81
Cardboard	6810	15.8	180	107
Coffee grounds	4900- 6500	11.4-15.1	-	-
Corn cobs	8000	18.6	300	178
Corrugated paper (loose)	7040	16.4	100	59
General wood wastes	4500- 6500	10.5-15.5	-	-
Grass (green)	2058	4.8	75	44
Hardboard	8170	19.0	900	534
Latex	10000	23.1	1200	712
Magazines	5250	12.2	945	561
Meat scraps	7623	17.7	400	237
Milk cartons (coated)	11330	26.3	80	47
Nut hulls	7700	17.9	-	-
Nylon	13620	31.7	200	119
Paraffin	18621	43.3	1400	830
Plastic-coated paper	7340	17.1	135	81
Polyethylene film	19780	46.0	20	12
Polypropylene	19860	46.2	100	59
Polystyrene	17700	41.1	175	104
Polyurethane (foamed)	17580	40.1	55	33
Resin-bonded fiberglass	19500	20.6	990	587
Rice hulls	5200- 6500	12.1-15.1	-	-
Rubber synthetics	14610	34.0	1200	712
Sawdust and shavings	4500- 7500	10.5-17.4	-	-
Shoe leather	7240	16.8	540	320
Tar paper	11500	26.7	450	267
Textile waste (nonsynthetic)	8000	18.6	280	166
Textile waste (synthetic)	15000	34.9	240	142
Vegetable food waste	1795	4.2	375	222
Wax paper	11500	26.7	150	89
Wood	9000	20.9	300	178

[17]Robert G.G. Schwieger, "Power from Waste, a Special Report," *Power*, (February, 1975), p. S.2.
[18]Kjell I. Erlandsson, "Using Solid Waste as a Fuel," *Plant Engineering*, (December 11, 1975), p. 133.

transportation. For instance, hydraulic debarking results in bark moisture content of 60–75%; drum debarking with wet handling, 45–63%; and drum debarking with dry handling, only 35–50%.[19] Sawmill wastes (sawdust, shaving, bark) and chips and paper mill wastes (mainly bark) can be burned on grates, in suspension or partially in suspension, with a fiscal burnout on a thin bed.

Bagasse, the remnants of sugar cane after the extraction of its liquid, consists of matted cellulose fibers and various particles. It has high moisture content (40–60%) and high ash content due to the silt picked up during the harvest.

Coffee grounds are usually pressed into cake form containing approximately 60% moisture. The residues from the instant coffee process are flushed into a tank from which they are taken in cake form to refractory cells to be burned by auxiliary fuel burners.

Industrial sludge is burned after premixing with dry sludge. The mixture is fed to a cage mill, where flash drying is initiated. The drying is completed by intermixing with boiler flue gases. A cyclone separator discharges the spent gases and the dry sludge is conveyed to the boiler.

4.12.3 *Municipal Waste*

The remainder of Section 4.12 deals mostly with municipal waste and Section 4.12.3, in particular, deals with mechanical stokers.

Stationary grates are seldom specified today. These are not cooled and they handle steam loads in the 5000–30,000 lb/hr (2300–13 600 kg/hr) range. Low-ash wastes, such as sawdust, can be burned on a flat or inclined stationary grate in a thin bed. Ash is blown off the grate by steam or air or may be raked manually. The fuel can be introduced by gravity hoppers or underfloor screw conveyors.

Dump grates, also seldom specified, are used in smaller boilers to handle wastes of high moisture content. The grate is divided into sections, each served by an individual feeder. Air and feed are suspended at one section and its fire is allowed to burn down before cleaning. The other sections continue normal operation. Each grate section is powered by steam operators and is tilted for cleaning and dumping ashes into the ashpit. During dumping, particulate emissions increase, requiring very efficient filtering equipment.

Reciprocating grates are versatile in three ways: First, they are able to accommodate a wide steam range of 5000 to 75,000 lb/hr (3200–34 000 kg/hr). Second, they can fit any length and width due to their modular construc-

tion. Third, they can burn waste products of almost any size, even unprocessed refuse. The assembly consists of alternate stationary and reciprocating rows, with the entire grate assembly being at a slope. Alternate rows are powered to reciprocate across the stationary rows. The movement tumbles and aerates the burning mass; gravity moves it toward an ashpit. Ash is dumped intermittently or continuously through a dumping grate or other provision. Underfire combustion air is supplied through ports in the grating. This design owes its origin to coal overfeed stokers, which were intended for caking bituminous coals that required agitation of the bed. This grate is especially suitable for larger, unshredded municipal refuse and is being used in the Harrisburg, Penna. and Chicago incinerators. One of the main advantages of this system is that the ash produced from raw refuse is nearly sterile. A major disadvantage of all mass burning grates, including the reciprocating type, is that the firing pattern of refuse is relatively uncontrollable and, therefore, the steam rate is similarly unpredictable.

Vibrating grates are similar to reciprocating grates in that their surfaces are vibrated intermittently to move the bed forward. A timing mechanism controls the frequency and duration of the vibration in proportion to load demand. The grates are water-cooled to prevent grate deterioration and the formation of clinkers.

Traveling grates are common for solid waste applications. The lower steam capacity limit is 50,000 lb/hr (22 700 kg/hr) with an upper limit in the neighborhood of 400,000 lb/hr (181 000 kg/hr). This type of grate can handle a variety of refuse sizes. Speed varies between 4–20 ft/hr (1.2–6.1 m/hr). Preheated air is supplied through slots. Traveling grates, in combination with spreader stokers, have greater capacities than reciprocating grates and cost less per ton of refuse burned.

Large systems have an overfire fan, which fulfills two purposes when it is associated with spreader stokers: it supplies air to the refuse feed nozzles and combustion air above the grates. In the reciprocating stoker, it supplies overfire air only. In addition, there is an underfire air fan, an oil burner fan and induced-draft fan that transports combustion gases from the boiler to atmosphere through a suitable air pollution control system. The induced draft fan maintains a slightly negative pressure in the boiler and helps prevent gases from escaping through the boiler housing.

Either electrostatic precipitators or bag houses are required to control particulates in the flue gases in most cases. The choice depends on an economic analysis and applicable regulations. Electrostatic precipitators are efficient and insure small pressure drops. Their operating

---

[19]Robert G. Schwieger, *op. cit.*, p. S.4.

costs are low; their first costs, high. Scrubbers are also used for dust collection, although to a lesser extent than precipitators or fabric filters in industrial boiler installations.

Pneumatic feeders are used almost exclusively for distributing shredded fuel to boilers. In some cases their distribution spouts act as spreader stokers. There is a wide variety of waste feed distribution systems, including overthrow motors and motorized dampers, to assure even fuel distribution throughout the furnace. In instances where the basic fuel is coal or oil, and a fixed percentage of municipal refuse is the supplementary fuel, a pneumatic refuse distribution system is required in addition to either coal spreader stokers or oil burners. The operation of overfeed type of spreader stokers promotes combustion of the finer particles in suspension and deposition of the coarser particles on the moving grate. Use of a spreader stoker presupposes shredding of the material. Only bulky material must be shredded when reciprocating grates are used.

Fan static pressure to overcome the resistance of the burning mass on the grate is difficult to predict because it depends on the size of the shredded particles, grate speed and type of grate. It can vary from 1-7 in. of water (0.24-1.74 kPa).

Moving grates provide continuous ash discharge and, generally, higher average ratings, increased efficiency, more uniform steam pressures and less attendant labor than stationary grate systems.

4.12.4 *Size Reduction, Separation, and Energy Resource Recovery*

Reduction in size of mixed municipal refuse and material separation may be thought of as two basic preprocessing steps for landfilling, incineration, pyrolysis or biodegradation. Size reduction alone shrinks volume up to 50%, an important advantage in landfilling. In a landfill operation, reduction and separation may become economically justifiable options, depending on land values, materials markets and the cost of ownership of a facility. An economic analysis of landfill costs versus refuse processing will establish the advantages of each. When separation is combined with size reduction, landfill requirements are shrunk even further. Credit may br obtained for the recovered materials, if a ready market exists. Economic recovery of costs renders any method of waste disposal more attractive.

When remains after all recoverable materials are removed can be dried and pulverized. The resultant product has a heating value on the order of 7500 Btu/lb (1.74×10^7 J/kg) and represents 60% of the original municipal refuse.[20]

Size reduction by shredding was used initially to increase combustion efficiency but there are other advantages:

1. Volume reduction by 50%,
2. Greater homogeneity of refuse,
3. Conveyor transporation, magnetic separation and air classification are facilitated,
4. More surface for pyrolysis and incineration becomes available by cutting up materials to 1-6 in. (25.4-152 mm) particles.

Primary shredders are of three types: vertical shaft, horizontal shaft and flail mills (not much used at present). Vertical and horizontal types, also known as "hammer-mills," reduce municipal refuse to a uniform mixture. The vertical mill discharges the shredded refuse through a bottom grate. The horizontal shaft mill can more readily reject objects that cannot easily be shredded. Secondary shredding is accomplished by horizontal and vertical shredders, disk mills, rasp mills, cage disintegrators, wet pulpers, grinders, pulverizers, etc., which can reduce material to 1-5 cm pieces. Extensive processing is required for further size reduction.

Separation, segregating the component materials, is an important step in processing municipal refuse. The economics of each separation technique depends on the market value of the salvaged material. The following are some practical methods used in separating materials:

Hand sorting is effective but subject to human error and escalating labor costs. One person can pick 0.45 to 0.68 metric tons of newsprint per hour.[21]

Screening of materials depends on the size that can be fed through a particular screen. The size of municipal refuse must first be reduced by some other method. Various materials can be separated by size. In practice, horizontal vibrating screens clog. Rotating screens are more effective and are self-cleaning.

Magnetic separators are in wide use. Separation depends on size reduction and the degree of segregation of ferrous from nonferrous materials. Magnetic separators incorporate either natural or electromagnets, and are generally either drum or belt type. Separation efficiencies vary with infeed conveyor system configuration, ferrous and contaminant content of feed, and may approach 90% recovery.

---

[20] C.J. Huang *et al.* "Energy Recovery from Solid Waste," Vol. 2- Technical Report,: Document N75-25292. Springfield, Va.: National Technical Information Service, 1975, p. 51.

[21] *Ibid*., p. 44.

Air classifiers separate materials according to density. A jet of air fluidizes lightweight materials and transports them out of the stream while the heavy materials fall to the bottom. Air classifier type depends on chute and air flow arrangement.

Optical sorting is based on light-reflecting properties of materials, so separation is by color of the material.

High intensity electrostatic separation is used to classify glass and aluminum. The process is sensitive to size. Feed material is exposed to an electric charge. The non-conductors, which retain the charge, adhere to an oppositely charged drum, while the conductors drop off. The process is repeated for various sizes at different stages until all material is sorted.

Froth flotation is a mineral processing technique used to separate glass. Glass particles are preconditioned with a hydrophobic reagent prior to entering the flotation circuit. Air bubbles attach themselves to the glass particles buoying them to the surface in the form of a froth, which is removed by skimmers. Rotors mix the glass slurry providing air-glass mixing.

Process economics may be drastically affected by recovered materials quality or by specifications that may be developed covering quality.

4.12.5 *Incineration*

Incineration is very effective in reducing refuse to as little as 20% by weight and 8% by volume of the original material.[22] The inert residue may be used in landfill or as a building material. From a power standpoint, incineration presently provides the main vehicle for steam generation. The greatest economic return is expected in the area of incineration because of the economic value of salvaged materials and the resultant volume reduction, followed, logically, by an equivalent reduction in landfill space requirements. The economic and environmental benefits of energy recovery from industrial and municipal waste are additional inducements. The following four processes constitute the base for current and future development of power generation from refuse:

1. Water-wall incinerators for steam generation
2. Supplemental fuel systems for steam generation
3. Incineration of prepared refuse in boilers for steam generation
4. CPU-400 process for electric generation.

Water-wall incinerators, with integral boiler, superheater, economizer or, alternately, *refractory-lined furnaces* with a waste heat boiler mounted in the flue

downstream of the furnace, constitute the main means for energy recovery and steam generation. Steam, in turn, may power steam turbines for electric generation. Some of the steam produced may be used to operate auxiliary services such as undergrate heaters, soot-blowers, building heating and shredder turbine drive. The remaining steam can be sold, if there is an available market, or used in the plant itself for power generation, process work, heating, cooling and testing.

Furnace temperatures must be maintained at 1400-1800F (760-982C) to destroy all odors without fusing fly ash to walls or tubes. The furnace walls are constructed of steel tubes with fins in an all-welded panel assembly that provides an air-tight enclosure for the furnace. A layer of wear blocks or refractory over the tubes near the grate prevents tube damage. The combustion gases lose heat in the boiler, superheater and economizer sections and are cooled from 2500F (1371C) to 450C (232C) before they enter the dust collectors. Thus, the boiler, in generating steam, serves to cool the gases as well. There are several North American plants using water-wall incinerators, including plants in Montreal, Chicago, Ill., Northwest Harrisburg, Penna., Nashville, Tenn. and Saugus, Mass. and far in excess of this number in Germany, Japan and elsewhere.

Boiler efficiency in refuse water-wall incinerators does not exceed 65% due to the nonhomogeneity of refuse and its high ash and moisture content. There are two potential problems in a mass-burning, suspension-firing refuse boiler: fouling and corrosion of heat transfer surfaces. Build-up of slag and fly ash deposit can be minimized by sizing the furnace adequately and arranging heat transfer surfaces properly: wide spacing of tubes and low velocities at the convection section. Proper furnace sizing involves (1) adequate volume and residence time for complete combustion, (2) tubes in contact with dry ash that have adequate heat absorption, and (3) a sufficiently high temperature to destroy all odors.

Particulate emissions entering the precipitator average 3.5 grains/SCF (8×10^{-3} kg/m^3) and sulfur oxides measure only 0.12% of the ultimate analysis gas, compared to 3% for coal. Refuse with a PVC content of 2% generates flue gases that contain 690 PPM of HCl and a 12% CO_2 level.[23]

The devices used for particulate collection are mechanical collectors, wet scrubbers, fabric filters and electrostatic precipitators.

Mechanical collectors, with 75-80% efficiency on suspension-fired systems, cannot meet current U.S. Environ-

---

[22]*Ibid.*, p. 57.

[23]*Ibid.*, p. 58.

mental Protection Agency (USEPA) criteria. Wet scrubbers require high power input for high efficiencies, which depend on turbulence, liquid droplet size and the amount of liquid used. Wet scrubbers require approximately the same floor space as mechanical collectors but two or three times the scrubber floor space for associated water and ash-handling equipment. In addition, elaborate and costly water-treating equipment is necessary to remove the pollutant from the water. Fabric filters (bag houses) exhibit efficiencies of 99% but operational costs are high; space requirements and pressure drop are disadvantages. Electrostatic precipitators capture 99.5% of the entrained fly-ash. While operating costs are low and space requirements are similar to those of fabric filters, first costs are considerable.

Supplemental fuel systems use refuse to supplement the basic boiler fuel, which most often is oil or coal. The percentage of supplemental fuel varies between 10-20% based on the heating value of coal. Successful experiments have taken place at 25%. In the St. Louis, Mo., Meramec plant, the refuse firing rate equaled approximately 20% (reduced later) of the total fuel heat input. The 20% limit seems to be practical for coal-firing plants, without overtaxing the dust-collecting capability of an existing or conventional boiler design. Experience at other, similar facilities in Munich, Germany, confirm this percentage. The German installation uses two chambers, one for coal or oil and one for refuse. The gases from the two chambers are mixed and are treated in one precipitator.

Gas or oil-fired boilers are not suitable for burning refuse as supplementary fuel without considerable and expensive modifications to handle bottom ash and emissions. Supplemental fuel incineration compares favorably with inexpensive landfill disposal. In general, refuse as a supplementary fuel has to establish a more successful record before it is given unqualified acceptance.

Incineration of prepared refuse constitutes an important variation of the supplemental fuel system. The refuse, treated and shredded in a central facility, is then transported to incinerators as a supplementary fuel. The operation improves the logistical basis for fuel availability, *i.e.*, secure delivery of the quantity of fuel needed for uninterrupted plant operation on a fixed percentage of refuse fuel.

4.12.6 *Pyrolysis*
Pyrolysis is a process in which organic materials decompose at elevated temperatures without complete combustion, *i.e.*, in an oxygen-free or oxygen-deficient atmosphere. Pyrolysis requires heat, an endothermic process. The chemical reactions of the pyrolytic process are complex, simultaneous and/or consecutive and are not understood in their entirety. Pyrolytic products are a mixture of combustible gases, liquids and solid residues.

The process has not been demonstrated on a wide, commercial scale yet and there are conflicting claims, making a technical and economic evaluation rather difficult. Pyrolysis is of interest, from a power standpoint, because it produces combustible gases and liquids of low calorific value. The gases can be expanded in a gas turbine, as in the CPU-400 process, or they can be piped a short distance to a boiler or a process. Refuse derived liquid fuels can be burned separately or they can be blended with oils.

Pyrolysis occurs in three types of reactors: (1) shaft (vertical and horizontal), (2) rotary kiln, and (3) fluidized beds.

In *vertical shaft reactors* the feed materials is introduced at the top and settles into the reactor by gravity. Pyrolytic gas passes to the top from where it is removed. The feed mechanisms include screw conveyors, rotary devices and rams. A residue discharge mechanism and gas collection manifold are necessary. *The horizontal shaft reactor* has a conveyor (a molten bed system or mechanical) through the reactor housing. Refuse is continuously pyrolized. Under these circumstances, the reliability of the conveyor at high temperatures is a problem.

The *rotary kiln* is a rotating cylinder inclined to the horizontal. Cylinder rotation and inclination help to mix and to move material from the feed to the discharge end. In order to withstand the high temperatures, the metal cylinder is lined with refactory. One of the problems is the sealing of feed and discharge ports.

Fluidized beds have received more attention than any other combustion process due to their simplicity and promise.

The reactor consists of a cylindrical tube containing sand (dolomite, in some processes). The sand is held on a porous plate under which there is a gas plenum. The gas flow can be controlled and, at low flow rates, the bed remains packed to the point where the exerted gas pressure is equal to the weight of the bed. As flow increases, the sand and solid particles remain in suspension, allowing free gas flow (fluidization). In this state, the bed obeys the law of buoyancy (Archimedes' principle), has viscosity and maintains a surface parallel to earth; in other words, it acts like a fluid. Also as flow increases, the entire mass assumes a boiling action and the particles act as a heat reservoir at a uniform temperature throughout. When a refuse particle enters the bed, it ignites and the heat of combustion is transmitted to the bed. This process maintains both the ignition of foreign materials and bed temperature, *i.e.*, the process is self-sustaining

The bed acts like a thermal flywheel, giving and receiving energy. Foreign particles entering the bed act as heat exchangers.

Some constraints exist in the operation of the bed. In practical terms, bed temperatures can not exceed 2000F (1090C) so as not to approach the softening point (of the bed medium),[24] gas velocity must range between excess sand carryover and one that can maintain fluidization.

Yield and product composition depend on bed temperature. At 3000F (1650C), the reactor products consist mostly of gas and slag phases. As temperature is lowered, the gas becomes richer in higher molecular hydrocarbons and a liquid phase may also be present. The solid phase becomes more heterogeneous at lower temperatures.

Indirect methods of heating the pyrolytic mass (through a heat transfer wall) are not prevalent, not only because of inefficiency (heavy refractory linings) but also because of slag and corrosion problems. However, in indirect heating there is less CO_2 and H_2O formation and reduced NO_x and N_2 content.

As in any other refuse-to-energy systems, feeds vary from untreated refuse to preprocessed feed. In general a dried, finely shredded feed with all solid inorganic materials removed is most suitable.

The atmospheric pressure fluidized bed has the following potential advantages:

1. All combustible wastes can be used.
2. Tall stacks are eliminated.
3. For steam generation, it is less expensive to own and operate than pulverized coal plants, equipped with stack gas scrubbers.
4. Oil shale can be burned directly. Efficient combustion of gas, oil and coal can take place at temperatures as low as 800-900C. The process is insensitive to ash content of the fuel, so that high ash coals and gas of low calorific value can be burned. Cheap, low-grade coal can be mixed with solid feed to increase heating value. Liquid fuel can be sprayed to maintain constant bed temperature.
5. Low combustion temperatures induce low partial pressures in the alkali metal salts, minimizing corrosion of walls in contact with the fluidized bed or with the gases above the bed. Outage and maintenance costs should be lower and soot blowers should not be necessary.
6. The heat transfer coefficient between fluidized solids and submerged pipes is high and the required

heat transfer surface is greatly reduced. In cogeneration applications, this fact should help maintain a high power-heat ratio. Most of the energy release takes place in the bed. Approximately 50-80% can be removed to generate steam. Excess air requirements are reduced when a tube bundle removes heat from the bed. Consequently, gas-handling equipment is also reduced in size and scope.

7. The low fouling and corrosion attributes, together with process additives such as lime, should make it possible to use high sulfur coals and a wider range of fuels with little additional equipment. By adding lime to the combustion bed a substantial portion of the sulfur can be retained, reducing atmospheric pollution. Limestone can also retain vanadium found in heavy oils. The removed sulfur is in dry form and might be used in other applications. Since solid waste is low in sulfur content, the primary source of air pollution would be particulate matter, which is easier to filter.
8. Due to lower combustion temperatures and the paucity of air in the combustion process, lower NO_x emissions are expected.
9. The reduction of excess air for complete combustion reduces the size and cost of gas-handling equipment and the extent of cleanup operations required to meet pollution standards.
10. Distillation products are consumed in the fluidized bed and reduce the amount of unburned hydrocarbons leaving the unit.
11. Fluidized beds have a high volumetric heat generation, leading to compact sizes.

Fluidized beds appear to have more advantages than disadvantages. However, the following disadvantages may be mentioned:

1. Most probably an auxiliary fuel supply is needed to maintain stable bed temperatures due to the varying feed quality. The auxiliary burners may be necessary when the quantity of feed is partially or totally interrupted for a variety of reasons.
2. Erosion and carryover problems associated with solid particles are of a rather persistent nature.
3. Improper transfer and separation of solids create problems in the bed or in the rest of the system.
4. Power control is difficult, since only gas volume can be manipulated in a limited way. A minimum gas velocity must be maintained to sustain pyrolysis. The production of gases or steam cannot be affected

[24] Richard C. Bailie, "Solid Waste Incineration in Fluidized Beds," Industrial Waste Engineering, (November, 1970), p. 25.

to a significant degree except within the limits of minimum and maximum gas velocities.

With fluidized beds, generation of heat, steam and utilization of the generated gases and liquids are possible. So far, the shaft and kiln reactors have found more practical applications. The fluidized bed is now beginning to move beyond the experimental stage and some commercial units are available.

Some of the high temperature pyrolytic processes produce gases of low calorific value, which cannot be piped economically for long distances but must be consumed locally. A gas with a minimum calorific value of 300 Btu/ft^3 (11.19 MJ/m^3) is considered practical; gases with lower calorific values require large pipes and enlargement of burners. Some gases derived from pyrolytic processes can be upgraded to methane. Still other low temperature processes yield pyrolytic oils, which can be blended successfully with some No. 6 oils or which can be burned by themselves in fluidized beds or in boilers.

4.12.7 *Eco-Fuel and CPU-400*

Eco-Fuel is a solid fuel sold to utilities by a New York City firm. Municipal refuse is shredded in two stages and undergoes two stages of classification to separate the heavy, noncombustible fraction from the lighter, combustible fraction, which is subjected to further size reduction in a mechanical separator.

A variation of this method, called Eco-Fuel^{TM} II, uses a magnetic separator and chemical catalyst to aid size reduction. The fuel can be delivered to the site by truck or rail since it is a solid. In some actual installations Eco-Fuel left a great deal of ash in the combustion process, requiring the use of electric precipitators.

The CPU-400 system is being proposed by Combustion Power Company. Its advanced design has been tested and is based on sound engineering principles. The combustion process is pyrolytic; however, it uses prepared refuse and the objective is to drive a turbine that generates electricity.

The process works as follows: trucks deposit refuse in a central facility where it is shredded and classified. The light materials are air-classified and stored for use in a fluidized bed. The heavy material drops out. Glass, stone and aluminum are separated and processed for sale.

The prepared refuse is injected pneumatically into a sand fluidized bed at 1400 F (760 C). The gas emanating from the bed, containing some ash and sand, is cleaned in three stages of cyclones. The first stage removes sand, stones and aluminum; the second and third stages remove sand and smaller particles down to a size of less than five microns. (Even with this amount of gas-cleaning, the gas turbine blades must be replaced every two years.) After leaving the cyclones, the cleaned gas enters a two-stage gas turbine. One stage drives the air compressor and the other the turbogenerator. This process has a very desirable result: gas directly from the pyrolytic process is piped to a gas turbine to be used as fuel, dispensing with intermediary equipment such as boilers, pumps, condensers, etc.

As of this writing, a complete pilot plant has been tested on a limited basis and has not succeeded, for several reasons: the gas must be cleaned to superior standards in order to avoid turbine erosion and the cleaning must be done at high temperatures. Molten aluminum solidifies throughout the system as aluminum oxide. The refuse feed is not of constant consistency and contributes to problems in maintaining the fluid nature of the bed. In basic terms, the energy process, fluidization and gas handling, are not advanced or consistent enough to fulfill the high requirements of power generation.

4.12.8 *Biodegradation and Methane Production*

Biodegradation is the reduction of refuse by either of two organic methods. One is biochemical, in which protein or glucose is produced from cellulose by hydrolysis using enzymes and fungi. However, this method is still experimental. The other method includes reduction of refuse by biological organisms, which includes aerobic or anaerobic conversion. Composting, which is an aspect of the aerobic biological method, is not dealt with here since it has no implications in the power field. However, methane, which is produced by anaerobic digestion of organic wastes, can play an important role in energy production.

Industrial production of methane is of interest to the power field as a fuel. Two main sources are sewage sludge and industrial wastes; the latter are derived not only from petrochemical plants but from food processing plants and agribusiness in the form of cattle manure.

The **Pfeffer-Dynatech** anaerobic digestion system produces both methane and carbon dioxide, which must be separated from it. This system is based on known technology but is still in the pilot stage. The economics are improved if a known market exists for CO_2 product and if credits can be obtained for the recovery of aluminum and glass. The system consumes 37% of its methane production for its own energy needs.

Another proposed methane recovery system, from which metals and glass have been removed, involves municipal refuse mixed with dry sludge and compacted in a specially prepared landfill. A leachate barrier is placed at the bottom and sides. Radial distribution tiles and vertical pipe shafts are built-in and the top surface is sealed from air and water. A pumping system pumps water throughout the landfill using the distribution field. The landfill is thus converted into an anaerobic digester. In approximately one year, methane production begins and reaches top production rates in about the second year and progressively decreases until it ceases about the tenth year, at which time the landfill can be dug and sold as compost. The site may be used as landfill or for any other purpose. Methods to speed decomposition are being studied. A recent Palos Verdes, Cal. (Los Angeles County) landfill experiment of one year's duration failed to show economic viability. In this experiment, the gas was cleaned to utility line purity.

Biochemical conversion of industrial products into fuel will have to compete with all other conversion methods. This area is still under development. One potential market is the conversion of cellulose to a liquid fuel, methanol.

BIBLIOGRAPHY

Actual Plants

Coad, William J. "Energy Economics: a Design Parameter," *Heating, Piping, Air Conditioning*, (June, 1973), 73–78.

Elliott, Thomas C. "New Airport Reflects Latest Technology," *Power*, (December, 1971), 34–38.

Heavey, Jr., William F. "Airports Face New Energy Demands," *Power*, (September, 1971), 73–75.

"How Coal-fired District Steam, HTW System Serves NYU Campus," *Heating, Piping & Air Conditioning*, (August, 1968), 68–71.

Kneen, Thomas B. "Trends in University Utility Expansion," *Proceedings, National District Heating Association* (now International), Vol. LVI, 1965, 117–120.

Michaud, Robert L. "Super Central Supplies Steam, CW to Minneapolis," *Heating, Piping & Air Conditioning*, (December, 1972), 43–51.

"Nashville Turns Solid Waste into District Steam and Chilled Water," *Power*, (December, 1974), 18–19.

"Piggybacked District Cooling for California's Capital Mall," *Heating, Piping & Air Conditioning*, (March, 1970), 75–82.

Prues, L.J. "Central Cooling for a Downtown Business Area," *ASHRAE Journal*, (May, 1971), 47–51.

Steinman, William R. "Case History: Total Energy at Work in Pennsylvania Shopping Mall," *ASHRAE Journal*, (September, 1971), 60–63.

Strauss, Sheldon D. "Heat Recovery Takes a Fresh Turn," *Power*, (January, 1976), 19–23.

Sullivan, F.P. "Co-op City Gets Unified Energy System," *Power*, (March, 1968), 72–77.

Whitaker, Lincoln E. "Campus Central Plant Heat and Cools," *Power*, (July, 1973), 28–29.

Coal and Coal-fired Equipment

"Ash Handling in Commercial and Industrial Plants," *Air Conditioning, Heating and Ventilating*, (April, 1961), 71–86.

"Coal Stoker Fired Steam Plants, Guide Specification: GS-4," Washington, D.C.:National Coal Association, 1963.

Coffin, B. Dwight. "Estimate the Cost of Your Next Coal-Fired Industrial Boiler Plant," *Power*, (October, 1977), 28–29.

"Combustion Controls for Industrial Power Boilers," Washington, D.C.: National Coal Association, 1965.

Corey, Richard C. and Ronald R. Bevan. "Burning Coal in CPI Boilers," *Chemical Engineering* (January 16, 1978), 110–128.

Dick, John L. "The Vibrating-Grate Stoker in Campus Heating Plants," *Proceedings, National District Heating Association* (now International), Vol. LII, 1961, 164–178.

"Estimating Coal Consumption from Steam Load," Section B-1.1. Washington, D.C.:National Coal Association, 1961.

Garvey, James R. "Designing Coal-Storage Bins for Positive Discharge," *Proceedings, National District Heating Association* (now International), Vol. L, 1959, 159–168.

Keith, R., *et al.* "A Special Report, Fuels," *Power*, (June, 1968), S.1-S.48.

Miller, Earl C. "Pressurized Firing with Spreader Stokers," *Proceedings, National District Heating Association* (now International), Vol. LII, 1961, 179–188.

"Modern Dust Collection for Coal-Fired Industrial Heating and Power Plants," Section F-2. Washington, D.C.: National Coal Association, 1961.

"Operation and Management for Coal-Fired Boilers," Section G-1. Washington, D.C.: National Coal Association, 1962.

Robertson, James E. "Selection and Sizing of Coal Burning Equipment," *Power Engineering*, (October, 1974), 44–45.

"Should You Convert to Coal?" *Power*, (July, 1976), 38–40.

Schwieger, Robert. "A Special Report, Power from Coal, Part I, Coal Selection and Handling," *Power*, (February, 1974), S.1-s.24, "Power from Coal, Part II, Coal Combustion, (March, 1974), S.25-S.48 and "Power from Coal, Part III, Combustion Pollution Controls," (April, 1974), S.49-S.64.

Condensers

Ferwerda, G.G.J. "Elaborations on What Happens Inside Steam Heated Condensers," *Heating, Piping, Air Conditioning*, (January, 1969), 140–148.

Harnish, J.R. "Controlling Condensing Temperature," *Air Conditioning, Heating and Ventilating*, (May, 1962), 80–81.

Lichtenstein, Joram and C.G. Siegfried. "Controlling Corrosion in Surface Condensers," *Power Engineering* (November, 1976), 54–57.

"Know Your Condenser," *Power Engineering* (reprint). Chicago:Technical Publishing Co., no date.

Miliaras, E.S. *Power Plants with Air-Cooled Condensing Systems*. Cambridge:MIT Press, 1974.

Pnueli, D. *Economic Comparison between Water and Air-Cooled Condensers*. Haifa: Israel Institute of Technology, 1959.

Smith, Ennis C. and Michael W. Larinoff. "Power Plant Siting, Performance and Economics with Dry Cooling Towers," *Proceedings of the American Power Conference*, Vol. 32, 1970, 544.

Spenser, Elliot. "Specifying Steam Surface Condensers," *Buildings Systems Design*, (July, 1970), 32–37.

Swift, Donald. "Starting and Running Your Condenser's Circulating Water System," *Power*, (October, 1957), 84–85, 192, 196, 198, 200, 212.

———— "Condensers: Surface and Jet Types Meet Most of Today's Needs," *Power*, (October, 1957), 84–85, 192, 196, 198, 200.

———— "Follow These Methods for Condenser Testing and Operation," *Power*, (March, 1958), 98–99, 202, 204, 210, 212.

Heat Balance

Baily, F.G., K.C. Cotton and R.C. Spenser. "Predicting the Performance of Large Steam Turbine-Generators Operating with Saturated and Low Superheat Steam Conditions," General Electric Company document GE-2454A.

Elston, C.W. and P.H. Knowlton. "Comparative Efficiencies of Central Station Reheat and Non-Reheat Steam Turbine-Generator Units," ASME paper 51-A-139.

Hegetschweiler, H. and R.L. Bartlett. "Predicting Performance of Large Steam Turbine-Generator Units for Central Stations," ASME paper 56-SA-52. General Electric Company document GER-1222.

Leung, Paul and Raymond E. Moore. "Thermodynamic and Economic Analyses of Closed Feedwater Heaters for Supercritical Pressure Steam Turbine Cycles," ASME paper 67-WA/PWR-5.

Newman, L.E. "Modern Extraction Turbines," Part I, *Power Plant Engineering* Part I, (January, 1945), 76–79 and Part II (February, 1945), 78–81, Part III (March, 1945), 82–86 and Part IV (April, 1945), 74–79.

Pollard, E.V. "Feedwater Heating Calculations for Steam Turbines," *Industry and Power*, (May, 1953), 60–66.

———— "Calculations of Comparable Heat Rate Correction Factors," *Industry and Power* (December, 1952), pp. 81–85 and (January, 1953), pp. 74–77.

Salisbury, J.K. "Optimization of Heater Design Conditions in Power Plant Cycles," ASME paper 68-WA/NE-12.

Salisbury, J. Kenneth., ed. Kent's Mechanical Engineers' Handbook; Power Volume. Section 8. "Steam Turbines and Engines." 12th ed. New York:John Wiley & Sons, Inc. 1954. pp. 8 (57) - 8(95).

Spenser, R.C., K.C. Cotton and C.N. Cannon (General Electric Company). "A Method for Predicting the Performance of Steam-Turbine-Generators 16,500 kW and Larger." General Electric Company document GER-2007c. Schenectady, N.Y., Rev. July 1974.

Willoughby, William W. "Steam Rate: Key to Turbine Selection," *Chemical Engineering* (September, 1978), 146–154.

Organic Fluid Rankine Cycle Systems

Angelino, G. and V. Moroni. "Perspectives for Waste Heat Recovery by Means of Organic Fluid Cycles," *1973 Transactions of the ASME, Journal of Engineering for Power*, (April, 1973), 75.

Barber, Robert E. "Rankine-Cycle Systems for Waste Heat Recovery," *Chemical Engineering*, (November 24, 1974), 101–104 & 106.

Curran, H.M., et al. (Hittman Associates). *Assessment of Solar-Powered Cooling of Buildings*. NTIS Document C00-2622-1. Sponsored by Energy Research and Development Administration, Washington, D.C. Springfield, Va.:National Technical Information Service, 1975.

Doyle, E., S. Helekar and R. Raymond. "Diesel-Organic Rankine Compound Engine Development," Waltham, Mass.:Thermo Electron Corp., 1977.

Morgan, D.T. and J.P. Davis. "High Efficiency Gas Turbine/Organic Rankine Cycle Combined Power Plant," ASME Paper 74-GT-35.

Morgan, Dean T. and Jerry P. Davis. "High Efficiency Decentralized Electrical Power Generation Utilizing Diesel Engines Coupled with Organic Working Fluid Rankine-Cycle Engines Operating on Diesel Reject Heat," Waltham, Mass.:Thermo Electron Corp., 1974.

"Organic Rankine Cycle Waste Heat Recovery/Power Generation Systems," Sarasota, Fla.: Kinetics Corp., no date.

Soini, H.E., P.S. Patel, et al. "Combined Diesel-Organic Rankine-Cycle Powerplant," Waltham, Mass.:Thermo Electron Corp., no date.

Sternlicht, Beno. "The Equipment Side of Low-Level Heat Recovery," *Power*, (June, 1975), 71–77.

_____ "Low-Level Heat Recovery Takes on Added Meaning as Fuel Cost Justify Investment," *Power*, (April, 1975), 84–87.

System Criteria and Design (See Chapter 2, Bibliography on Power Generation, Central Plant and Process Heat Overview.)

Alhart, H.E. "How Meter System Monitors Lube Oil," *Power*, (November, 1971), 61–62.

Bolafio, Robert E. "Flash Tank Selection and Application," *Air Conditioning, Heating and Ventilating*, (June, 1962), Reference Data, no page.

_____ "Sizing Vertical Flash Tanks," *Air Conditioning, Heating and Ventilating*, (July, 1964), Reference Data, no page.

Booser, E.R. and D.A. Smeaton. "Piping for Circulating-Oil Lube Systems," *Power*, (August, 1961), 91–93.

Booser, E.R. and D.A. Smeaton. "Instrumentation and Control of Circulating-Oil Lube Systems," *Power*, (May, 1962), 74–76.

Crosthwait Jr., D.N. "A System Approach to Flash Tank Applications," *Heating, Piping, Air Conditioning*, (January, 1971), 137–139.

Dawson, Richard S. "Compression Tank Selection Chart," *Air Conditioning, Heating and Ventilating*, (June, 1957), 83–84.

Diamant, R.M.E. "District Heating with Combined Heat and Power Generation," *The Heating & Air Conditioning Journal*, (January, 1977), 16–18 and (February, 1977), 28–32.

Ediss, B.G. "Steam Injection Can Improve Gas Turbines," *Power*, (June, 1970), 82–84.

Ecabert, E. and L. Stutz. "Economics of Combined Gas-Steam Cycle Plants," *Sulzer Technical Review*, (November, 1965), 81–85.

Elmendorf, R.G. "Today's Soot-Blower Design," *Power*, (December, 1957), 107–109, 214, 216, 218, 220.

Elmenius, L. "The Application of By-Product Power Rate," Tappi Magazine (May, 1972), 713–718.

Ferwerda, G.G.J. "Flash Tank Design and Steam Usage," *Heating, Piping, Air Conditioning*, (July, 1969), 99–103.

Fischer, P., J.W. Suitor and R.B. Ritter. "Fouling Measurement Techniques," *Chemical Engineering Progress*, (July, 1975), 66–72.

Gore, Eugene and A. Salim Qureshi. "Design Guidelines for Total and Selective Energy Systems," *Heating, Piping, Air Conditioning*, (September, 1977), 57–64.

Gore, Eugene and A. Salim Qureshi. "Design Guidelines for Continuous Duty Standby Systems," *Heating, Piping, Air Conditioning*, (September, 1977), 69–73.

Hicks, Tyler G. "Analysis of Boiler Air Ducts and Gas Uptakes," *Standard Handbook of Engineering Calculations*, 1972, Chapter 3, 253–257.

Fleming, W.S. "Total Energy Concepts for Refrigeration and Air Conditioning," *ASHRAE Journal*, Part I, (November, 1970), 49–55 and Part II, (December, 1970), 32–39.

Hafer, A.A. and W.B. Wilson. "Gas Turbine Exhaust-Heat Recovery," ASME paper 54-A-194.

Hichley, Peter. "How to Avoid Problems of Waste-Heat," *Chemical Engineering*, (September, 1975), 94–98.

Kates, Edgar J. "A Special Report, Oil and Gas Engines," *Power*, (December, 1964), S.1-S.32.

Klug, Harold H. "Problems in Steam Design," *ASHRAE Journal*, (April, 1964), 43–49.

Krieg, Tom E. "How to Size Condensate Receivers for Gravity Return Systems," *Heating, Piping, Air Conditioning*, (June, 1968), 124–127.

Lockhart, H.A. and G.F. Carlson. "Compression Tank Selection for Hot Water Heating Systems," ASHVE Journal Section, *Heating, Piping, Air Conditioning*, (April, 1953), 132–139.

Mangan, J.C. and R.C. Pettit. "Combined Cycle with Unfired Boiler Has High Efficiency," *Power Engineering*, Part I, (August, 1963), 59–61 and Part II, (September, 1963), 47–49.

Mathur, Jimmy. "Performance of Steam Heat-Exchangers," *Chemical Engineering*, (September 3, 1973), 103–104.

McConnell, John E. and Lars Elmenius. "A Systematic Approach to the Economic Selection of Design Parameters for an Integrated Industrial Power Plant," *Combustion*, (December, 1976), 7–12.

McConnell, John E. and Keith C. Wein. "Combined Gas-Steam Cycles Offer Savings to Industry," *Power*, (May, 1963), 68–71.

Meckler, Milton. "Options for On-Site Power," *Power*, (March, 1976), 33–35.

Miller, Carl E. "Standardized Industrial Boilers for District Heating," *Proceedings, National District Heating Association* (now International), Vol. XLIV, 1953, 198.

Nagib, M.M. "Analysis of a Combined Gas Turbine and Absorption Refrigeration Cycle," *Transaction of the American Society of Mechanical Engineers* (ASME), Series A, Vol. 93, 1971, 29.

Neal, Gordon W. "Advantages of Combined Power/Process Generating Plants," *Power Engineering*, (February, 1977), 56–59.

North American Mfg. Co., *Combustion Handbook*, 2nd Ed. Cleveland, Ohio, 1978.

O'Keefe, William. "Air Dryers, Aftercoolers and Filters," *Power*, (December, 1972), 21–28.

Patterson, M.M., E.V. Pollard and W.B. Wilson. "Economics of Higher Pressures and Temperatures for Steam Turbines and Industrial Plants," *Proceedings of the American Power Conference*, Vol. XIX, 1957, 222.

"Picking Compressors for Air Soot Blowing," *Power*, (September, 1956), 84–86.

Pirille, Walter. "Precipitator Performance Hinges on Control," *Power*, (January, 1975), 23–26.

Phillips, E.J. "M-d Turbine Lubricating-Oil Systems," *Power*, (June, 1959), 91–93.

Reed, J.M., *et al.* "Fuel Oil for Engineered Systems," *Air Conditioning, Heating and Ventilating*, (July, 1963), 42–71.

Samuels, G. and J.T. Meador (Oak Ridge National Laboratory). MIUS Technology Evaluation - Prime Movers. NTIS Document ORNL/HUP/MIUS-11. Springfield, Va.:National Technical Information Service, 1974.

Schwieger, Robert G. "A Special Report, Heat Exchangers," *Power*, (June, 1970), 33–48.

———. "Future Brightens for Combined-Cycle Plants," *Power*, (October, 1971), 105–109.

———. "Plant Design Today, a New Challenge," *Power*, (November, 1976), 37–60.

———. "A Special Report, Industrial Boilers, What's Happening Today," *Power*, (February, 1977), S.1-S.24.

Schletty, R.G. "Pressure Tank System Control," *Power*, (November, 1957), 118–120.

Schneider, Gilbert G., *et al.* "Selecting and Specifying Electrostatic Precipitators," *Chemical Engineering*, (May 26, 1975), 94–107.

Sculthorpe, William L. *Design of High Pressure Steam and High Temperature Water Plants.* New York:Industrial Press, Inc., 1972.

Shore, James T. "Total Energy," Chapter 14, vol. 3, Maintenance & Basic Fundamentals, *Sawyer's Gas Turbine Engineering Handbook.* 2nd Ed. Ed. J.W. Sawyer. Stamford, Conn.:Gas Turbine Publications, Inc. 1972.

Sullivan, F.P. "A Special Report, Electric Generators," *Power*, (March, 1966), S.1-S.24.

Swift, D.C. "Burners, Stokers and Combustion Air Affect Your Boiler's Rating," *Power*, (October, 1955), 106–107, 210, 214, 216, 218.

"The ABC of Engine-driven Generators and Their Control," Bulletin 2200-PRD-224, Minneapolis:Electric Machinery Mfg. Co., 1955.

The 1976 Energy Management Guidebook, published by the editors of *Power* magazine.

Trocolli, J.E. "What you should know about Flash Tanks," *Actual Specifying Engineer*, (now *Specifying Engineer*), (January, 1971), 71–73.

Whirl, S.F. "District Heating Using Electric Boilers and Steam Accumulators," *Proceedings, International District Heating Association*, Vol. LVIII, 1967, 62.

Wilson, W.B. "The Role of Turbines in Industrial Energy Systems," *Combustion*, (October, 1969), 39–44.

Wilson, W.B. and W.J. Hefner. "Economic Selection of Plant Cycles and Fuels for Gas Turbines," ASME Publication 74-GT-84.

Young, Norman W. "Keeping New Boiler Capital at a Minimum for District Heating Systems," *Proceedings, National District Heating Association* (now International), Vol. XIVI, 1955, 141.

Solid Waste

Alvarez, Ronald J. "Study of Conversion of Solid Waste to Energy in North America," *Proceedings of the National Waste Processing Conference.* Published by ASME, Solid Waste Div., 1976, 163.

Black, Dugald O. *"Energy Recovery from Solid Waste: A Review of Current Technology."* NTIS Document PB 230-633. Springfield, Va.:National Technical Information Service, 1976.

Bump, Robert L. "Handling Ash from Bark-fired Boilers," *Power*, (February, 1977), 95–96.

Cheremisinoff, Paul N. and Angelo C. Morresi. *Energy from Solid Wastes.* Vol. 1. New York:Marcel Dekker, Inc., 1976.

Elliott, E.D. "Possible Use of Fluid-Bed Combustion and Heat Transfer," *Proceedings, Total Energy Conference,* Vol. 36, 1974, 663.

Fisus, D.E., P.G. Gorman, *et al. St. Louis Demonstration: Refuse Processing Plant Equipment, Facilities and Environmental Evaluations.* NTIS Document PB-272 757/GWE. Springfield, Va.: National Technical Information Service, 1977.

Gore, Eugene, *et al.* "Design Guidelines for Solid Waste Boiler Systems," *Heating, Piping, Air Conditioning,* (August, 1977), 65–66.

Jones, Jerry. "Converting Solid Wastes and Residues To Fuel," *Chemical Engineering*, (January 2, 1978), 87–94.

Kispert, R.G., *et al.* "An Evaluation of Methane Production from Solid Waste," *Resource Recovery and Conservation* (No. 1, 1976), 245–255.

Klumb, David. "Union Electric Company's Solid Waste Utilization System," *Journal of the Washington Academy of Sciences*, (March, 1976), 217–239.

Metcalf and Eddy, Inc. and City of Lynn, Mass. *Generation of Steam from Solid Wastes.* Document PB-214-166. Sponsored by U.S. Environmental Protection Agency, Washington, D.C. Springfield, Va.:National Technical Information Service, 1972.

Neville, Charles B. "How District Heating/Cooling and Solid Waste Disposal Became Part of a Downtown Urban Renewal Project," *Specifying Engineer*, (February, 1976), 50–53.

Papamarcos, John. "Power from Solid Waste," *Power Engineering*, (September, 1974), 46–55.

Pennsylvania Department of Community Affairs. *Reclamation of Energy from Solid Waste:Theory and Practice. A Selected, Annotated Bibliography for Pennsylvania Local Government Officials.* Document PB-267 800/IWE. Springfield, Va.:National Technical Information Service, 1976.

Reese, T.G. and R.C. Waddle. *General Survey of Solid-Waste Management.* Document N74-27523/1. Springfield, Va.: National Technical Information Service, 1974.

Schultz, Helmut, W. "Energy from Municipal Refuse: a Comparison of Ten Processes," *Professional Engineer*, (November, 1974), 20–24.

———— "Cost/Benefits of Solid Waste Reuse," *Environmental Science and Technology*, (May, 1975), 423–427.

Sheng, Henry P. "Energy Recovery from Municipal Solid Waste and Method of Comparing Refuse-Derived Fuels," *Resource Recovery and Conservation*, (No.1, 1975), 85–93.

Trethaway, William. "Energy Recovery and Thermal Disposal of Wastes Utilizing Fluidized Bed Reactor Systems," *Proceedings of the National Waste Processing Conference*, Published by ASME, Solid Waste Div., 1976, 117.

Wilson, Maurice J. "Heat Energy from Waste:Cash for Trash," *Heating, Piping, Air Conditioning*, (April, 1974), 51–56.

———— "The Markets and the Economics of Heat Energy from Solid Waste Incinerators," *Journal of the Washington Academy of Sciences*, (March, 1976), 197–206.

CHAPTER **5**

Drives

5.1 Mechanical Drive Steam Turbines

5.1.1 *Classification*

Mechanical drive steam turbines, so called because they are ideally suited to drive mechanical equipment (refrigeration compressors, pumps, fans and other plant auxiliaries), are efficient and economical. They require no intervening linkages when converting the thermal energy of steam directly to the kinetic energy of a rotating shaft.

There is no single, definitive classification of steam turbines. One way, illustrated in Fig. 5-1, types them by operating characteristics. In the lower pressure range, as in Table 5-1, the materials of construction, whether cast iron, cast steel, etc., provide another item of differentiation, which is typical. Above 600 psig, each manufacturer has his own practice. Economic considerations tend to group turbines in terms of pressure, temperature and steam flow, as indicated in the following pressure and temperature classes, which have no official basis, and in Table 5-2, which is a NEMA standard.

Unit Size	Psig	Deg F[1]
Small	150– 400	500– 750
Medium	400– 600	750– 825
Large	600– 900	750– 950
Large	900–2000	825–1050

5.1.2 *Back Pressure Turbines*

Most mechanical drive turbines in the process industries are of the back pressure or reducing valve type. An assured demand for low pressure steam in a central plant

---

[1] *De Laval Engineering Handbook,* 3rd ed., Ed. Hans Gartmann, New York: McGraw-Hill Book Co., 1970.

Table 5.1
Turbine Material Classification

Material Class	1	2	3
Steam Pressure	Up to 250 psig (1720 kPa)	Up to 600 psig (4140 kPa)	Up to 600 psig
Steam Temperature	Up to 500 F (260 C)	Up to 600 F (316 C)	Up to 750 F (399 C)
Exhaust Pressure	Up to 75 psig (517 kPa)	Up to 75 psig	Up to 75 psig
Steam Chest	Turbine Quality Cast Iron	Cast Steel	Cast Steel
Casing	Turbine Quality Cast Iron	Cast Iron	Cast Steel
Nozzles and Blading	Stainless Steel	Stainless Steel	Stainless Steel
Inlet Valve and Cage	Carbon Steel Stellited	Carbon Steel Stellited	Carbon Steel Stellited

can render the back pressure turbine very useful. Figures 5-3 and 5-4 suggest the magnitude of power generation by using a topping turbine to produce low cost power and increase plant efficiency. If self-generation is part of the overall plan, the costs of high pressure boilers may be justified by using a high pressure, multistage turbine to generate, exhausting into a medium pressure turbine drive for a centrifugal refrigeration machine and exhausting this turbine into an absorption machine at low pressure in a piggy back arrangement.

125

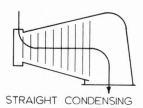

STRAIGHT CONDENSING

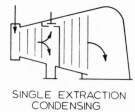

SINGLE EXTRACTION
CONDENSING

DOUBLE EXTRACTION
CONDENSING

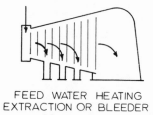

FEED WATER HEATING
EXTRACTION OR BLEEDER
CONDENSING

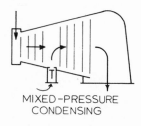

MIXED-PRESSURE
CONDENSING

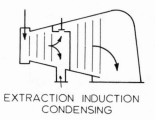

EXTRACTION INDUCTION
CONDENSING

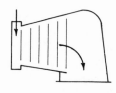

LOW-PRESSURE CONDENSING

REHEAT

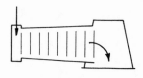

TOPPING OR BACK-PRESSURE
NON-CONDENSING

From Theory and Design of Steam and Gas Turbines by John F. Lee. Used with permission of McGraw-Hill Book Co.

Figure 5-1. Condensing and noncondensing turbine arrangements.

Table 5-2
NEMA Pressure and Temperature Classification of Steam Turbines [2,3]

	Rated Initial Steam Pressure, Psig	Rated Initial Steam Temperature					
		500F and Less	501F to 600F	601F to 750F	751F to 825F	826F to 900F	901F to 950F
Multistage / Single stage	101 to 250	1	..	..	..	..	..
	251 to 400	..	2	3	4	..	..
	401 to 600	..	5	6	7	8	..
	601 to 900	..	..	9	10	11	..
	901 to 1250	..	..	12	13	14	15
	1251 to 1500	..	..	..	16	17	18

[2] *NEMA Publication SM 21-1970,* Multistage Steam Turbines for Mechanical Drive Service, National Electrical Manufacturers Association, Table 4-1, Part 4, p. 2.

[3] *NEMA Publication SM 22-2970,* Single-State Steam Turbines for Mechanical Drive Service, National Electrical Manufacturers Association, Table 4-1, Part 4, p. 1.

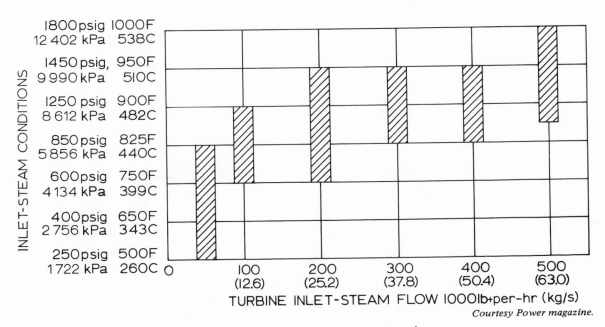

Courtesy Power magazine.

Figure 5-2. Usual throttle steam conditions vs. steam flows.[4]

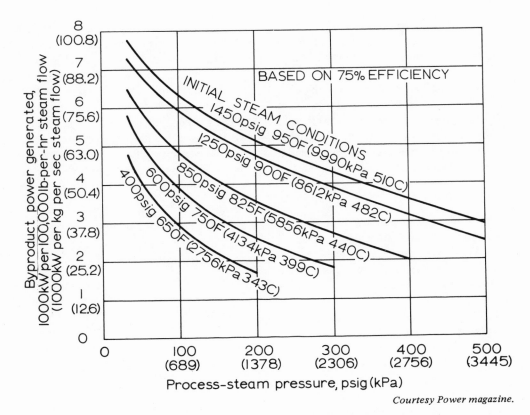

Courtesy Power magazine.

Figure 5-3. By-product power generation vs. initial steam conditions.[4]

[4] W. B. Wilson, "How to Use Steam Turbines in Refining, Petrochemical and Chemical Industries." *Power*, February, 1960, p. 78.

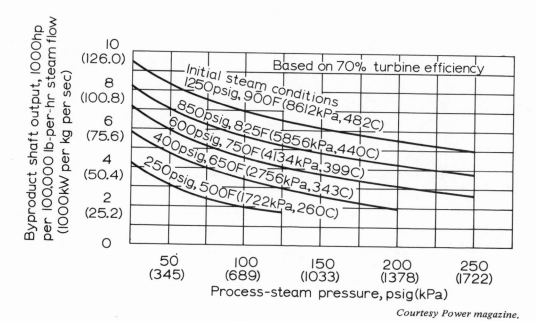

Figure 5-4. By-product shaft horsepower vs. initial steam conditions.[5]

The term "topping turbine" owes itself to the practice of fitting a high pressure turbine exhaust into existing lower pressure boilers or turbines, hence the term "topping."

Because back pressure turbines have high steam rates, they must be very efficient in order to fit into the heat balance of an existing plant. Therefore, multistage machines are required. A back-pressure turbine can be provided with an exhaust pressure regulator, when constant exhaust pressure is desirable.

5.1.3 *Condensing Turbines*

In the condensing turbine, steam expands from throttle conditions to condensing pressure. Its main advantage is a low steam rate. The higher the pressure at inlet conditions, the lower the steam rate for the same condensing pressure. Throttle conditions are invariably determined by economics, except in some process industries where there is an abundance of waste steam. The cost of higher pressure boilers, piping and turbines must be weighed against the savings because of lower steam rates over the life of the equipment. To compensate for the high waste of heat in condenser heat removal, use of the highly efficient multistage machine may be considered. A single stage machine is economically attractive in the smaller

sizes; however, additional equipment is necessary to implement the condensing cycle: surface condenser, hot well pumps, steam ejectors or vacuum pumps to maintain condenser vacuum, a cooling water system involving ponds, lakes or cooling towers, and other accessories which increase initial costs. Condensing turbines are applicable when process steam is not a concurrent demand with power generation or shaft output, since steam is then solely a waste product. When waste steam available to the condensing turbine cannot be used, a bypass may be used to conduct it directly to the surface condenser.

5.1.4 *Extraction Turbines*

An extraction turbine consists of two turbines in series, enclosed within the same casing. Extraction steam serves process needs or feeds to auxiliary equipment at one pressure; the second turbine, if non-condensing, may serve other process needs at another pressure. A second condensing turbine usually drives an electric generator. An extraction turbine may be condensing or non-condensing. Usually, extraction steam flow is controlled automatically by a built-in extraction valve, which controls the extraction pressure.

If extraction is uncontrolled, the turbine may be known as a "bleeder" turbine. A hole in the casing is adequate, but the extraction steam pressure is bound to vary considerably, depending on the proportion of steam that travels to the condenser. In an extraction turbine, all

――――――
[5]Wilson, "How to Use Steam Turbines in Refining, Petrochemical and Chemical Industries," *op. cit.*, p. 78.
[6]*Ibid.*

of the steam does work up to the extraction point. This portion of the work is done at a high steam rate, whereas the portion which proceeds to the condenser is the only part of the steam that has done work at a low steam rate.

Addition of a reducing valve to an uncontrolled extraction line to hold steam pressure steady is often called "reducing valve extraction."

Double extraction points are sometimes used in industrial work and triple extraction points for feedwater heating in power plants. In central energy plants, it is possible to use the non-condensing extraction turbine for power generation with an extraction point for driving a centrifugal compressor and exhaust steam for hot water heating, absorption refrigeration or any other project need. In the case of condensing turbines, an extraction point can feed refrigeration machines when needed; when not, the steam proceeds to the condensing section to generate power, if project needs call for this type of operation. An extraction turbine is not built for less than 500 hp (373 kW).[7]

The mixed pressure turbine and the extraction-induction turbine are variations of the extraction turbine. In the mixed pressure turbine, steam expands to a suitable intermediate pressure, at which point low pressure steam emanating from various processes, waste heat boilers or accumulators is admitted into the turbine to augment flow. Low pressure steam is considered the primary steam source and maximum use is made of it, as long as it is available, before supplementary high pressure steam is brought into the turbine.

The extraction-induction (or "mixed-flow") turbine can function as a mixed pressure or an extraction turbine, *i.e.*, it can extract steam at the required pressure or admit it at the same pressure. It is geared to provide a convenient solution to the problem of irregular steam supply and demand. Process steam, when in abundance, can be induced into the turbine to drive a generator or centrifugal compressor, but when there is a heavy process steam demand, the machine is operated in its extraction mode, with generator or compressor load satisfied only partially. A solution of this type, naturally, requires extremely careful load analysis.

5.1.5 *Low Pressure Turbines*

Substantial energy is extracted from steam when it is expanded from a pressure at, or just above, atmospheric to high vacuum and subsequently condensed. Turbines operating in this low pressure range generate power from

steam that otherwise would be wasted. Plants with a dependable source of low pressure steam can generate power or drive machines or auxiliaries while improving plant efficiency. If low pressure steam supply is erratic, a mixed pressure turbine may be employed to meet power demand.

5.2 Steam Turbine Performance

Quite often it becomes necessary to estimate approximate steam rates and flows in connection with preliminary estimates for new plants, expansion or modernization of plants, when making comparisons between steam-driven and electrically driven auxiliaries, in tentative equipment selection for space arrangement, or for energy studies. Of course, the most dependable method is to obtain such information from a manufacturer. Turbine equipment purchase must be based on precise selections with turbine performance specified. All manufacturers have performance test curves for all their turbines and it is appropriate to let manufacturers offer their best machine for a specific task. However, the length of time it sometimes takes to get such information is such that some approximate calculations or shortcut methods may be permissible.[8,9,10] The charts and figures that follow will result in a maximum error of 10%.

Small and medium size turbines belong to three basic categories:

1. Turbogenerators for condensing, noncondensing and low pressure operation.
2. Mechanical drive, single-stage turbines, condensing and noncondensing.
3. Mechanical drive, multi-stage turbines, condensing and noncondensing.

Data for extraction and induction type turbines have not been included here because these require special study based on anticipated power load and steam demand. Such data should be obtained from manufacturers but are also available in the literature.[11,12,13] Each extraction point adds, roughly, about 15% to the price of the turbine.

[7]*Prime Movers*, Ed. William Staniar, New York: McGraw Hill Book Company, 1966, pp. 12–225.

[8]Richard Shifler, "How to Predict Turbine Performance," *Power*, September, 1959, pp. 80–82.
[9]*De Laval Engineering Handbook, op. cit.*, pp. 5–7 to 5–11.
[10]J. Kenneth Salisbury, *Steam Turbines and Their Cycles*, New York: John Wiley and Sons, 1950, Chapter 25, Estimating Steam Rates, pp. 562–578.
[11]Richard Shifler, "Condensing Automatic-Extraction Turbines," *Power*, December, 1959, pp. 96–98.
[12]Richard Shifler, "Noncondensing Automatic-Extraction Turbines," *Power*, January, 1960, pp. 73–75.
[13]L. E. Newman and Others, *Modern Turbines*, New York: John Wiley and Sons Inc., 1944, pp. 77–91.

Tubogenerators: Figures 5-5, 5-6, and 5-7, with theoretical steam rate factors for noncondensing, condensing and low pressure turbines, result in steam rate estimates within an error margin of 5-10%.[14]

The following example will illustrate the use of these charts:

Given: A 2000 kW condensing turbogenerator, at 0.80 power factor, throttle steam at 200 psia (1378 kPa) and 100 F (56 C) superheat, exhausting at 2 in. Hg abs (6.76 kPa).

Find: Throttle flow and steam rate at 100%, 75%, 50% and 25% load.

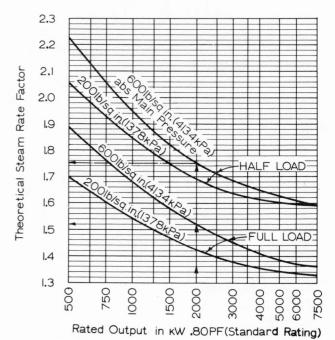

Rated Output in kW .80PF (Standard Rating)

Reprinted by permission of General Electric Co. © 1944 by General Electric Co.

Figure 5-6. Theoretical steam rate factor for condensing turbines.[15]

We now construct Table 5-3 and enter the results of the above calculations on lines 2 and 4 (50% and 100% load). To find throttle flow at 25% and 75%, we plot the steam rates and throttle flows of lines 2 and 4 on Fig. 5-8.

Through the two points representing throttle flow, since it is nearly true that throttle flow is linear, we draw

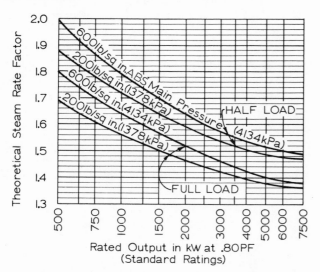

Rated Output in kW at .80PF (Standard Ratings)

Reprinted by permission of General Electric Co. © 1944 by General Electric Co.

Figure 5-5. Theoretical steam rate factor for noncondensing turbines.[14]

From the ASME theoretical steam rate tables, the theoretical steam rate equals 9.55 lb/kW-hr (0.012 kg/W-s). From Fig. 5-6, theoretical steam rate factors at full and half loads are 1.46 and 1.59, respectively.

Full load steam rate = 9.55 × 1.46
= 13.94 lb/kW-hr (0.0019 kg/kW-s)
Half load steam rate = 9.55 × 1.59
= 15.18 lb/kW-hr (0.0019 kg/kW-s)
Full load flow = 2000 kW × 13.94 lb/kW-hr
= 27880 lb/hr (3.51 kg/s)
Half load flow = 1000 kW × 15.18 lb/kW-hr
= 15180 lb/hr (1.91 kg/s)

[14]*Ibid*., p. 31.

[15]Newman, *op, cit*., p. 37.

[16]Newman, *op. cit*., p. 33.

[17]Newman, *op cit*., p. 55.

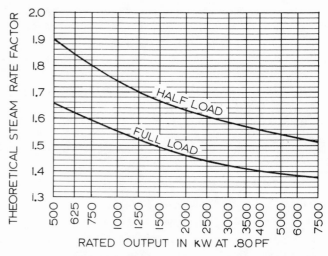

RATED OUTPUT IN KW AT .80PF

Reprinted by permission of General Electric Co. © 1944 by General Electric Co.

Figure 5-7. Theoretical steam rate for low pressure condensing turbines.[16]

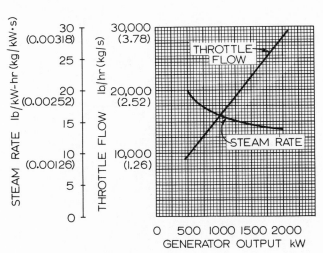

Figure 5-8. A plot of the data of Table 5-3.

a straight line, extending the line down to 25% load (500 kW) and read flow of 22,400 lb/hr (2.82 kg/s) at 75% load (1500 kW) and 9750 lb/hr (1.22 kg/s) at 25% load (500 kW), entering the data in Table 5-3.

Table 5-3

Point	Percent Load	Load kW	Steam rate lb/kw-hr (kg/W-s)	Throttle flow lb/hr (kg/s)
1	25	500	19.50 (0.0024)	9750 (1.22)
2	50	1000	15.18 (0.0019)	15180 (1.91)
3	75	1500	14.93 (0.0018)	22400 (2.82)
4	100	2000	13.94 (0.0017)	27880 (3.51)

To find the steam rate at the 75% and 25% load points, the newly found flows at these points can be divided by the corresponding kW. Hence

$$\text{Steam rate at 75\% load} = \frac{22{,}400 \text{ lb/hr}}{1500 \text{ kW}}$$

$$= 14.93 \frac{\text{lb}}{\text{kW-hr}} \ (0.0018 \text{ kg/kW-s})$$

$$\text{Steam rate at 25\% load} = \frac{9750 \text{ lb/hr}}{500 \text{ kW}}$$

$$= 19.50 \frac{\text{lb}}{\text{kW-hr}} \ (0.0014 \text{ kg/kW-s})$$

These data, too, are entered in Table 5-3 and plotted in Fig. 5-8, for the record.

Mechanical Drive, Single Stage Turbines: A single stage turbine is one of the most basic prime movers in industrial plant design, the workhorse of industry, because it is one of the least expensive drives and is flexible enough to be designed, depending on first cost, at intermediate or low efficiency. Because of the performance variations for design factors among manufacturers, such as throttle steam pressure and temperature, initial superhead speed and exhaust pressure, it is impossible to present one set of curves that would satisfy every situation.

The single stage turbines' relatively high steam rate makes it very suitable for noncondensing service and its exhaust steam has many uses in process work. In central plant work, on account of its wide range, 15-3500 hp (11-261 kW), it can be used successfully for auxiliary pumps, fans, etc. and as a centrifugal refrigeration compressor driver in a combination centrifugal-absorption arrangement.

As a condensing turbine, the single stage machine is at a disadvantage vis-à-vis the multistage turbine which, due to its higher efficiencies, can keep steam rates low and, therefore, steam plant size within reasonable limits.

Figures 5-9, 5-10 and 5-11 indicate typical steam rates for condensing and noncondensing service. Single stage turbine Fig. 5-12, a curve derived from the data of two manufacturers, is useful in determining water rates at partial loads for condensing and noncondensing single stage turbines. These curves are accurate within 5%, but are without superheat and horsepower losses which can change the picture to some extent. The manufacturers can supply them on request.

Example 1: From the ASME theoretical steam rate tables, the steam rate equals 10.04 lb/kW-hr (0.00126 kg/kW-s) at 225 psig (1550 kPa) and 450 F (232 C). From Fig. 5-11, at a theoretical steam rate of 10.04 lb/kW-hr and 6000 rpm, the estimated base rate is 17.1 lb/hp-hr. From Fig. 5-12, at 50% load, the theoretical steam rate factor is 1.22. Therefore, the estimated steam rate at 50% load is 1.22 × 17.1 = 20.86 lb/hp-hr (0.0026 kg/kW-s). Note that the theoretical steam rate factor or the steam consumption factor, as it is sometimes known, is the inverse of efficiency.

Mechanical Drive, Multistage Turbines: Many more design variations are possible with multistage than are with single-stage turbines. Many standard components can be combined in a number of ways to obtain a wide range of efficiencies and steam rates. Figures 5-13 through 5-19 define the performance of multi-valve, multi-stage

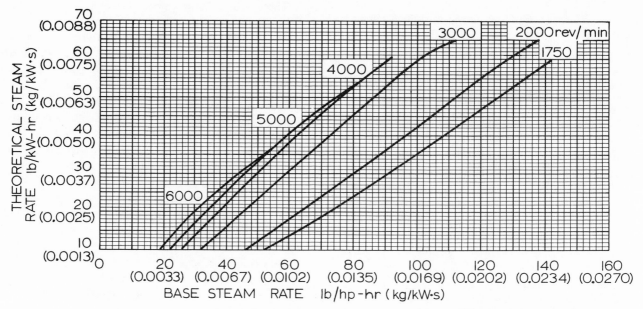

Figure 5-9. Single stage turbine performance. Range up to 1400 hp (1044 kW).

turbines from 400 to 50,000 kW. In general, multistage turbines can be designed to take advantage of high pressures and temperatures. They can operate at high speeds, in excess of 11,000 rpm. Some of the benefits are better speed and flow control, smaller losses in general and less

steam leakage losses in particular, more rugged thrust and journal bearings and, of course, high efficiencies. Because multistage turbines are expensive, their use and the conditions at which they must operate must be judiciously weighed.[18,19]

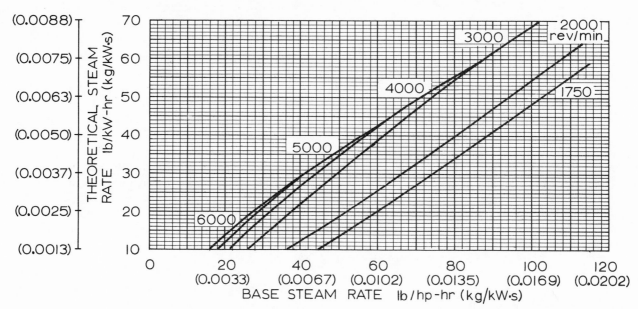

Figure 5-10. Single stage turbine performance. Range up to 2500 hp (1865 kW).

[18,19] Shifler, "How to Predict Turbine Performance," *op. cit.*, p. 81.

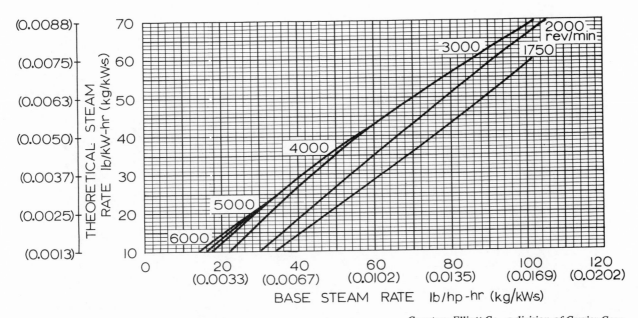

Figure 5-11. Single stage turbine performance. Range up to 3500 hp (2611 kW).

The theoretical steam rate factors of Fig. 5-19, derived by grouping a number of turbines of one manufacturer, do not necessarily fit all turbines.

The generalized expression for an approximate steam rate is as follows:

Approx. steam rate, lb/hp-hr

$$= \frac{\text{Theoretical steam rate, lb/kW-hr} \times 0.746 \text{ kW/hp}}{\text{Basic efficiency} \times \text{Correction factors}}$$

Partial load steam rate = Full load steam rate × theoretical steam rate factor

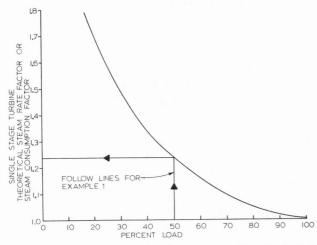

Figure 5-12. Mechanical drive single stage turbine theoretical steam rates for condensing and noncondensing service.

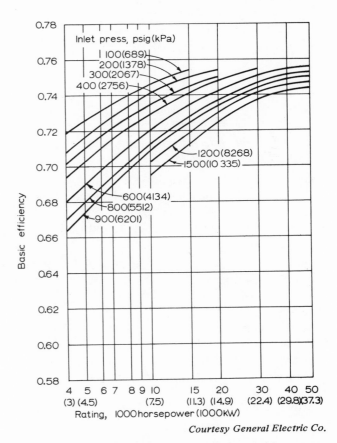

Figure 5-13. Basic efficiency of multi-valve, multi-stage condensing turbine.

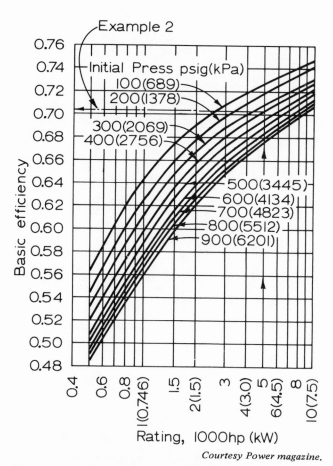

Figure 5-14. Basic efficiency of multi-stage condensing turbine of up to 10,000 hp (7460 kW).

Courtesy Power magazine.

Estimating steam rate for condensing turbine at full load and part load–Example 2.

Find: steam rate at 100% and 50% load.

Given: rated load, 5000 hp; speed, 6000 rpm; inlet steam at 400 psig (2756 kPa) and 600F (315 C) or 151F (84 C) superheat; exhaust at 3″ Hg abs (10.1 kPa).

Theoretical steam rate (ASME = 8.5 lb/kW-hr
tables) (0.0010 kg/kW-s)
Basic efficiency (Fig. 5-14 @
5000 hp and 400 psig) = 0.705
Superheat factor at 151 F
superheat (Fig. 5-17) = 1.013
Speed factor (Fig. 5-16 @
6000 rpm and 5000 hp,
3″ Hg abs) = 0.977
Theoretical steam rate
factor at 50% (Fig. 5-19) = 1.076

Estimated full load steam rate

$$= \frac{8.5 \times 0.746}{0.705 \times 1.013 \times 0.977} = 9.087 \text{ lb/hp-hr}$$
$$(0.0015 \text{ kg/kW-s})$$

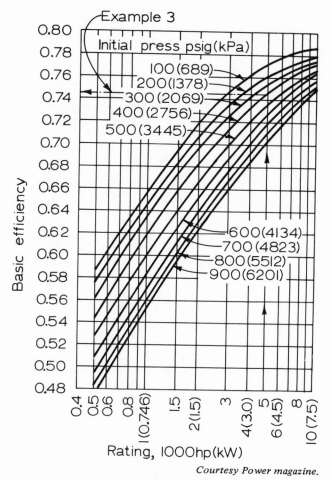

Figure 5-15. Basic efficiency of multi-stage noncondensing turbine.

Courtesy Power magazine.

Estimated steam rate at 50% load = 9.087 × 1.076
= 9.777 lb/hp-hr (0.0016 kg/kW-s)

Estimating steam rate for noncondensing turbine at full load and part load–Example 3.

Find: steam rate at 100% and 50% load.

Given: rated load 5000 hp; speed, 6000 rpm; inlet steam at 400 psig (2765 kPa) and 600F (315 C) or 151F (84 C) superheat; exhaust at 15 psig.

Theoretical steam rate = 15.24 lb/kW-hr
(ASME tables) (0.0019 kg/kW-s)
Basic efficiency (Fig. 5-15 @
5000 hp and 400 psig) = 0.745
Superheat factor at 151 F
superheat (Fig. 5-17) = 1.013
Speed factor (Fig. 5-16 @
6000 rpm and 5000 hp,
3″ Hg abs) = 0.977

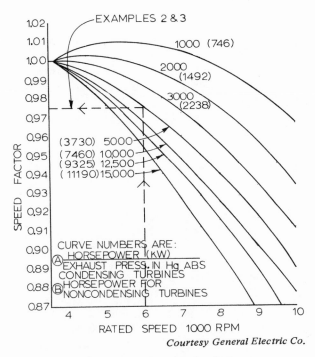

Figure 5-16. Speed correction factor.

Pressure ratio factor @ $\dfrac{P_e}{P_i} = \dfrac{15 + 14.7}{400 + 14.7} = 0.072,$

from Fig. 5-18 = 1.0
Theoretical steam rate factor at 50% (Fig. 5-19) = 1.151

Estimated full load steam rate

$$= \dfrac{15.24 \times 0.746}{0.745 \times 1.013 \times 0.977 \times 1.0} = 15.41 \text{ lb/hp-hr}$$

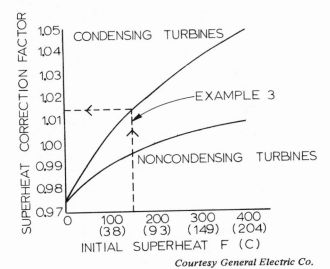

Figure 5-17. Superheat correction factor.

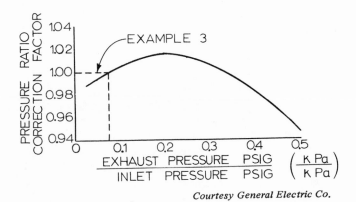

Figure 5-18. Pressure ratio correction factor for noncondensing turbines.

Estimated steam rate at 50% load = 15.41 × 1.151
= 17.74 lb/hp-hr (0.0029 kg/kW·s)

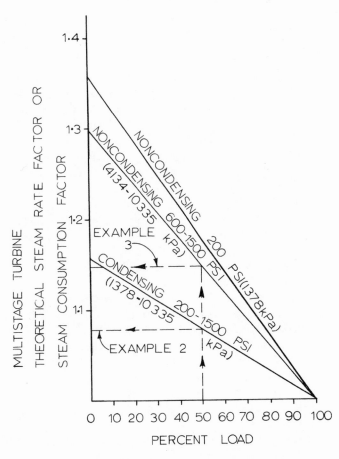

Figure 5-19. Theoretical steam rate factor for multistage turbines (for turbines over 1000 hp (746 kW)).

5.3 Variable Speed Drive

Variable speed drive can be used in central plants to drive centrifugal refrigeration compressors and the various auxiliaries such as draft fans, and feedwater, circulating, river cooling water and condenser water pumps.

The gas turbine and, even more so, the steam turbine, have proven to be the ideal variable speed drive for centrifugal compressors. Fluid or eddy current variable couplings, if applied to multi-thousand horsepower compressors, would prove too expensive for the savings they might offer in energy consumption. Because of energy losses or high costs, or size limitations, hydrokinetic, hydroviscous or magnetic couplings will remain marginal in large refrigeration compressor applications.

In the 300-500 ton centrifugal compressor range, natural gas and, sometimes, diesel engines do provide a variable speed drive. However, vibration, expansion and alignment problems constitute cause for a second look. In total energy applications, gas or gasoline engine drive must receive serious consideration as drives for auxiliaries. Motor drive coupled with variable speed in total energy systems entails a too substantial investment to supply the auxiliaries.

Variable speed drive for the auxiliaries, especially pumps, is quite often a matter of design necessity. There are quite a few drives to satisfy this need. Besides the hydrokinetic, hydroviscous, variable frequency-variable voltage and eddy current couplings, there are wound rotor a-c and direct current motors and gear trains. Some manufacturers combine a hydroviscous system with gears to obtain high reduction ratios but at the risk of some complications. Slip drives, in general, are inefficient, especially in the low speeds.

Wound rotor motors have resistance added to the rotor circuit to change speed. Inrush current characteristics are excellent, never exceeding 150% of full load current but the major disadvantages still are speed regulation in steps, a great deal of power wasted in the external resistors, requiring a large ventilation system to remove generated heat, large space requirement for the resistors and the control drum and a substantial first cost. Wound rotor or saturable reactor motors are rarely used in central plants. Figure 5-20 represents an interesting control for wound rotor motors. The line sensor feels line pressure and operates the air compressor to keep the electrode water at the appropriate level. The pump discharge serves as an electrolyte cooler.

Direct current motors provide excellent speed control but direct current power is not ordinarily available at central plants. However, most operators are versed in their maintenance, which all commutating machines incur. The tendency is to use direct current motors only for special applications.

Gear trains are dependable and fairly inexpensive but provide fixed reduction ratios and, therefore, find limited applications.

Variable frequency, variable voltage (VFVV) drives are part of recent solid state development and offer marvelous flexibility and good torque characteristics. Their current upper limit is 200 hp, although frequency or voltage drives can reach 600 hp or even higher. Very high costs tend to limit their use.

Finally, there are variable slip type drives. The most common of this kind are: 1. Fluid couplings (hydrokinetic). 2. Hydroviscous couplings. 3. Eddy current couplings (magnetic). All three can handle: (1) constant-torque machines, such as reciprocating compressors and gear pumps, (2) variable torque devices, such as centrifugal fans, pumps and compressors, power generators and cooling towers geared to maintain constant temperature water, (3) constant horsepower loads, which normally do not exist in central plants but are found in some industrial processes such as wiredrawing.

In general, variable speed drives add considerable cost to a project. Their use can be based partially on energy saving grounds and on strict project requirements.

1. *Fluid Coupling (Hydrokinetic)*. The most representative type of fluid coupling is the automobile automatic transmission. Four basic components constitute the fluid drive: (1) the impeller (driving rotor), (2) the runner (driven rotor), (3) the casing (rotor envelope) and, (4) a hydraulic system that varies the amount of fluid (usually, oil) between impeller and runner.

A circulating oil pump, driven by an external motor or by the input shaft at constant speed, pumps oil from the sump to an oil cooler (if used) and subsequently to the rotors. Oil entering the rotating casing is thrown against its periphery by centrifugal force. Subsequently, the oil enters the space between impeller and runner in the form of a liquid ring (vortex). Energy is transferred from the input to the output shaft as follows: the impeller pumps oil by accelerating oil particles radially; the particles are decelerated in the runner where the kinetic energy of the oil is absorbed and transmitted to the output shaft as horsepower.

In variable speed applications, oil is injected or is withdrawn from the casing by various automatic and extremely sensitive mechanisms that vary vortex size (oil mass), thus, providing infinitely variable speed control. (See Fig. 5-21.)

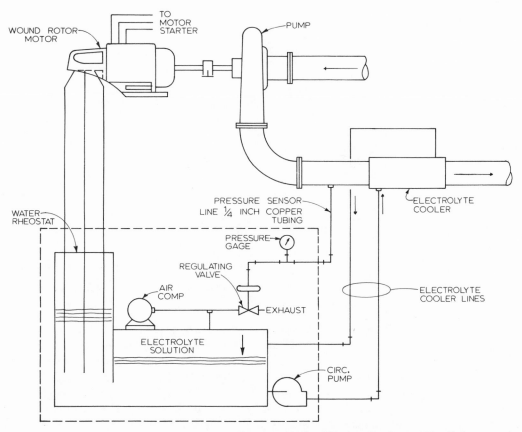

Courtesy Air Conditioning, Heating and Ventilating magazine.

Figure 5-20. Wound rotor motor using a water rheostat.

Fluid drive has the following main advantages: (1) controlled torque and no load starting, (2) controlled acceleration, (3) infinite speed control, (4) conservation

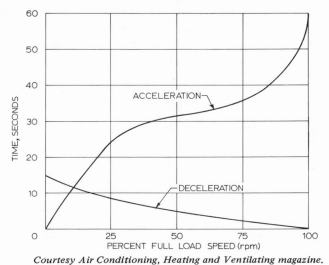

Courtesy Air Conditioning, Heating and Ventilating magazine.

Figure 5-21. Typical speed response curve for fluid coupling drive.

of energy for centrifugal fans and pumps by operating at required flow, and (5) reduced maintenance at lower speeds.

High initial cost and unfamiliarity of maintenance crews with repair and maintenance procedures are distinct disadvantages. An additional disadvantage is slip of 1–3% at full speed and 30–35% at half speed, entailing power losses.

Horsepower range is 1-40,000 (1-29 840 kW) in various classes of fluid couplings.

2. *Hydroviscous Coupling.* Shearing of an oil film between two rotating disks, one belonging to the input and the other to the output shaft, is the principle of transmitting torque. Torque control is accomplished by applying a clamping force to the disks.

In schematic form, as shown in Fig. 5-22, a disk attached to the motor (input) shaft revolves between two disks attached to the driven (output) shaft. The oil pump forces a continuously running thin film of oil between the driving and driven disks, carrying away any heat

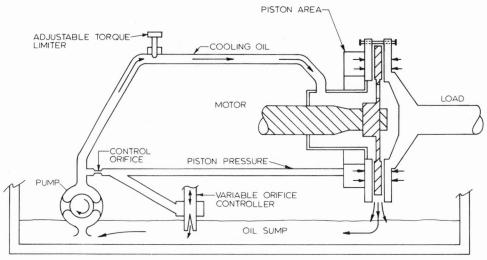

Courtesy U.S. Electrical Motors, Division of Emerson Electric Co.

Figure 5-22. Schematic arrangement of hydroviscous coupling. "InfiDrive" 3–200 hp range.

that is produced. The disks are suitably grooved and sufaced to allow variable speed control. The input disk accelerates oil particles tangentially while the centrifugal force accelerates them outwardly. The oil film is continuously replaced by being discharged to the oil sump and subsequently pump through an oil cooler (not shown in Fig. 5-22) located on the pump discharge to the center of the rotating disks.

The amount of transmitted torque depends on the shearing action taking place and the area of disk surface. To increase the area of the working surfaces, a number of disks are combined in a disk "pack." Disks are alternately splined to the input and output shafts with complete freedom to slide axially.

A clamp force is applied at one end of the disk pack by means of a stationary hydraulic piston, which transmits the applied force to all disks due to their freedom to slide. A varying clamp force controls piston pressure in response to output shaft speed, which feeds back a signal through a proportional hydraulic or pneumatic signal. Control pressure can be manual or automatic.

Unlike fluid and eddy current couplings, which exhibit slip losses of 2–5% and output speeds of 95–98% of input speeds, hydroviscous couplings can transmit torque up to 100% of input speed. Therefore, additional power savings of roughly 2.5% are possible. Loss of oil film can damage a shearing surface. Proper alignment is essential to maintain the presence of an oil film and close clearances between rotating members. Response to speed and load changes is rapid and proper controls can incorporate torque-limited features. Input speed is up to 3600 rpm

and the range is 1-20,000 hp (1-14 920 kw) in different models by various manufacturers. Hydroviscous couplings have wide applications in various industries such as lumber, cement and paper mills.

3. *Eddy Current Coupling (magnetic).* An eddy current coupling is a rugged device that provides speed control for boiler draft fans, centrifugal pumps, adjustable frequency motor generator sets, etc. The basic control circuit (see Fig. 5-23) has a d–c rectifier to provide current for rotor excitation through two output shaft slip rings (see Fig. 5-24). There is an automatic controller with feedback from the tachometer generator.

Eddy currents are introduced in the outer solid metal ring which, in rotating, cut across the magnetic flux set up by the electromagnets across the air gap. A great deal of heat is generated, as with wound rotor or saturable reactor motors, and the coupling must be ventilated.

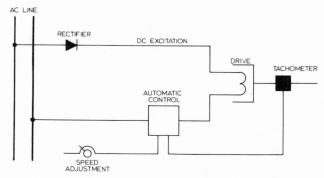

Courtesy Power magazine.

Figure 5-23. Basic control circuit of eddy current coupling.

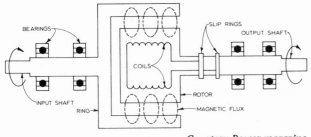

Figure 5-24. Schematic arrangement of eddy current coupling.

Speed response is very rapid, almost instantaneous. Applications involve constant and variable torque, although magnetic couplings are best suited for variable torque. However, for small loads and limited speed ranges, they can be used for constant torque. Speed regulation is within ± 2%. Efficiency drops rapidly with decreasing speed. The couplings range in size from a few to 3000 hp (2238 kW).

Variable Frequency, Variable Voltage (VFVV) Drive. Variable speed motors can be regulated either by variable frequency or by variable voltage and often by a combination of both. At low speed, variable voltage as such does not develop adequate torque.

On the other hand, variable frequency, without voltage variation, is limited to a narrow speed range. Variable frequency can be accomplished with a number of designs as a more economical drive, if voltage variation is not considered. Variable frequency at constant voltage produces substantial power losses in the form of heat and is not energy-saving. Exclusive use of either voltage or frequency variation results in a specially designed motor.

A combined VFVV design, in spite of its high cost, has much to recommend it.

1. A standard induction motor may be used and, therefore, any existing squirrel cage motor may be backfitted.
2. A well-designed system incurs only 5% heat dissipation.
3. Starting current characteristics are excellent, not exceeding 150% of full load current.
4. Power loss decreases with decreasing speed, unlike the eddy current coupling.
5. The drive may be installed remotely from the motor.

The basic principle of a VFVV design is that frequency changes, which regulate speed, are accompanied by proportional changes in voltage to maintain motor flux at a constant level. This solid state drive, in general outline, works as follows: A rectifier converts a-c voltage into adjustable d-c voltage, which becomes the input to a three-phase inverter, producing a waveform voltage. Adjustment of the wave voltage frequency produces motor speed variation. The control logic maintains output voltage. Figure 5-25 indicates that frequency is governed by the speed input signal and that a ratio between voltage and frequency is maintained and it further shows that gate-drive signals and good starting are ensured.

Very roughly, the cost ratios of a constant speed motor starter to the eddy current and VFVV drives, respectively, are 1 : 2 : 4.

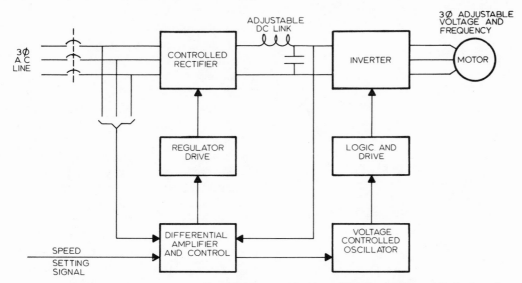

Courtesy Electric Machinery Mfg. Co., Division of Turbodyne Corp.

Figure 5-25. Typical block diagram of a variable frequency, variable voltage drive.

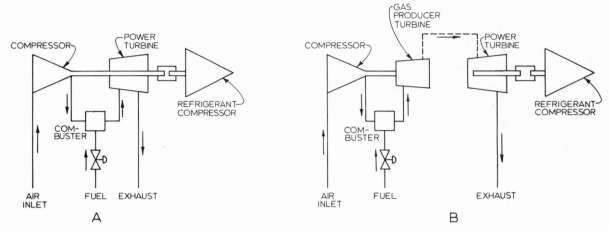

Courtesy York Division of Borg-Warner Corp.

Figure 5-26. Typical shaft arrangements for gas turbines: A. single-shaft; B. two-shaft.

5.4 Gas Turbines as Prime Movers for Centrifugal Refrigeration Machines

The factors that favor gas turbines for refrigeration machine drive applications are many. They include: light-weight, with minimum foundation requirements; rapid startup capability; ability to operate automatically with minimum supervision; less floor space requirement than for equivalent horsepower diesel engines; negligible lubricating oil requirements; fairly clean exhaust gases, which can be utilized for substantial heat recovery; and the ability to use both gas and oil fuel and be integrally pre-packaged and mounted on skids.

The unfavorable factors are: low thermal efficiency, which is less than that of reciprocating engines, except for the latest gas turbine generation, rapidly falling efficiency

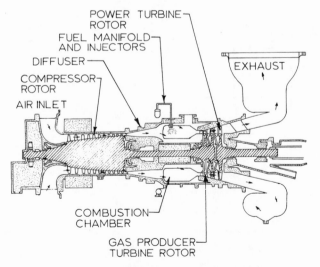

Figure 5-27. Cross-section of a two-shaft turbine.

at part loads for the single shaft arrangement, substantial loss of horsepower output with higher ambient space temperature; and high initial cost in $/hp, though installed costs may be comparable.

Gas turbines fall into two categories based on shaft arrangement. The "single shaft" design has the power turbine, air compressor and centrifugal refrigeration compressor mounted on one, interconnected shaft as illustrated in Fig. 5-26A. This is a simple cycle, not incorporating intercooling, reheating, exhaust heat recovery or other improvements.

Figure 5-26B, which illustrates the two-shaft or "free turbine" or "split shaft" arrangement, as it is sometimes known, shows the output-load shaft (power turbine) mechanically independent of the gas producer. That portion of the turbine required to drive the air compressor is called the gas producer turbine. The power turbine, which is mounted on a second shaft, drives the centrifugal refrigeration compressor through a reducing gear box. See, also, Fig. 5-27.

In general, single shaft turbines are more suitable for constant speed applications such as generators. The mechanical simplicity of a single shaft turbine reduces its cost somewhat. The ability of a single shaft turbine to handle speed variations is excellent and fuel control can be simple and direct, not unlike that of an automobile. (See Section 6.12, "Gas Turbine Controls" and Section 6.13, "Governors".) Starter power and torque requirements are greater for single-shaft turbines, since the entire mass of the rotating components must be accelerated as well as the refrigeration compressor. The resulting high starting torque usually dictates use of a piping bypass around the compressor or throttling of compressor suction. When the refrigeration system uses hydrocarbon refrigerants, it is possible to blow the contents of the system to a flare system to reduce gas density and, consequently, reduce the starting torque. In the case of halocarbon refrigerants, the same results can be obtained by blowing down the system. In conclusion, additional equipment is necessary, involving expense and complication, to reduce the starting torque of a single-shaft turbine.

The two-shaft turbine permits a wide range of speeds at full power, which contributes to better part-load economy. It also exhibits an inherent lag in response between the gas producer and power turbines, despite the fact that each can be equipped with a speed governor. The starter is required to accelerate the gas producer turbine only, which pressurizes the system. Gases leaving the gas producer turbine flow through the free power turbine forming a fluid coupling. The remaining energy is absorbed by the power turbine and is transferred to the output shaft which drives the load of the centrifugal refrigeration compressor. Although one of the characteristics of the two-shaft turbine is high stall torque, this does not necessarily require a high horsepower starter. For these reasons, a two-shaft turbine is, in general, better suited to centrifugal refrigeration machine applications.

Turbine speeds vary from 40,000 to 60,000 rpm for 50-200 hp (37-149 kW) and as low as 3600 rpm in the 20,000-30,000 hp (14 920-22 380 kW) range. Speed reducers range between 1200-6000 rpm.[20] Figures 5-28 and 5-29 are typical two-shaft gas turbine performance curves.

Starter systems are of various types. Compressed air (150-200 psig) and gas starters seem to enjoy a great degree of popularity, followed by a-c and d-c motors. Other types are batteries, chargers, motor-generator sets, reciprocating engines, high pressure impingement, hydraulic, and hand-crank.

Fuels for gas turbines can be natural gas or liquefied petroleum gas such as propane and butane, which gasify at room temperatures. Some turbines use JP-1, a form of kerosene, and JP-4, a blend of gasoline and diesel oils. Residual oils and even pulverized coal, though only experimentally, are also used to fuel industrial turbines. Some turbines burn two fuels, usually gas and oil, incorporating two complete fuel systems which can be switched from one to the other while in operation.

Sources of losses, other than mechanical, which must be accounted for, occur in the inlet and outlet ducts and gearbox, or are due to altitude and elevated inlet air temperature. Refer to Figs. 5-30 and 5-31.

Entering air temperature has a pronounced effect on turbine performance. As the air temperature increases air mass flow and compression ratio decrease. Higher temperature air entering the compressor means air at higher temperature leaving the compressor and, therefore, less energy can be imparted to this air in the combustor to produce a desired turbine inlet air temperature. In a typical gas turbine, if the compressor inlet temperature were increased from 50 to 115 F, the power output would decrease by 30%.[21] For every degree Celsius above 59 F (15 C), the turbine is derated by approximately 1%, and for every 1000 feet (300 m) above sea level, the turbine is derated 3½%.[22] Hence, some thought may be given to precooling the air during the summer through chilled water coils. An economic balance can be struck between cooling coil installation and operating the necessary refrigeration on one hand and the unavailability of installed turbine on the other.

Turbine exhaust heat recovery is a very important consideration in reducing the amount of fuel per output horsepower. Exhaust gas can be processed through a waste heat boiler to produce steam for process use or it can be used as combustion air for plant boilers.

Figures 4-33 and 4-34 of the previous chapter describe performance of simple cycles and variations thereof; *i.e.*, regenerative cycle, regenerative with reheating cycle, regenerative with intercooling cycle and regenerative with reheating and intercooling cycle. The following explanatory notes will help to define these cycles. The objective of these variations is to increase the thermal efficiency of the cycle and explore other side effects.

Basic cycle (A). This cycle is illustrated in Fig. 5-26 and is the subject of Sec. 5.4 "Gas Turbines as Prime Movers for Centrifugal Refrigeration Machines."

Regenerative cycle (B). Regeneration consists of heating compressor discharge air before it enters the combustion chamber by using a counterflow heat exchanger. On one side of the heat exchanger flow the hot, turbine exhaust gases and on the other side the relatively cool air discharged from the compressor. The net effect is to increase the efficiency of the cycle and to reduce the compression ratio for the same turbine-inlet gas temperature of the simple cycle (A). It is apparent by simple comparison that the gains of this cycle surpass those of cycles (C), (D) and (E). These gains are obtained inexpensively by investing in a simple heat exchanger.

[20] American Gas Association, *Gas Turbine Manual*, New York: The Industrial Press, 1965, p. 26.

[21] *Ibid.*, p. 27
[22] R.M.E. Diamant, *Total Energy*, Oxford: Pergamon Press, 1970, p. 33.

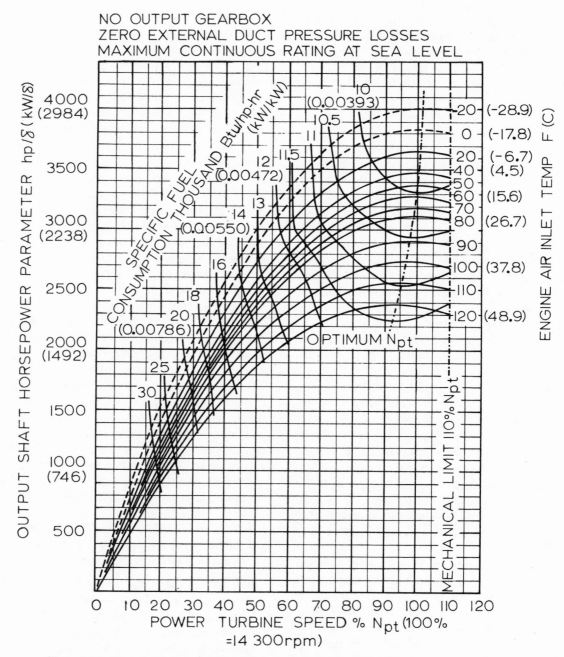

Figure 5-28. Typical two-shaft, 3000 hp (2238 kW) gas turbine performance of output vs. speed.

Regenerative cycle with reheating (C). This cycle, besides regeneration, includes reheating of the exhaust gases from the first turbine in a second combustor chamber before they enter a second (reheat) turbine. Efficiency is increased even further than in cycle (B), although the pressure ratio tends to rise at higher efficiencies. Work optimization indicates that work should be divided approximately in half between the two turbines.

Regenerative cycle with intercooling (D). Cycle (B) efficiency can be increased by introducing intercooling in two stage compressors, thus reducing compressor horsepower. The pressure ratio tends to increase.

Regenerative cycle with reheating and intercooling (E). This cycle combines all previous features and increases cycle efficiency even more than all previous cycles. The efficiency curves flatten out over a wide range of compression ratios and the compression ratios are pushed even higher than in the other cycles.[23-26]

Efficiencies for different cycle turbines, as illustrated in Figs. 5-32, 5-33, and 5-34, are in the following ranges:[27]

Simple plants	15-23%
Plants with regenerative heat exchange	21-27%

Two shaft turbines with intercoolers, regeneration and reheat   32-33%

Analyses performed by this author on turbine plants, with waste heat boilers and continuous use of the by-product steam, showed plant efficiencies in 50-60% range.

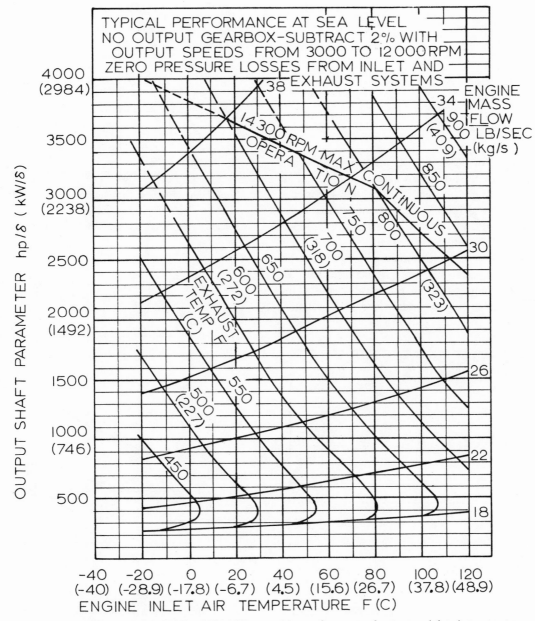

Figure 5-29. Typical two-shaft, 3000 hp (2238 kW) gas turbine performance of output vs. inlet air temperature.

[23]*Prime Movers.* Ed. William Staniar, *op. cit.*, pp. 12-193 and 12-194.

[24]*Ibid.*, p. 12-193.

[25]*Ibid.*, p. 12-200.

[26]*Ibid.*, p. 12-197.

[27]Diamant, Total Energy, *op. cit.*, p. 33.

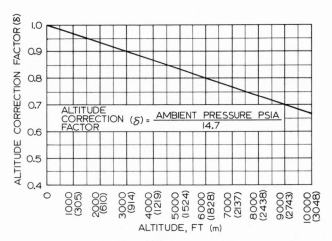

Figure 5-30. Gas turbine performance correction for altitude.

Operating costs for diesel engines are well-defined and documented by ASME and various manufacturers.[28] This is not the case with gas turbines. Not enough experience has been amassed and reliable data are not available for the engineer to make an independent evaluation of operating expenses. However, some authentic data are available.[29,30] Reblading at the manufacturer's plant is necessary on a periodic basis and so is turbine overhaul. To make an independent evaluation of operating costs, the engineer must rely on the experience and data available.

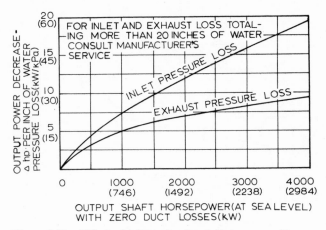

Figure 5-31. Effect of inlet and exhaust duct pressure losses on power output.

[28]1972 Report on Diesel and Gas Engines, Power Costs, the American Society of Mechanical Engineers.

[29]GTI Editorial Staff. "Gas Turbine Maintenance Costs," *Gas Turbine International*, January-February, 1977, pp. 12–15.

[30]*Sawyer's Gas Turbine Engineering Handbooks, vol. III, Maintenance & Basic Fundamentals.* Ed. J.W. Sawyer, Stamford, Conn.: Gas Turbine Publications, Inc., 1972.

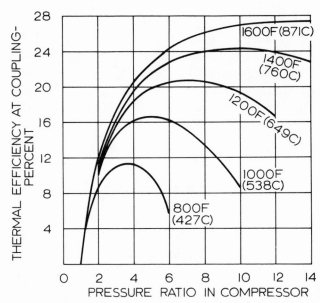

From Prime Movers by William Staniar. Used with permission of McGraw- Hill Book Co.

Figure 5-32. Basic gas turbine cycle thermal efficiency.

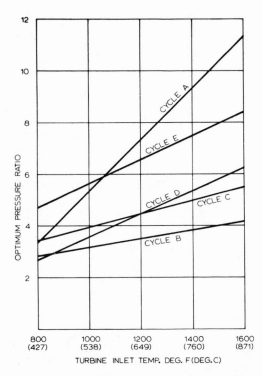

From Prime Movers by William Staniar. Used with permission of McGraw Hill-Book Co.

Figure 5-33. Optimum pressure ratios for various gas turbine cycles.

Fuel consumption, time between overhauls, types of over-haul, cost of overhauls, cost of spare parts, gas turbine manufacturer's service policies, existence of independent service shops and independent spare part providers constitute items to be evaluated before purchasing a gas turbine. (See Fig. 5-35.)

After deciding to use gas turbines, the engineer must decide whether a single or a double-shaft turbine is more suitable and whether the airplane or the industrial gas turbine is preferable. He must make an economic comparison and analysis and be convinced that a gas turbine is the appropriate drive in terms of plant objectives and cycle adaptability.

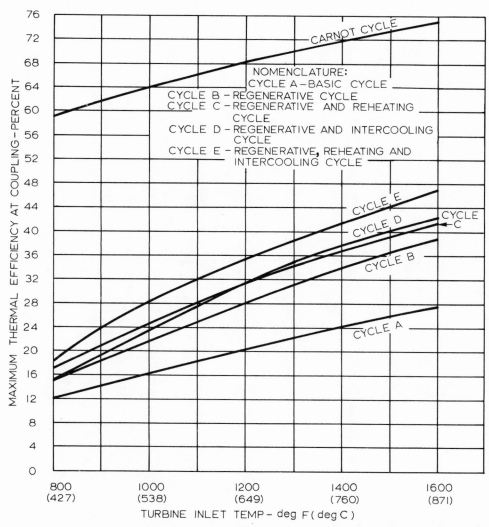

From Prime Movers by William Staniar. Used with permission of McGraw-Hill Book Co.

Figure 5-34. Maximum thermal efficiencies for various gas turbine cycles.

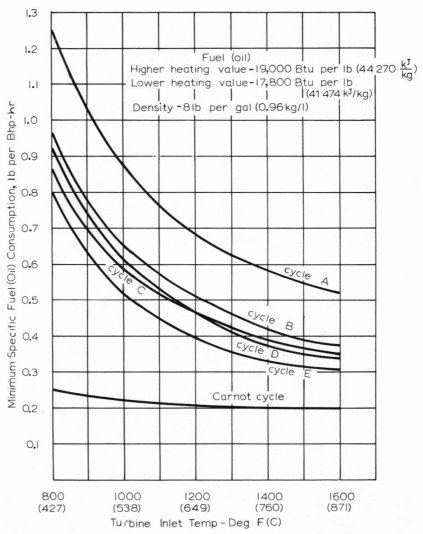

From Prime Movers by William Staniar. Used with permission of McGraw-Hill Book Co.

Figure 5-35. Specific fuel oil consumption.

BIBLIOGRAPHY

Drives

ASHRAE Handbook and Product Directory, 1975 Equipment, *Engine and Turbine Drives*, Chapter 32.

Thornton, Althea. "A Special Report, Belts and Chains," *Power*, (May, 1957), 76–94.

Wood's V-Belt Drive Engineering Handbook. Chambersburg, Penna: T.B. Wood's Sons Company, 1971.

Gas Turbines and Engines

Apitz, C.R. "The Economic Application of Gas Turbines to Large-Tonnage Air Conditioning Plants," ASME paper 61-WA-202.

Barrangon, M. "The Gas Engines—On Site Movers," *ASHRAE Journal*, (February, 1970), 50–53.

Blankenbaker, John. "Engine Drives for Compressors Come of Age," *Air Conditioning, Heating and Ventilating*, (May, 1967), G17–G20.

Caplow, S. David and Sidney A. Bresler. "Economics of Gas Turbine Drives," *Chemical Engineering*, (May 27, 1967), 103–10.

Congress, A. Eugene. "Can Gas Turbines Compete as Air Conditioning Power Sources?" *Heating, Piping, Air Conditioning*, (December, 1963), 87–91.

Gas Turbines Principles and Practice. Consulting Ed. Sir Harold Roxbee Cox. New York: D. Van Nostrand Company, Inc., 1955.

Gore, Eugene and A. Salim Qureshi. "Design Guidelines for Engine-driven Chillers and Heat Pump Systems," *Heating, Piping, Air Conditioning*, (August, 1977), 59–64.

Hambleton, W.V. "General Design Considerations for Gas Turbine Waste Generators," ASME paper 68-GT-44.

Hawkes, R.W. "Gas Turbines for Air Conditioning," *Heating, Piping, Air Conditioning*, (December, 1964), 141–44.

Jarvis, G. "Maintaining Gas Turbines in Pipeline Service," *Power*, (February, 1972), 56–57.

Jennings, Burgess H. and Willard L. Rogers. *Gas Turbine Analysis and Practice.* New York: Dover Publications, 1953.

Judge, Arthur W. *Modern Gas Turbines.* 2nd Edition. London: Chapman & Hall, Ltd, 1950.

Knight, C.E. "Here's What You Need to Know Before You Specify Gas Turbines," *Power Engineering*, (June, 1965), 47–49.

Koubek, J.H. and J.M. Kovacik. "Evaluating Gas Turbines for Process Applications, Economic Guides for the Decision Maker," ASME paper 71-GT-50.

Landman, William J. "The Basics of Choosing Prime Movers for Chillers," *Actual Specifying Engineer* (now *Specifying Engineer*), (August, 1974), 94–97.

Moftan, M. "Gas Turbine Performance Under Varying Ambient Temperatures." ASME paper 71-GT-57.

Ritchie, P.J.S. "Engines other than Gas-Turbines," *Proceedings, Total Energy Conference,* London, England. Held in Brighton, England, November 29–December 1971, Paper 9, 155.

Shearer, B. "Reciprocating Engines," *Proceedings, Total Energy Conference*, London, England. Held in Brighton, England, November 29–December, 1971, 155.

Skrotzki, B.G.A. "A Special Report, Gas Turbines," *Power*, (December, 1963), S1–S24.

Viera, S. "Gas Turbine's Application in Central Chilled Water and Heating Plants," ASME paper 69-GT-53.

Steam Turbines

Appendix A to Test Code for Steam Turbines (with 1968 Addenda) (ASME PTC 6-1964).

Coffey, John J. "Is There a Steam Turbine in Your Future?" *Air Conditioning, Heating and Ventilating*, (April, 1968), 43–47.

Direct-Connected Steam Turbine Synchronous Generator Units, Air Cooled (NEMA Standard Publication SM-12-1967).

Direct-Connected Steam Turbine Synchronous Generator Units, Hydrogen Cooled (20,000 to 30,000 kW, inclusive), (NEMA Standard Publication SM 13-1967).

General-Purpose Steam Turbines for Refinery Services (American Petroleum Institute, API Standard 611, 1st Edition, 1969).

Marks, R.H. "A Special Report, Steam Turbines," *Power*, (June, 1962), s.1–s.40.

Multistage Steam Turbines for Mechanical Drive Service (NEMA Standard Publication SM 21-1970).

Simplified Procedures for Routine Performance Tests of Steam Turbines (ASME PTC 6S Report-1970).

Single-Stage Steam Turbines for Mechanical Drive Service (NEMA Standard Publication SM 22-1970).

Speed Governing and Pressure Control of Steam-Turbine Generator Units (NEMA Standard Publication 46-112, 1971).

Steam Turbines (with 1968 Addenda) (ASME Performance Test Code, PTC 6-1964).

Variable Speed Drives

Alcock, D.N. "Understanding Solid-State, Variable Frequency Power Supplies for Variable Speed Pumping," *Power*, (January, 1977), 28–32.

Cook, Ernest F. "Adjustable Speed Drives," *Chemical Engineering*, (September 6, 1971), 70–82.

Goss, B.L. "Developments in A–C Adjustable-Speed Drive Systems," *Proceedings of American Power Conference*, Vol. 33, 1971, 369–376.

Liu, John K. "Energy Efficient Variable Speed Drives for Pipeline Pumping," Instrument Society of America (ISA), paper 78-815, 1978.

McGinness, Jr., John E. "Drivers for Variable-Speed Systems," *Air Conditioning, Heating and Ventilating*, (May, 1963), 75–79.

Peach, Norman. "Control Speed with Eddy-Current Couplings," *Power*, (April, 1970), 78–79.

CHAPTER 6

Controls

Aspects of control discussed in this chapter are those, which are basic to central plant design, that is, essential for controlling the processes and directing the plant's general operations.

First, compressor safeguards and internal capacity controls are examined, distinguishing between internal controls (low pressure cutouts, hot gas bypass, etc.) and external controls, such as a leaving water temperature controller.

The role of the refrigerant condenser deserves our closest attention because of its influence on machine capacity and stability. Machine load capacity and the unwelcome phenomenon of compressor surge come under scrutiny here. Herein we discuss different complex control loops required to coordinate a number of refrigeration machines and attempt to rate each according to its inherent disadvantages and how is promotes economical operation.

Steam and gas turbine and engine controls are examined in detail because turbines appear both in turbine-generator combinations and as mechanical drivers in a variety of situations.

Automation is an integral part of today's technology. Here we are not so much concerned about hardware or software, which continually changes, but about the criteria for its general application in sophisticated plant routines.

6.1 Compressor Controls

Compressor controls belong to three basic types. Safety controls are the first type, whose function is to prevent damage to the refrigeration machine by restricting a condition to a predetermined magnitude. They consist of high or low safety switches that operate in an "on-off" fashion, ordinarily requiring resetting in case of failure.

The second type may be called operational controls because they restrict a dangerous condition for a limited length of time until the condition is overcome or disappears. At other times, operational controls allow another part of the system to compensate for the deficiencies of the part of the system where a dangerous condition exists. Thus, operational controls manage to isolate, postpone or compensate for abnormal conditions. In this category we include monitoring or alarm functions because they are a preventive nature.

The third type are capacity controls. Their main function is to make the compressor responsive to load and rate of load changes while maintaining system stability. There are two aspects to capacity control. The first is what may be called compressor capacity controls. In the second, control of the operation of multiple machines, we must consider a variety of factors, such as type of prime movers and refrigeration machine size, type, and arrangement.

6.2 Compressor Safety Controls

Safety controls consist of controllers with on-off operation and sensors located in the measured variable. They can be pneumatic, electronic or electric and some may include characteristics of two types. They usually are snap-acting switches, either normally open or normally closed, depending on function. They can be coupled with indicating lights or buzzers, local or remote, and must be reset manually after the malfunction is corrected. Although a great many are here, only a few may be considered indispensable, among which are high pressure and low chilled water cutouts. These are common to both hermetic and open centrifugal machines.

High pressure cutout: Usually located in the condenser shell, the control's normally closed contacts open on an increase in discharge pressure, which may be due to reduced condenser water flow, excessive condenser tube scaling or the presence of air in the condenser. The switch setting which depends on the application, must not exceed the leak test pressure of the high side, which covers the compressor discharge piping and the condenser, including condenser float or orifices to the evaporator.

High temperature cutout: A switch with a normally closed contact, it opens on excessive increase in discharge gas temperature to halt compressor operation. It can substitute for the high pressure cutout or may work in parallel with it.

Low pressure cutout: Interchangeable with the low refrigerant temperature cutout, this switch is actuated by suction pressure and its normally closed contacts open to stop compressor operation.

Low refrigerant temperature cutout: A low limit temperature switch, its bulb senses evaporator temperature. Its setting is 3F below design suction temperature or 34F, whichever is lower. When evaporator temperature reaches the setpoint the switch contacts open, stopping the compressor.

Low oil pressure cutout: This safety switch senses pressure difference between the oil reservoir and the oil leaving the filter. A typical setting of 12 psi differential will close the normally open contacts and a differential of 6 psi will open them. If the pressure falls below a safe minimum (6 psi in the above illustration), the switch opens the control circuit to stop the compressor.

High oil temperature cutout: The sensor of this switch opens the normally closed contacts of the control circuit to stop the compressor if oil temperature becomes excessive.

Low chilled water or brine temperature cutout and recycle switch: When the leaving chilled water temperature drops approximately 5F (8.8C) below normal design temperature, or reaches 36F (2.2C), whichever is higher, compressor operation stops. However, the chilled water pump continues to run. When water temperature has risen approximately 10F (5.6C) the switch contacts close and the compressor restarts. The sensor is usually located in the last evaporator pass and the controller acts both as low limit protection and recycler. This switch must cut out ahead of the low refrigerant temperature cutout setting in order to maintain recycling capability. The recycling capability may be eliminated, if a delay in restarting the compressor is not of importance. Recycling, rather than stopping the compressor, assumes that lowering of the chilled water temperature is the result of an accidental but harmless malfunction and that restarting of the compressor will, therefore, be uneventful.

Chilled water flow switch: This paddle type switch is mounted on the discharge side of the chilled water pump. Its normally open contacts close when chilled water flow is established, thus allowing the compressor to go on. Proof of minimally acceptable flow is necessary for automatic operation. Some engineers prefer pressure switches for this function since paddles are subject to damage.

Condenser water flow switch: Its operation is similar to that of the chilled water system.

Compressor start-stop push buttons: These are momentary contact push buttons, with green and red indicating lights. They start and stop the compressor manually either from the refrigeration machine control panel, the motor control center, the control room, or from any of these locations. If the compressor shuts down on recycle, the green light stays on.

Low voltage reset: A relay is wired into the compressor starter circuit to disconnect the motor from a low voltage source. In case of low voltage outage, the relay must be manually reset.

Additional controls: Solenoids may be installed to provide safety for unusual circumstances. Whether to employ such items as level indicator and alarm in the evaporator, solenoids to cut off flow of coolant water to oil coolers when the compressor is down, etc., is left to the judgment of the designer.

6.3 Operational Controls

Compressor automatic-manual switch: Some manufacturers provide this control with their machines. When the machine is on "automatic," it is under the control of the leaving chilled water temperature controller. On "manual," the capacity of the machine is controlled by high and low capacity push buttons, which preset the prerotation vanes. The motor overload can override the manual setting to protect the motor.

Cycle timer: Some machines are provided with a time delay usually 20-30 minutes, between repeated starts of the compressor. A motor-driven timer provides a normally closed contact for one minute and keeps the contact open for the remainder of the period.

First level alarm lights: Each safety control can have an additional set of normally open contacts which may operate an amber or blinking light to warn the operator that a dangerous level is approaching and that the compressor is about to be shut down by the safety cutout unless corrective action is taken.

Prerotation vane control: Several functions are involved in the operation of the prerotation vanes as they affect not only capacity, which depends on their angular position, but also the safety of the machine and the manner of starting the compressor.

The vane-positioning mechanism is directed by an electronic or pneumatic controller. In some designs, inlet guide vanes remain closed when the compressor is started in order to limit motor inrush current. The compressor is allowed to come up to full speed before its motor is connected across the line at full voltage. Whether it is a pneumatic or hydraulic vane operator, it is important that vanes close on compressor shutdown or control failure.

In the pneumatic system, a 3-way air-solenoid valve dumps air from the pneumatic vane cylinder operator causing the vanes to close. When starting the compressor, the solenoid allows air to pass through to open the vanes. In the hydraulic system (Fig. 6-1) 3-way solenoid oil valves bleed oil from one side of the operating cylinder while keeping pressure on the other side of the piston to maintain the vanes in the closed position. An interlocking circuit, actuated through a piston-operated limit switch, insures that the vanes remain closed before the compressor starts. Some machines have manual switches for operating the vanes. In addition, a remote vane position indicator is available. This same indicator can be converted to a load indicator through the use of a calibrated

dial. Motor overload control also controls the position of the prerotation vanes.

Motor overload control: Motor overload is monitored by a current transformer and resistor in the motor starter circuit, which performs a double function. Some manufacturers include a current limiting function, which can limit the current drawn by the compressor motor to as little as 40% of full load current, thus acting as an electrical demand controller.

More importantly, in its second function, the motor overload controller can override the chilled water temperature controller in order to enforce minimum vane opening during motor start. In case of overcurrent, the overload controller can force the vanes to close entirely, reducing compressor load until the current drops to just below the 100% position, when the chilled water temperature controller can take charge again.

Figure 6-3 indicates variable and fixed ratio demand limiters as designed by one manufacturer. Variable ratio current transformers (CT) are used with low voltage installations. The CT is installed on one of the compressor motor power legs. By connecting the load limiting relay (LLR) to the various taps of the CT, one can vary the secondary current through the LLR to limit the motor

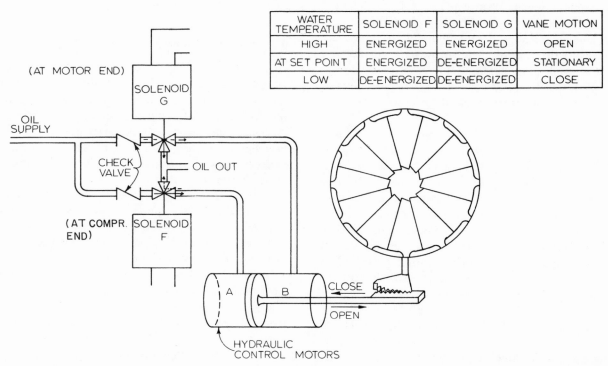

WATER TEMPERATURE	SOLENOID F	SOLENOID G	VANE MOTION
HIGH	ENERGIZED	ENERGIZED	OPEN
AT SET POINT	ENERGIZED	DE-ENERGIZED	STATIONARY
LOW	DE-ENERGIZED	DE-ENERGIZED	CLOSE

Courtesy Carrier Corp. © 1977, Carrier Corp.

Figure 6-1. Hydraulic vane operator (schematic). They can control capacity down to 15–20% for large machines without the benefit of a hot gas bypass.

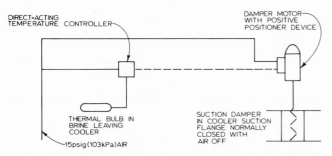

Courtesy Carrier Corp., from Handbook of Air Conditioning Design. © McGraw-Hill Book Co.

Figure 6-2. Suction damper control system. Used in older machines, limiting minimum load to approximately 40% without resorting to hot gas bypass.

load to any desired percentage up to 100% of full load. Readjustment of the tapped percentage is accomplished through switch S4. A fixed ratio autotransformer is used for high voltage installations, *i.e.*, over 600 volts. In Fig. 6-3 a 5000-volt primary rating is used.

Oil heaters: The sump electric oil heater, energized when the compressor is down, evaporates any refrigerant absorbed in the lubricating oil and prevents further absorption during shutdown. The heater is essential where miscibility between oil and refrigerant exists. Thermostatic control keeps the oil in the sump between 150-170F (66-77C), depending on installation requirements.

Vibration detector: For certain seismic or nuclear applications, motor bearing vibration detectors for open compressors may be specified. If vibration limits set by NEMA are exceeded, a standby unit may be brought on line. A remote alarm in the control room may be deemed desirable.

Bearing high temperature detector: For all large motors except, perhaps, those having anti-friction bearings, bearing thermocouples may be provided to annunciate bearing hot spots.

Stator winding temperature detector: In order not to exceed NEMA standards concerning stator winding temperatures, it is appropriate for large motors to be provided with a thermocouple to annunciate excessive temperatures.

6.4 Centrifugal Compressor Performance

Any treatment of centrifugal compressor capacity controls must be preceded by a general understanding of compressor behavior as it affects capacity controls. In addition, process refrigeration controls vary markedly from those for central heating and cooling plants. One of the basic reasons for this difference is that a process

heating and cooling system is usually tailored to a particular process, so that, although some generalizations are applicable, for the most part combinations of components are geared to serve a particular situation and the controls must be designed accordingly.

In contrast, manufacturers of refrigeration machinery for the non-process heating and cooling central plant have achieved extensive standardization of their components, which are assembled in packages of predictable performance. Capacity controls have been simplified to a remarkable degree and all exotic features eliminated, to achieve not only simplicity for simplicity's sake but the ability to computerize plant functions and reduce maintenance attention for today's scantily manned plants.

In the following, the terms "inlet guide vanes" and "prerotation vanes" are fully interchangeable, since they are both commonly used.

Constant speed, rather than variable speed compressor operation is the usual mode of operation, because most compressors are driven by a-c motors.

Ordinarily, variable speed drive is associated with steam or gas turbines and, occasionally, with gas engine drives. Fluid drives of various types can and have been successfully used with electric motors. Automatic speed of wound rotor motors is rather cumbersome and expensive, the main reason it is so rarely employed.

Figure 6-4 shows typical compressor performance at constant speed. The heavy upper curve represents the maximum head capability of the compressor. High head capability decreases along with flow. The area below the line is one of operational stability and the area above it is characterized by surging. The heavy curve thus represents the locus of surge points.

Table 6-1
Minimum Capacity Expressed as Percentage of Full Load

No. of Stages	Inlet Vane Control		Speed Control	
	With Reduced Head	With Constant Head	With Reduced Head	With Constant Head
1	15	25	30	40
2	25	35	40	50
3	35	49	50	60
4	45	60	60	70
5	55	70	70	80

Figure 6-4 shows that a centrifugal compressor has a maximum head and variable volume characteristic. It balances variable discharges with large volume changes. When maximum head is more than required, it may be absorbed by the closing vanes, thus trimming flow to

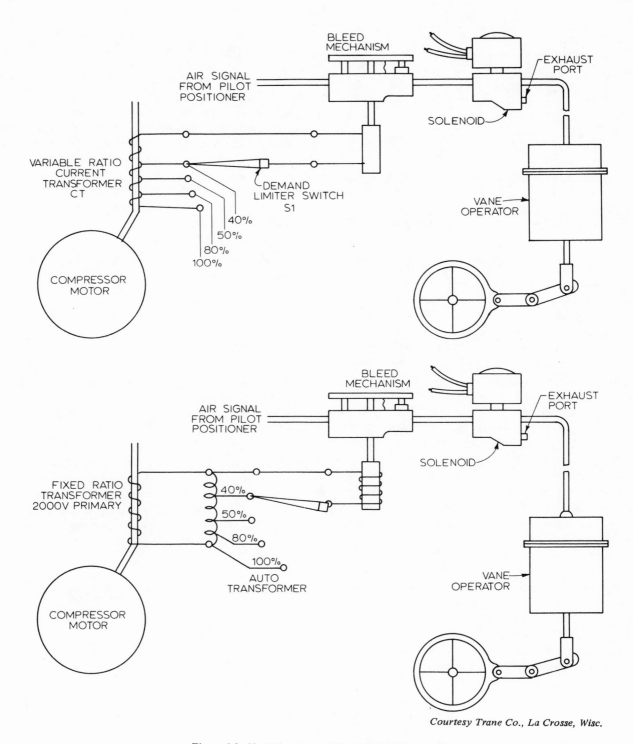

Courtesy Trane Co., La Crosse, Wisc.

Figure 6-3. Variable ratio and fixed ratio limiters.

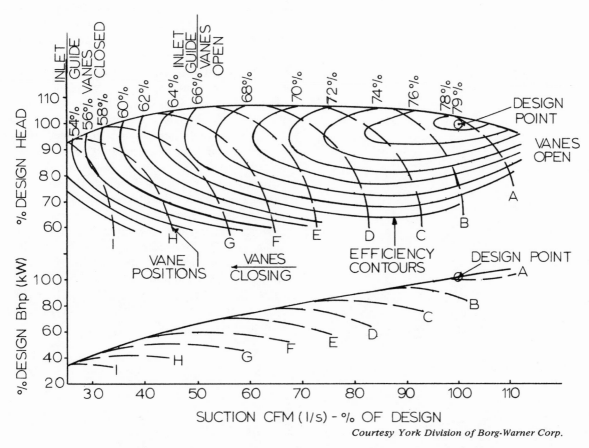

Figure 6-4. Typical compressor performance with inlet vane guide control.

system demand. The selected operating speed establishes the design head. By adjusting inlet guide vanes or speed, the compressor can operate at any point below the curve.

Inlet guide vanes can assume an infinity of positions and thus produce a series of compressor performance curves, (*A* to *I* in Fig. 6-4) to match system demand. Some centrifugal machines, usually of an older vintage and of the open type, have suction dampers, which are inefficient when compared to inlet guide vanes (Fig. 6-2). Inlet guide vanes extend the operating range to a lower flow region than either damper or speed control are capable of reaching. Table 6-1 indicates that lower discharge pressures extend the maximum operating range during part load conditions. In comparing Fig. 6-5 to 6-4 we see that inlet guide control has a wider operating range than speed control. Figure 6-5 indicates that different performance exists for each speed at which adequate head develops to satisfy conditions at that speed. Speed control has the advantage of requiring less horsepower approximately between 70-100% load.

Figure 6-6 indicates the effect of condenser water temperature on compressor speed and kW input for a compressor driven by a wound rotor motor. Figure 6-7 shows the wound rotor motor performing inefficiently in comparison with other drives because of its inherent disadvantages.

The inlet guide vanes, located at the inlet of the compressor, throttle gas to meet refrigeration machine load. By imparting a rotational movement to the incoming gas, they increase compressor efficiency. When throttling reduces flow capacity, head capability drops due to the friction incurred across the inlet guide vanes.

6.5 Compressor Surge

Surge is characteristic of centrifugal machines, be they compressors or fans, at low loads. It may be attributed to compressor inability to handle low flow at high head. A centrifugal compressor can compress a minimum suction volume at a specific head pressure, below which

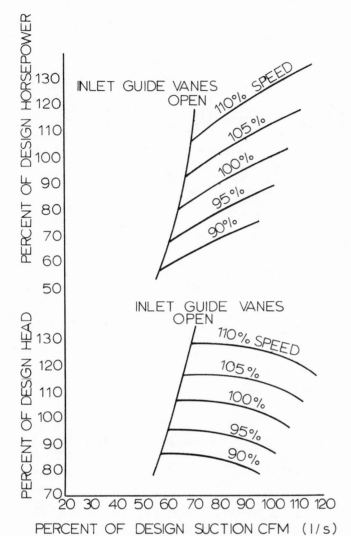

PERCENT OF DESIGN SUCTION CFM (l/s)

Courtesy York Division of Borg-Warner Corp.

Figure 6-5. Typical compressor performance with variable speed.

instability is likely to occur. Expressed differently, an event known as surge occurs when the system head pressure requirement exceeds compressor head capability for a given flow. During surge, there is a reversal of flow in the impeller which distorts compressor performance significantly, usually by an uneven, roaring and sometimes screaming noise. Flow reversal is due to the fact that the impeller cannot maintain condenser pressure. Flow reversal then lowers condenser pressure to the point where flow reverts to the normal direction. When pressure in the condenser builds up once more, surge recurs and the cycle is repeated. As a result of fluctuating pressure the prime mover is loaded and unloaded alternately.

In Figs. 6-8 and 6-9 are plotted the locus of surge points with vane and speed control. Curves of constant pressure

indicate instrument air pressure versus relative vane setting or speed percentage. Curve *A* is that of a heating system or heat pump cycle, Curve *B* is typical of a cooling cycle. Both *A* and *B* assume constant condenser and chilled water flow. It may be noticed in Fig. 6-8 that *A* enters the surge region at approximately the 50% load point (Point *A*) whereas Curve *B* does so near the 0% point with the vanes virtually closed (Point *B*). Curve *B* is desirable insofar as load range extension is concerned and one may expect the machine operating on Curve *A* to remain in surge condition for some time unless corrective action is taken. Figures 6-8 and 6-9 indicate relative opening and closing positions for the hot gas bypass. The hot gas bypass extends the stable control range below minimum CFM capacity, which occurs at approximately 8 psig. The minimum valve position corresponds to 9 psig. If control pressure drops, the bypass valve opens to allow enough gas to satisfy the minimum CFM requirement. The valve opens fully at 3 psig and, if sized adequately, will keep the compressor operating at zero load. The indicated pressures are only illustrative.

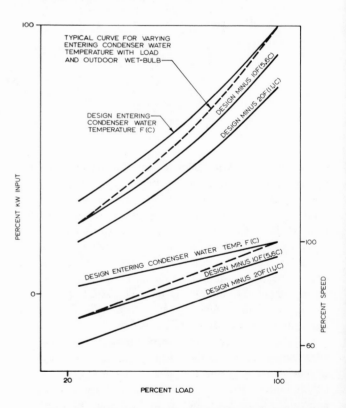

Courtesy Carrier Corp., from Handbook of Air Conditioning Design. © McGraw-Hill Book Co. Used with permission of McGraw-Hill Book Co.

Figure 6-6. Performance of a typical refrigeration machine driven by a wound rotor motor.

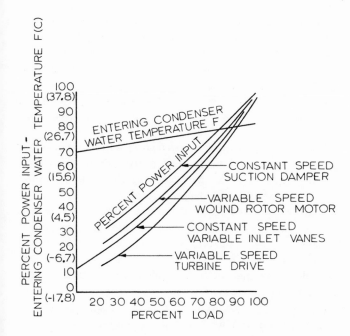

Figure 6-7. Comparative performance of centrifugal compressor capacity control methods.

Comparison of Fig. 6-8 with Fig. 6-9, vane *vs.* speed control, reveals that the locus of surge points with speed control is quite steep and that small changes in speed result in substantial head reduction with but modest flow control. In consequence, it is generally agreed that "an advantage of speed control is higher compressor operating efficiencies in the range of from 60%–100% of design full load conditions."[1,2]

Proposals to reduce or eliminate surging include addition of heat transfer surface to the condenser, change of refrigerant or modification of impeller size. Although these remedies are theoretically feasible, they cannot be applied in everyday practice because they cannot satisfy all load conditions and because economic considerations have forced manufacturers to standardize refrigeration machine sizes, refrigerants and impellers.

In order to predict surge conditions in an actual design, it is advisable to plot system conditions on the performance curve of a selected compressor. This approach will lead to practical control measures with greater assurance

[1] D.L. Cooksey, "Optimizing Compressor Operation," *Air Conditioning, Heating and Ventilating*, (September, 1967), p. 46.

[2] "Part I Feasibility Study Information," *Centrifugal Compression Systems Applications*, York, Penna: York Corp., 1968, pp. 1–2.

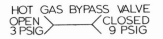

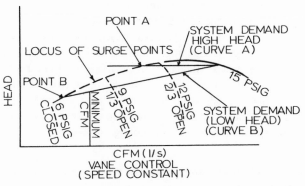

Figure 6-8. Centrifugal compressor, variable system demand, vane control with constant speed. Hypothetical instrument air pressures are shown for opening and closing hot gas bypass valve to impart greater compressor operating stability at low loads. (See Section 6-7.)

than by following the rule of thumb, that is, assuming that no surging exists above 50% of design load.

Whether for heat pump or strictly cooling applications, the following are acceptable ways to extend operation below the surge point:

1. When surge occurs, use a smaller refrigeration machine, if one is available, so that what is partial load on the larger machine becomes a substantially large

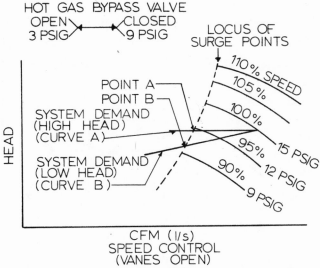

Figure 6-9. Centrifugal compressor, variable system demand speed control with vanes open. (See Section 6-7.)

load for the smaller one. This approach calls for proper planning during the design stage.

2. Decrease compressor head at partial load by allowing chilled water leaving temperature to rise or lowering leaving condenser water temperature.

3. Close inlet guide vanes to a minimum position and lower head by applying automatic hot gas bypass.

4. Reduce speed and apply automatic hot gas bypass.

5. Reduce speed manually or automatically and keep inlet guide vanes under temperature control of the leaving chilled water to conserve horsepower. If surge occurs, increase speed manually until surge disappears, without changing inlet guide control.

6.6 Variable Speed and Condenser Water Temperature at Reduced Loads

Full load performance defines suction and condensing temperature design parameters. At reduced loads, as less heat must be transferred, condensing temperature tends to decrease and suction temperature to rise, unless there is suction temperature control.

Rising construction and energy costs sometimes dictate reductions in the amounts of condenser water in order to save on piping and operating costs. Designers tend to use fewer gallons per ton of refrigeration than has been customary. The resulting higher condenser temperature may bring a particular compressor into the surge region at a higher load percentage, a risk that ought to be examined during design. Any savings due to lower water rates ought to be balanced against cost of larger condenser surface.

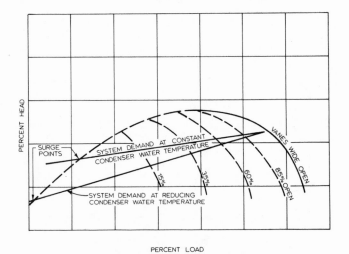

Courtesy York Division of Borg-Warner Corp.

Figure 6-10. Effect of condenser water temperature on compressor performance at part load and at constant speed.

Curves *B* of Figs. 6-8 and 6-9 indicate a decreasing lift characteristic, even at constant water flow and temperature. In Fig. 6-10 lower heads and therefore lower lifts are accentuated with decreasing water temperature at constant compressor speed. The overall result is immensely beneficial in avoiding surge even at very low loads by removing the operating point farther away from the surge line.

Figure 6-11 indicates the interrelationship between compressor speed, inlet guide vane position and condenser water temperature. Modern centrifugal machines can operate at an entering condenser water temperature slightly higher or even the same as the returning chilled water temperature. However, a word of caution is necessary because, at low condenser water temperature flow of refrigerant coolant to motor, oil-refrigerant miscibility and metering of refrigerant from condenser to evaporator can be adversely affected.

Figure 6-12 shows that there can be considerable horsepower savings at low partial loads with proper use of speed reset, *i.e.*, changing speed by means of a controller acting on a signal emanating from a sensor measuring chilled water temperature, condenser water pressure or another system variable. One can expect an extremely economical and efficient operation using low condenser water temperature and speed reduction reset either by load or condenser entering water with inlet guide vanes practically open.

Lowering chiller condenser water temperature should be encouraged. Two factors promote lower condenser water temperatures: lower atmospheric wet-bulb temperatures and reduction of tower load. A reduction of atmospheric wet-bulb improves tower performance by supplying cooler water to the refrigeration cycle; ordinarily, an unpredictable reduction of building cooling load also occurs. If, on the other hand, there is a reduction of load which is more or less independent of external conditions, there is a corresponding drop of condenser water temperature to the chiller at constant flow, provided atmospheric wet-bulb is low enough to permit this water temperature depression. Prevailing wet-bulb temperature at a particular location, the nature of the load, *i.e.*, high internal gains, high solar gain, high outside air loads, etc., and other variables affect the degree of water temperature reduction. A rule of thumb assumes that a 5 F deg (2.8 C) reduction in condenser water temperature corresponds to a 25% reduction in tower load.[3]

Instead of automatic adjustment, manual reset may be used to vary the speed as major incremental load reduc-

[3]Cooksey, *op. cit.*, p. 45.

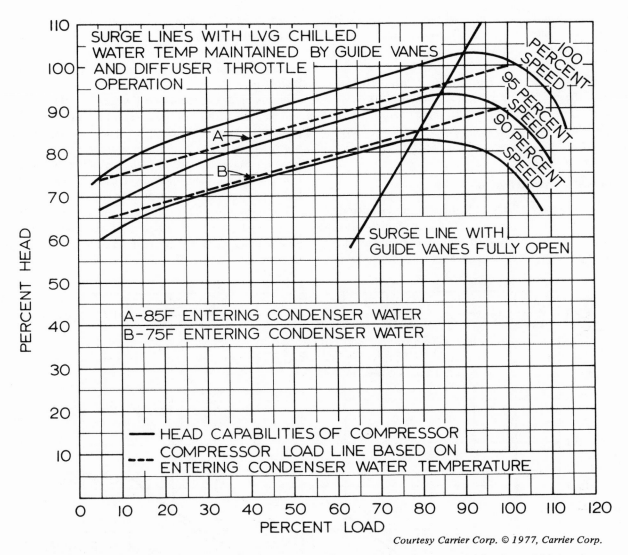

Figure 6-11. Effect of condenser water temperature on compressor performance at part load and at variable speed.

tions occur. The inlet guide vanes must be practically in the wide open position with each resetting. Under these circumstances, the overall lift and brake horsepower requirement per ton are greatly reduced. Constant water flow, both chilled and condenser, must be maintained. Neither condensing temperature nor leaving condenser water temperature must be the measured variable as they are not indicative of load conditions and each state point can represent a combination of factors, which can only indicate true load by sheer accident. It is preferable to control speed from condenser water rather than load as it is somewhat less complicated from an instrumentation point of view, although less dependable and accurate.

On the other hand, Fig. 6-13 indicates that, at constant speed and partial load, low condenser water temperature is of very little benefit. Dependence on straight vane control at very low loads can result in high frictional losses through the vanes, thus increasing substantially the horsepower per ton requirement of the compressor. Figure 6-15, which is similar to Fig. 6-13 but by another manufacturer, shows limited hp gains at lower condenser water temperatures at constant compressor speed. Figure 6-14 shows slightly more extensive hp gains for a machine specifically designed to take advantage of lower condenser water temperatures.

One manufacturer has developed a method for compressor part load control. Figure 6-16 shows a single stage compressor equipped with prerotation vanes, which are not shown. In addition, a diffuser ring is placed around the periphery of the impeller. Through mechanical linkages and cams, the ring is sequenced with the inlet guide vanes by the vane operator. The cam programming schedules the diffuser ring not to operate above 50% load so that all control is done only by vanes. Below 50%, the diffuser ring and vanes operate simultaneously. As load is reduced, the ring moves in to block part of the passage and restrict flow, so that it provides the compressor with increased head potential to avoid reentry of the gas into the impeller and to continue surge-free operation.

6.7 Automatic Hot Gas Bypass

In order to prevent a large centrifugal compressor from surging, and to match extensive partial load requirements, some manufacturers continue to apply the hot gas bypass method. Gas from the discharge side of the compressor is shunted to the suction side to satisfy compressor minimum flow requirements. During the hot gas bypass operation, the refrigeration load remains constant at low levels and compressor gas flow remains artificially constant at high levels.

Ideally, hot gas should be bypassed to the evaporator and sprayed with refrigerant liquid to desuperheat the hot gas before it enters the compressor so as to minimize compressor horsepower requirements. See Fig. 6-17. Thorough mixing is essential and, if an evaporator connection is not available, then the sprayed gas can be led through a liquid suction trap as a second best choice.

The operation is normally quiet and automatic although highly inefficient. If off-peak conditions continue for long periods, hot gas bypass can be an inordinately expensive method.

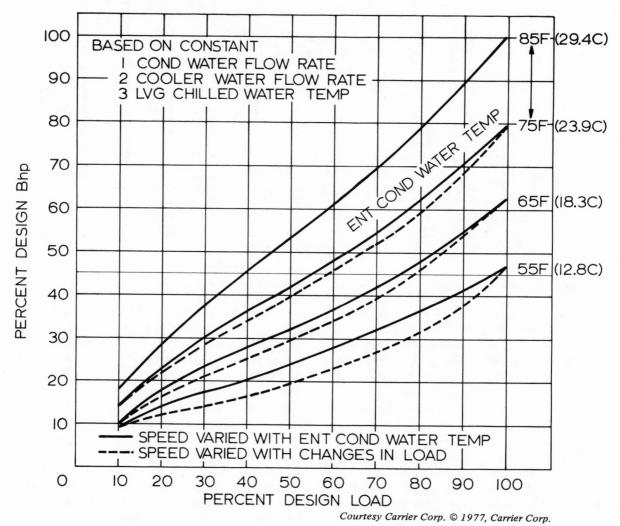

Figure 6-12. Effect of condenser water temperature on compressor brake horsepower at variable speed.

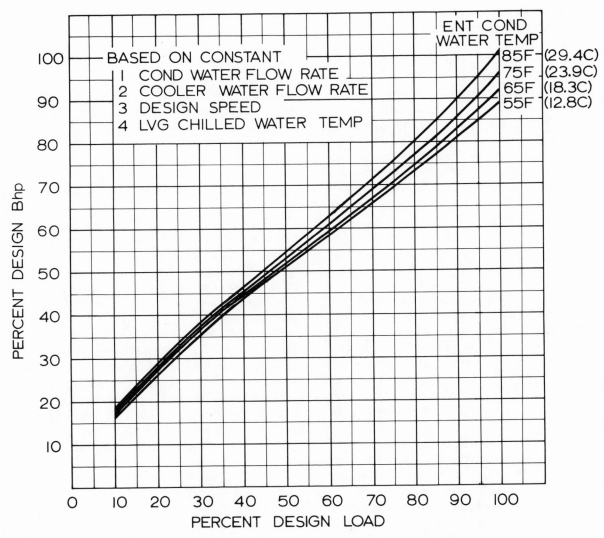

Figure 6-13. Effect of condenser water temperature on compressor brake horsepower at constant speed.

Another condition that can create high compressor head, reduction of suction pressure, can also be alleviated by bypassing hot gas from the discharge to the suction side.

Figures 6-18 and 6-19 show hot gas bypass directly across the discharge and suction side of the compressor in schematic form for illustrative purposes only. In interpreting Fig. 6-18 reference should be made to Figs. 6-8 and 6-9 indicating compressor performance with variable speed and variable vane control.

Referring to Fig. 6-8, it is observed that, at 15 psig (103 kPa), the (inlet guide) vanes are fully open and, at 6 psig (41 kPa), fully closed. Working with Curve *B*, the minimum flow rate before surge occurs is at about 8 psig

(55 kPa). The hot gas bypass can be set to start opening at 9 psig (62 kPa), which can also be the air pressure corresponding to the minimum vane opening position. As the pressure continues to drop, the hot gas bypass valve open further to satisfy minimum compressor CFM requirement. At 3 psig (20 kPa) it is fully open and can ensure stable operation at minimal loads. In Fig. 6-8 note that, depending on state point, a surge condition can exist between Points *A* and *B* on the surge line.

If the bypass is set for Point *A*, it will be bypassing unnecessarily during periods of low head. On the other hand, if it is set at *B*, it will be undergoing surge at times of high head operation. On most air conditioning applications this may not be a problem. Resetting the air pressure at which the bypass valve will open is one way of resolving such a problem.

In Fig. 6-19, pressure controller PC-1 positions inlet guide vanes and hot gas bypass to maintain constant compressor suction pressure. Discharge pressure controller PC-2 resets PC-1 as it senses changes in discharge pressure. Vane control is extended and gas bypass is delayed during low head periods.

Another less frequently used method is to measure suction-discharge pressure ratio to reset the bypass valve.

In general, when hot gas bypass is used with variable vane or speed control, the control circuit includes a minimum vane or speed condition before the hot gas bypass valve opens.

6.8 Pneumatic Control Loops

A number of pneumatic loops are needed for a full complement of operating refrigeration machine controls. Figure 6-20 includes four basic loops for different control needs, not all present at any one time. The inlet guide vane control loop is part of every large centrifugal machine but is occasionally replaced by a suction line damper. The other illustrated loops are a hot gas bypass valve (A), a subcooler level control (C), and a turbine speed control (D).

The inlet guide vane loop (B) is composed of three parts: a temperature controller output signal, an instrument air to valve positioner and instrument air to the underside of the piston to close the guide vanes.

A 3-15 psig (20-103 kPa) air-operated controller output regulates the position of the inlet guide vanes. The signal

passes through the 3-way hand valve, minimum output cumulator, or motor overload controller, low refrigerant pressure override controller, solenoid valve and adjustable restrictor. The vanes are fully open at 15 psig (103 kPa) and close at 3 to 6 psig (20-41 kPa), depending on make. When the compressor shuts down, the vanes come to the fully closed position, preventing reversal of flow from the high pressure region of the condenser to the low pressure region of the evaporator. When the compressor is started, the guide vanes are closed, so that the starting torque requirements remain low; a time delay relay in the control circuit energizes the solenoid valve and the current limiter at a predetermined interval after starting the compressor, allowing control air to flow to the vane operator and putting it directly under the control of the temperature controller. An adjustable restrictor dampens the sensitivity of the current limiter and prolongs the time for the vane to travel from the fully open to the fully closed position. One manufacturer uses an air accumulator in lieu of a restrictor.

The instrument air line maintains a constant 40 psig (276 kPa) on the underside of the piston, which will close the vanes on compressor shutdown by bleeding air from the positioner through the side port of the guide vane solenoid valve, which becomes deenergized. As already mentioned, the 40 psig (276 kPa) air pressure will keep the guide vanes closed when the compressor starts, protecting the compressor from overload.

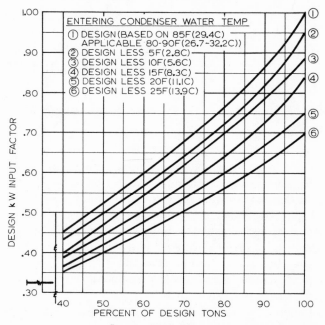

Courtesy York Division of Borg-Warner Corp.

Figure 6-15. Reduction of power input with reductions in condenser water temperature at constant speed (typical).

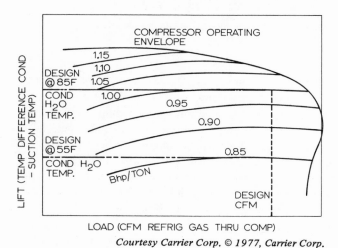

Courtesy Carrier Corp. © 1977, Carrier Corp.

Figure 6-14. Typical horsepower map at variable condenser water temperatures at constant speed.

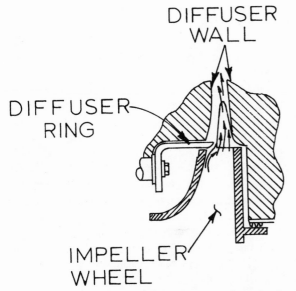

DIFFUSER WALL

DIFFUSER RING

IMPELLER WHEEL

Courtesy Carrier Corp. © 1977, Carrier Corp.

Figure 6-16. Surge control by diffuser ring.

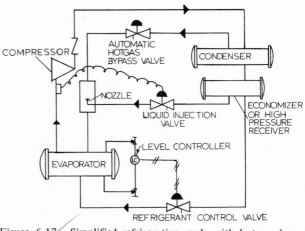

Figure 6-17. Simplified refrigeration cycle with hot gas bypass and liquid injection.

The current limiter is an override control that reduces the pneumatic signal in order to close the vanes when current exceeds full-load amperage, regardless of the magnitude of the controller signal or the high setting of the cumulator, which maintains the vanes at a minimum opening position. If the hot gas bypass is used, a split signal is employed, so that the vane operator can employ a 6-15 psig (41-103 kPa) range and a minimum opening position at 9 psig, so that the vanes remain partially open to accommodate the hot gas flow. The hot gas bypass valve is reverse-acting so that it opens at 3 psig (21 kPa) and closes at 9 psig (62 kPa). The hot gas solenoid is de-energized when the compressor stops, opening the side port and bleeding the air line to the valve, causing the valve to open. An open hot gas valve is desirable in order to equalize the high and low sides of the compressor at shutdown for systems operating above 32F (0C) suction temperature.

Another useful override device, provided only by one manufacturer, is the low refrigerant pressure override controller, which senses evaporator pressure and bleeds air from the vane operator, closing vanes and reducing capacity.

The subcooler control loop (C) is used on very large machines, which do not have external thermal or flash economizers but which have a subcooler bundle at the bottom of the condenser shell. The level controller (as in Fig. 6-17) keeps the subcooler tubes flooded and releases liquid refrigerant to the evaporator as needed to keep the liquid level in the evaporator appropriate to the magnitude of the load. Entering condenser water first comes in contact with the refrigerant liquid surrounding the sub-cooler tubes, subcooling the condensed liquid below its saturation temperature, which prevents flashing in the liquid line and improves cycle efficiency. The overall results are beneficial because subcooling helps minimize liquid valve and pipe size and reduces friction and insulation thickness. Note that high pressure control is used, while in some industrial applications, evaporator control is used in conjunction with a high pressure receiver.

The subcooler level controller, by using proportional action, releases more refrigerant to the evaporator as load decreases, assuring flooded evaporator tubes. This is necessary because, as a result of violent boiling activity at full load, less refrigerant is required in the evaporator while, at low loads, as boiling decreases, the heat exchanger performs best if most of its tubes are flooded.

Smaller centrifugal machines have high pressure floats or orifices and external economizers. The economics of the smaller machines are such that they favor this arrangement.

The turbine governor control loop (D) is discussed in Section 6.12, "Gas Turbine Controls" and Section 6.13, "Governors." Figure 6-12 illustrates the operating economies in controlling speed by using the entering condenser water temperature.

6.9 Control of Leaving Chilled Water Temperature

Chillers in Series: Figure 6-21 represents evaporators in series but condensers in series or in parallel. Each chiller has an individual temperature controller. Both units reduce capacity simultaneously in response to leaving water temperature. The return chilled water temperature controller cycles each unit off the line, Unit *A* first, when the return water drops to a predetermined temperature.

Leaving water temperature controller *TC-A & B* in Fig. 6-22 operates Unit *B* on a branch pressure of 3-11 psi (21-76 kPa). If Unit *B* cannot maintain temperature, Unit *A* comes on line at 10 psi (69 kPa) and is modulated to full load at 18 psi (124 kPa). This is the preferred and more economical control arrangement because only one instrument is necessary instead of three, and coordination of multiple instrument settings is avoided entirely.

Chillers in Parallel: Each chiller has its own temperature controller in Fig. 6-23. Capacity modulation of both chillers is simultaneous. The return water controller cycles the units off and on as required to maintain cycle efficiency as programmed. However, if one pump is in operation and one unit is down, water continues to circulate through the inactive unit, requiring overchilling through the active unit to maintain a constant water temperature. Over chilling may cause a freeze-up condition and each controller can act as a low limit temperature controller to shut down the machine.

The arrangement shown in Fig. 6-24 may be used for water temperatures exceeding 44 F (6.7 C). It guarantees the same leaving temperature from both chillers but does not offer individual freeze-up protection, which must be provided through separate low-limit controllers. Even with overchilling, it is highly unlikely that a freeze-up can occur, even without the low limit controller. However, it is necessary to check one chiller operation before omitting the low limit controllers.

Figure 6-25 is an economical and often-used arrangement, utilizing a split range controller. Branch pressure of 3-11 psig (20-76 kPa) operates Unit *A* and 10-18 psig (69-124 kPa) operates Unit *B*.

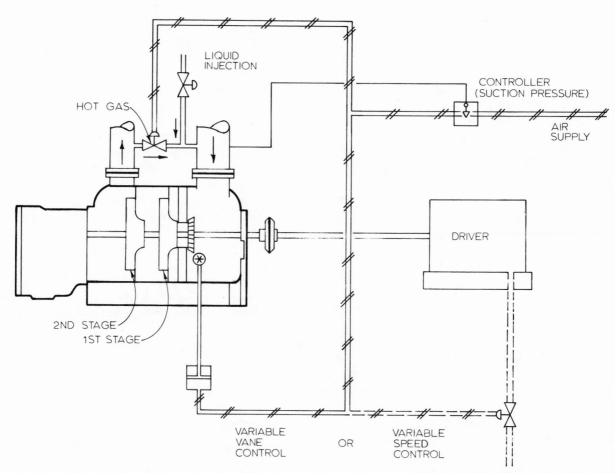

Courtesy York Division of Borg-Warner Corp.

Figure 6-18. Schematic representation of speed or vane control in combination with hot gas bypass for low load range controlled by suction pressure controller.

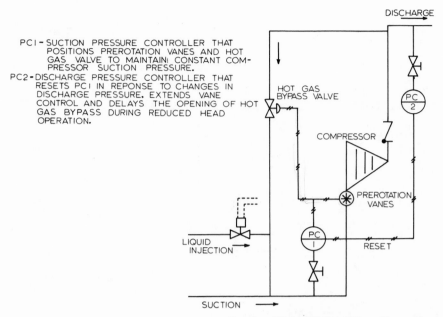

PC1 - SUCTION PRESSURE CONTROLLER THAT POSITIONS PREROTATION VANES AND HOT GAS VALVE TO MAINTAIN CONSTANT COMPRESSOR SUCTION PRESSURE.

PC2 - DISCHARGE PRESSURE CONTROLLER THAT RESETS PC1 IN REPONSE TO CHANGES IN DISCHARGE PRESSURE. EXTENDS VANE CONTROL AND DELAYS THE OPENING OF HOT GAS BYPASS DURING REDUCED HEAD OPERATION.

Courtesy York Division of Borg-Warner Corp.

Figure 6-19. Schematic control diagram indicating suction pressure controller reset by variations in discharge pressure.

6.10 Piggyback Arrangement in Series, Combination of Centrifugal and Absorption Machines

Leaving chilled water temperature in Fig. 6-26 is maintained by temperature controller *TC-1* which, at full load, opens the inlet guide vanes to the wide open position. The steam turbine speed controller opens main steam admission valve *V-1* to the wide open position for full load. Steam is expanded through the turbine and is exhausted at 15 psig which, after incurring a 3-psi drop through valves *V-2* and *V-3*, enters the absorption machines at 12 psig.

On reduction of load, the turbine's speed controller, in order to maintain constant speed, closes *V-1* to reduce steam flow. Steam pressure to the absorption machines may be expected to drop. The capacities of both centrifugal and absorption machines diminish. If chilled water temperature begins to rise, *TC-1* opens compressor inlet guide vanes and the speed controller modulates *V-1*. The capacity of all machines then increases, lowering the leaving water temperature. This cycle repeats itself. Absorption unit valves *V-2* and *V-3* are either fully open or fully closed, depending on whether the absorption units are working or not.

As load decreases, some of the units can be automatically shut down. Temperature Controller *TC-2* is located in the chilled water return and, since the return water tem-

perature is indicative of refrigeration load size, *TC-2* will shut down Absorption Unit 2 when branch air pressure drops to a point at which Pneumatic-Electric relay *PE-1* is activated, thus, bleeding valves *V-2* and *V-5* and engaging the starting interlock in the Absorption Unit 2 control panel. *V-2* shuts steam flow and *V-5* bypasses condenser water around Absorption Unit 2.

On further reduction of load, branch air pressure will drop further and *PE-2* deactivates solenoid valve *S-1*, bleeding *V-1*, stopping the turbine and activating *S-2*, admitting air to *V-4* and sending steam directly to Absorption Unit 2. An arrangement can be made whereby Absorption Unit 2 can be operated along with Absorption Unit 1, provided *S-3* can be opened manually or remotely by a switch on the control board and *V-4* can be sized to accommodate both machines. *TC-3* is set for 44F (6.7 C), a setting low enough not to satisfy either of the absorption machines. Essentially, the absorption machines operate at maximum output at all times as long as they are coupled with the centrifugal machine. When the centrifugal machine shuts down, *TC-3* operates Absorption Unit 1 and regulates the leaving water temperature to maintain 44F (6.7 C).

To obtain greater compressor efficiency at lower loads, it is possible to reset turbine speed by using *TC-2* branch air pressure to reset the turbine speed controller. How-

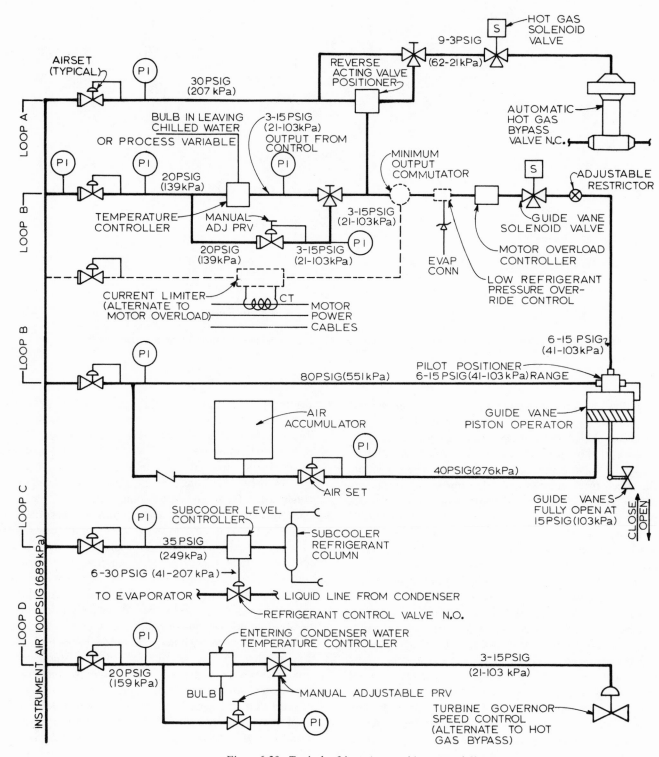

Figure 6-20. Typical refrigeration machine control diagram.

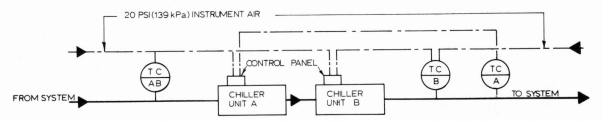

Figure 6-21. Chillers in series with individual controllers, *TC-A* and *TC-B*. Return water controller *TC-AB* takes Unit *A* off line first.

ever, compressor surge can occur and must be guarded against.

6.11 Piggyback Arrangement in Parallel, Combination of Centrifugal and Absorption Machines

The controls of this arrangement are almost identical to those of the series arrangement discussed in the previous section. Some of the parameters for the controls change. Most notably, evaporator temperatures for the absorption machines decrease to a considerable degree. This is because there are no high and low machines as in the series arrangement. In a sense, all machines in parallel are "low" machines, since they all receive the coldest condenser water available.

Any overchilling required to maintain leaving water temperature, if one of the absorption machines is shut down, could cause the active unit to freeze-up. Ordinarily, absorption machines, in addition to their size limitation of 2000 tons (7040 kW), are not capable of operation at evaporator temperatures as low as the centrifugals. For centrifugals, 36F (2.2 C) leaving water is the low limit, whereas 39-40F (3.9-4.4 C) is the limit for absorption machines. This 3-4F (1.7-2.2 C) temperature difference may be supremely important in some central installations, to the point where absorption machines may not be applicable.

Figure 6-27 shows *TC-3* on the leaving side of all machines. The centrifugal and the two absorption machines, with their individual pumps and check valves, are in parallel. Otherwise the control system resembles that of the series arrangement.

The main idea behind combining centrifugal and absorption machines is the economical steam rate that results.

As long as the combined steam rate is smaller than the sum of individual steam rates, it becomes necessary to maintain the combination. When the combined steam rate exceeds the individual steam rate, the combination ceases to be useful and individual machine operation is indicated.

The purpose of the high pressure controller is to prevent steam pressure in the turbine exhaust from exceeding a predetermined value, usually 15 psig (103 kPa). If pressure exceeds 15 psig, the pressure controller overrides *TC-1* and starts closing the inlet guide vanes to reduce compressor load, which, in turn, forces the speed controller to close steam admission valve *V-1*, thus, reducing exhaust pressure. Another method involves having the pressure regulator reset *TC-1* to a higher temperature, thus, biasing the centrifugal machine to reduce its load. Reducing load by closing the inlet vanes or resetting *TC-1* will result in higher leaving water temperatures.

A load-sharing relationship exists between the centrifugal and absorption machines. The centrifugal machine acts as a perfect throttling steam control valve for the absorption machine. The absorption machine, in turn, must fulfill the role of a perfect condenser for the centrifugal machine. If these functions are not well matched, a pressure imbalance exists in the exhaust pipe. To restore this balance fully, both the centrifugal and absorption machines as well as their respective chilled water flows must be rebalanced. This would involve extremely cumbersome controls without an analogous economic benefit. The percent load carried by each type of machine is fairly constant down to about 30% of total load. Load imbalance is improbable, except at very low loads. At such low loads, it is advisable to switch to single unit operation at full load efficiency.

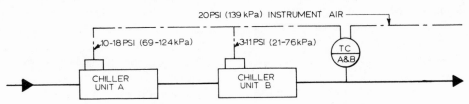

Figure 6-22. Chillers in series with split range controller, *TC-A & B*.

6.12 Gas Turbine Controls

There are a number of problems peculiar to gas turbines in connection with the air-fuel ratio.

High exhaust gas temperatures may be incurred since air is introduced by the air compressor, which is independent of the governor, whose function is to control fuel admission. Air is introduced into the turbine not only to sustain combustion, but to keep exhaust gases at design temperature. If the mix of air introduced into the turbine is inadequate, then the combustion process will overheat the air. Limitations on exhaust gas temperature and on the rate of fuel increase are necessary to prevent such temperature excursions.

Compressor stall during acceleration, a result of uncontrolled maximum fuel injection or a high rate of fuel increase (rich mixture), must be avoided by limiting maximum fuel flow or the rate of fuel admission.

A *flame-out* may result from a sudden decrease in fuel. Since a gas turbine depends on a continuous combustion process, limitations must be imposed either on the deceleration rate or in the rate of fuel decrease.

Some additional problems lie in the following areas: single versus two shaft gas turbines, small versus large turbines and matching gas turbines to refrigeration centrifugal compressors, taking into account how both the compressor and turbine affect each other.

The following two-column tabulation indicates factors, for both single and split shaft turbines, which influence governor selection and control mode operation:

Single Shaft

1. Dynamic system response is easier to calculate and constant speed control response is fast.

2. Only one governor is needed.

3. It is more suitable to constant speed applications such as generator drives, pumps and compressors (if started in unloaded condition).

4. Costs are reduced not only due to the simplicity of a single shaft turbine but, also, due to less complicated controls.

5. Starter power and starting torque requirements are high because the entire rotating mass must be accelerated.

6. Centrifugal compressor surge has less of an effect due to the larger rotating mass of the turbine.

Split Shaft

1. Dynamic system analysis is complicated, less predictable than it is for single shaft units, and is hard to calculate.

2. Usually both shafts need governors.

3. There exists a time lag in combustor response to load change.

4. Split shaft turbines exhibit high breakaway torque and can handle startup loads without resorting to high horsepower starting equipment.

5. Speed flexibility characterizes these turbines as they can match the speed-load requirements of many types of driven equipment, including pumps and compressors.

6. On surge, *i.e.*, temporary loss of compressor load, the compressor tends to run away, tripping overspeed safety, before the operation can be stabilized.

7. The possibility of reverse rotation, damaging the compressor, is less likely.

Turbine size heavily influences the type of controls used. However, there are no valid generalizations; each case must be analyzed separately. It is safe to say, however, that: relayed governors are better suited for small gas turbines (50-150 hp or 37-112 kW) if constant speed is not required, while mechanical-hydraulic governors are preferable for large gas turbines. Using electric governors on large turbines involves high cost but offers special design attributes that mechanical-hydraulic governors can not accommodate satisfactorily.

Matching gas turbine to centrifugal refrigerant compressor performance requires cooperation between the two respective manufacturers in addition to that of the control and governor manufacturers for a fully integrated system.

The possibility of resetting gas turbine speed from the temperature of the condenser water leaving the cooling tower was discussed in Section 6.8 (see Figs. 6-12 and 6-20). In throttling the turbine fuel valve, one must meet refrigerant compressor horsepower and speed requirements. Suction pressure or chilled water supply temperature are two other variables that can be used to control gas turbine drivers. A hot gas bypass, as part of the control scheme, can begin to open as the gas turbine reaches a minimum maintainable speed.

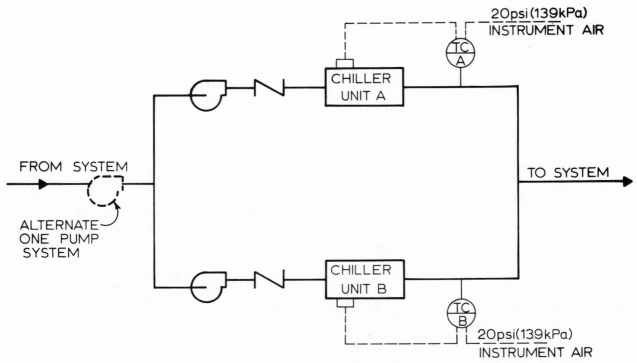

Figure 6-23. Chillers in parallel with individual controllers, offering a measure of freeze-up protection for water temperature around 40 F (4.4 C).

Two-shaft turbines can be controlled by measuring suction pressure, refrigerant compressor discharge flow or condenser and chilled water temperature, if a variable speed gas turbine driver is to be used. For constant speed applications, sensing turbine shaft speed is adequate.

Variable turbine speed (to regulate compressor capacity) is achieved by modulating the fuel valve in response to system load. This type of operation depends on a thorough understanding of the dynamic responses of both compressor and gas turbine and a realistic approach to controls that are suitable for a given set of components.

6.13 Governors

A governor regulates the rate at which a prime mover converts energy into mechanical power. The suitability of a governor depends on the type and size of prime mover and the type of application.

A governor should not be used as a safety device, such as an overspeed trip, which function must be delegated to a completely separate device. In other words, a governor must serve normal operating purposes, not emergency functions.

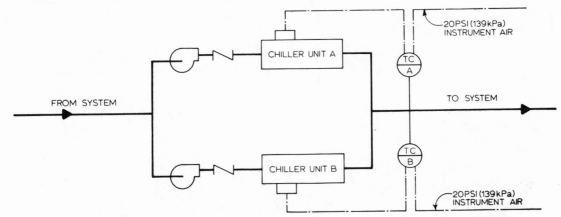

Figure 6-24. Chillers in parallel with individual controllers in the common line. Suitable for water temperatures above 44 F (6.7 C).

Under the term "governing," there are five groups of components:

1. Speed governor (mechanical, hydraulic, electrical or a combination).
2. Speed control mechanism (relays, servos, pressure or power amplifiers, etc.).
3. Governor-controlled valves.
4. Speed changers.
5. External control devices.

Speed is one of the most important variables in controlling engines or turbines. The main function of the governor is to maintain constant speed, in constant speed applications, despite variations in load or minor variations in fuel or steam supply. Speed governors measure shaft speed so as to control steam or fuel admission by positioning valves, which keep speed within speed regulation limits. A steady state condition is reached when developed power equals load power demand.

Some of the prime mover characteristics that affect control accuracy are:

1. Manifold or steam chest lag. Steam chest space determines not only time lag but, also, the relationship between valve movement and steam flow when a new power load demand arises.
2. Dead time between steam or fuel admission or charging and conversion to energy.
3. In the case of internal combustion engines, time for all cylinders to develop torque at new required levels.
4. Total rotating mass, which affects transient performance.
5. Damping and torque versus speed characteristics of a system.

6. Associated control devices, such as voltage regulators, pressure reducing valves, etc.

The conclusion is that the prime mover characteristics are fixed by design and, therefore, all other components must be carefully coordinated and selected to obtain optimum system performance. The governor and the control system are only two aspects of a multi-faceted problem. One must consider the system in its entirety as well as its desired behavior in order to satisfy all engine-load combinations. Mismatching of an expensive and responsive governor with a highly inert system is a waste. By the same token, a sluggish governor grafted on to a reactive system is an invitation to instability.

Variable speed is another mode of operation for prime movers. There are governors which can accept external signals and adjust speed accordingly. For instance, a centrifugal refrigeration machine can operate in the variable speed mode, when a constant leaving water temperature is maintained by a leaving water controller as load changes. Other examples are those of maintaining constant air compressor discharge pressure despite load variation or changes in cooling water temperature and of maintaining back pressure in a topping turbine, including the piggyback refrigeration machine arrangement.

6.13.1 *Governor Nomenclature and Definitions*

A definition of some of the most important terms used in governor technology may be helpful. Some of these terms are defined in NEMA publications.

Speed changer is a device, usually in the form of a spring, for changing the setting of the speed-governing system within the specified speed range while the turbine is operating.

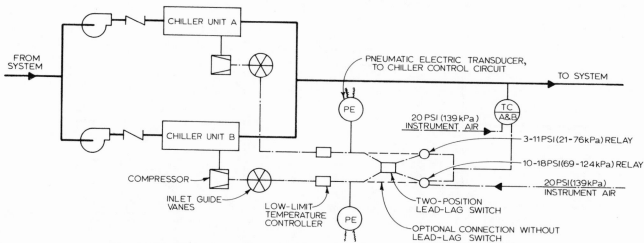

Figure 6-25. Chillers in parallel with split range controller, *TC-A & B*, and lead-lag switch.

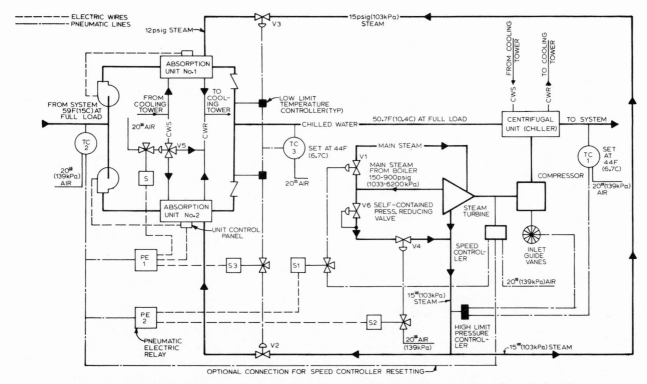

Figure 6-26. Combination of steam turbine driven centrifugal machines and absorption units in series.

Speed range is all the operating speeds, with upper and lower limits, expressed as a percentage of rated speed.

Speed regulation is the change in speed (decrease) between no load and full load speed.

Percent regulation

$$= \frac{\text{Speed at zero power} - \text{speed at rated power}}{\text{speed at rated power}} \times 100$$

Speed droop has come to mean speed regulation, which is based on power, whereas droop is expressed as a percentage of governed speed. Droop is reflected mechanically by the full to zero servo speed position. The difference is of no importance. A fuller explanation of droop is given in Section 6.13.3.

Dead band reflects the insensitivity of the speed-governing mechanism and is the total change in speed for which there is no corresponding change in the position of the governor-controlled valve.

Trip speed is the speed at which the engine or turbine protective device comes into play.

Speed stability is the capability of the speed-governing system to maintain steady state speed within specified limits.

Oscillation is represented by the limits of "hunting" under steady state conditions. Stability or instability refers to the magnitude of the oscillation.

Maximum speed rise is the momentary rise in speed experienced by the engine or turbine when the load is reduced suddenly from rated to zero.

Percent speed rise

$$= \frac{\text{max. speed at zero power output} - \text{rated speed}}{\text{rated speed}} \times 100$$

6.13.2 *Speed Governor System Classification*

NEMA *B, C* and *D* governor systems for steam turbines are sensitive to speed through special speed-sensing elements which, through relays and servomotors, can reposition the turbine steam valves. NEMA class *A* systems usually actuate the valve directly (see Table 6-2). The Institute of Electrical and Electronic Engineers (IEEE) publishes some standards concerning speed-governing of steam turbines, internal combustion and gas turbine generator units.[4,5,6] The American Society of Mechanical

[4] AIEE Publication No. 600 *Recommended Specification for Speed-Governing of Steam Turbines Intended to Drive Electric Generators Rated at 500 kW and Larger*, American Institute of Electrical Engineers (IEEE), 1959.

[5] AIEE Publication No. 606 *Recommended Specification for Speed Governing of Internal Combustion Engine-Generator Units*, American Institute of Electrical Engineers (IEEE), 1959.

[6] IEEE Publication No. 282, *Proposed IEEE Recommended Specification for Speed Governing and Temperature Protection of Gas Turbines Intended to Drive Electric Generators*, American Institute of Electrical and Electronic Engineers, 1968.

Table 6-2
Speed Govenor System Classification[8,9]

Class of Governor System		Speed Range, Percent (As Specified)	Maximum Speed Regulation, Percent	Maximum Speed Variation, Percent, ±	Maximum Speed Rise, Percent	Trip Speed Setting, Percent
Single Stage	A	10, 20, 30, 50, 65	10	0.76	13	115
	B	10, 20, 30, 50, 65, 80	6	0.50	7	110
	C (Multi Stage)	10, 20, 30, 50, 65, 80	4	0.25	7	110
	D	10, 30, 50, 65, 80, 85, 90	0.50	0.25	7	110

All values are in percent of rated speed.

Engineers (ASME) publishes a standard on gas turbine power plants.[7] Table 6-2 is a performance criterion for speed regulation. However, multistage steam turbines have only *C* and *D* governor systems in order to safeguard the improved efficiency of multistage operation. NEMA *A* governor system can be used where speed regulation is of no importance. Therefore, for steam turbines, the least expensive arrangement is the single admission valve, reinforced by a number of manual valves to improve part load efficiency (see Fig. 6-32, Section 6.13.8, "Hand Valve").

For larger turbines, or more precise control of smaller steam turbines where maximum efficiency is a goal, multiple automatic nozzle control, by using hydraulic servo motors, is employed at considerably increased cost.

6.13.3 *Speed Changer Versus Droop*

Speed changers affect the speed setting of a governing system by changing speeder spring tension (see Fig. 6-29 for speeder spring application). The speed changer upper limit should not exceed rated turbine speed unless special provisions are made to test overspeed trips. The governor is only a controlling device and must not be used to limit the safety of operating conditions. A governor shutdown device should only be used for normal and not for emergency purposes. The speeder spring is part of the speed

[7]ASME Power Test Code ASME PTC 22-1966 and ANSI PTC 22-1974, *Gas Turbine Power Plants*, American Society of Mechanical Engineers.

[8]NEMA Publication SM-22-1970, *Single-stage Steam Turbines for Mechanical Drive Service*, National Electrical Manufacturers Association, part 3, p. 2.

[9]NEMA Publication SM 21-1970, *Multistage Steam Turbines for Mechanical Drive Service*, National Electrical Manufacturers Association, part 3, p. 2.

changer mechanism. Operation of a speeder spring is described in Section 6.13.4.

External control devices can operate the steam valve directly and restrict governor function to speed-limiting. The external devices can amplify the power of their signal by regulating a pilot valve and having these signals maintain set speed by continually resetting the governor.

The speed changer, by being reset manually or automatically, can maintain constant turbine speed, as indicated by loci *A,B,C* and *D* in Fig. 6-28. By changing spring tension it is possible to eliminate the 6% speed regulation shown in Fig. 6-28. Rated speed can be maintained for all power outputs.

As pointed out in Section 6.13.9 "Governors and Parallel Operation," droop is essential in paralleling operations when using mechanical hydraulic governors.

6 13.4 *Relayed Governors*

This type of governor is known under other names such as "direct acting governor" or "simple governor." Figure 6-29 indicates its basic operation.

The flyweights, or ballarms in this case, raise or lower the plunger assembly. Whenever, the control land is pushed down by the speeder spring force, the gear pumps act directly on the power piston, raising it. In doing so, the terminal shaft lever is raised, lifting the speed droop lever pin and relaxing the speeder spring tension, lifting the floating lever and creating droop by decreasing speed.

When the load is decreased, the engine speeds up, raising the plunger and draining the oil under the piston to the sump, rotating the lever to the "decrease fuel" position and reducing speed.

This type of governor operates on speed regulation (droop) which helps to stabilize an engine or divide load between parallel units. It is suitable where isochronous speed, constant speed or zero percentage speed regulation, is not required and is applicable to small diesel, gas or gasoline engines.

6.13.5 *Electrical Governors*

Under this title one may include both electric and solid state governors.

The basic components of an electric governor are:

1. A *control system*, whose primary signal is either a speed sensor, *i.e.*, electrical impulses emanating from a magnetic pickup near a rotating gear or a tachometer generator or an alternate method of generating a signal by using sensors to generate frequency and power. In either case, the signal, which is converted to a d-c proportional sigínal, is compared to the reference speed, frequency or power. The amplifier, the main part of the control system, in association with auxiliary elements in the control box, regulates the output signal to increase, decrease or hold steady the fuel or steam supply to maintain speed.

2. An *operator* (actuator), capable of moving and positioning the fuel or steam mechanism. A flyweight governor is unnecessary since the signal originates in rotating gear as explained above. Occasionally, depending on the application, a ballhead governor may be used as a governor redundancy.

3. A *speed changer*, which can be outside the control box, in the form of potentiometer. Speed can also be easily changed or biased by electrical signals from a process controller, permitting precise control of process temperatures, liquid levels, pressures or other variables. Engine-driven or turbine-driven boiler feed pumps, chilled or condenser water pumps feeding large circuits, and centrifugal compressors vary speed to maintain operating conditions by having the speed changer reset.

The main reasons for using electrical governors are as follows:

1. They have a faster response under transient conditions. It is claimed that, in some cases, response time is about half that of a hydraulic operator.

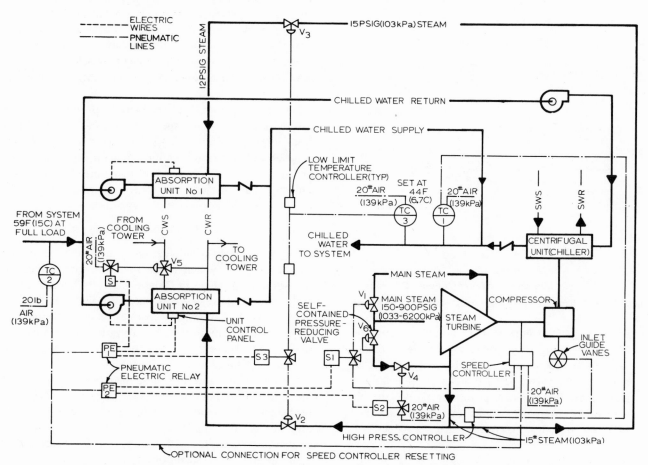

Figure 6-27. Combination of steam turbine driven centrifugal and two absorption machines in parallel.

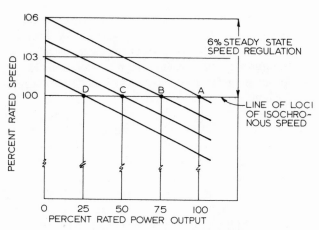

Figure 6-28. Speed regulation curves (slanted lines).

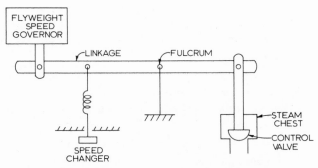

Figure 6-30. Schematic diagram of mechanical governor.

Governors having frequency and power sensors will always be used on generator set applications, while those sensing speed from magnetic pickups can either be used on generator or mechanical drive applications. Electric governors are used mainly on systems which have complex requirements or require automatic operation.

6.13.6 Mechanical Governors

This type of governor is antiquated. It has been used in constant speed, constant load applications. The movement of the flyweights through the linkage is translated into valve motion so as to reduce or increase steam flow into the turbine. Flyweights, flyarms, or ballarms are used interchangeably and are weights which move and assume position in accordance with the speed of rotation. See Fig. 6-30.

The regulation of the mechanical governor is in the 6-10% range. The hand speed changer has an adjustment from 100% to 80% of rated speed. This is a NEMA class A governor and is not recommended for centrifugal compressor applications, on account of its inability to handle sudden load changes, nor for compressors in parallel, which require a high degree of regulation.

6.13.7 Mechanical-Hydraulic Governors

The term mechanical-hydraulic is indicative of the flyweight governor, which is the speed-sensitive element of the governor mechanism. There is a truly hydraulic governor, whose speed sensor is a centrifugal oil pump driven off the engine shaft, which operates on the principle that oil pressure, the motivating signal, varies as the square of shaft speed. However, we use the term hydraulic here to mean a mechanical-hydraulic device.

The hydraulic governor described herein has a great many variations and operates isochronously, provided the load does not exceed engine or turbine capacity. However, for mostly paralleling operations, this type of governor can be provided with droop, be it for diesel, gas, or dual fuel engines, or steam turbines.

2. Machines or turbines (steam or gas) can be paralleled at isochronous speeds (zero regulation), whenever automatic paralleling, load sharing or frequency regulation are important. Nevertheless, droop may be introduced, if required.

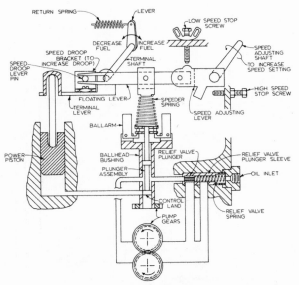

Courtesy Woodward Governor Company.

Figure 6-29. Schematic diagram of typical SG (simple governor).

3. By adding such components as automatic synchronizers, paralleling phase switches, transfer switches, etc., automation or semi-automation may be effected, relieving operators of decision-making responsibilities.
4. Many more functions are possible with electric than with other governors, such as time control, temperature controls, signal converters, overspeed controls, etc.
5. Speed change can be achieved easily.

The main hydraulic governor applications encompass prime movers for pumps, compressors, variable speed d-c generators, and alternators.

Hydraulic governors are very flexible and can be designed to include special characteristics and to perform such special functions as engine load limitation, engine shutdown in case of lubricating oil failure and other desirable objectives.

The characteristic features of the hydraulic governor are:

1. An *hydraulic system*, including an oil sump and gear oil pump, an hydraulic oil tubing system, a space inside the control system to store oil under pressure and the ability to relieve excessive pressure.
2. A *centrifugal flyweight governor* directly actuates a pilot valve that regulates circulation of oil under pressure to and from the power cylinder.
3. A *power cylinder*, otherwise known as a servo motor or amplifier piston which, by moving in either direction, repositions an engine fuel feed mechanism or turbine steam valve.
4. A *compensatory (feedback) system* that stabilizes the operation.
5. A *speed adjustment mechanism*, local or remote.

There are two types of power cylinders and pistons: (1) A spring-loaded piston, whose spring force acts on one side of the piston at all times, while the other side is subjected to variable hydraulic pressure, moving the piston in either direction. (2) A differential power piston, subjected to hydraulic pressure on both sides of the piston; the side with the smallest area receiving full oil pump pressure at all times while pressure on the opposite side varies.

Figure 6-31 is a composite picture of an Elliott Company steam turbine and a Woodward Governor Company hydraulic governor system. The governor, type PG (pressure compensated), consists of the five basic elements mentioned previously. However, the turbine manufacturer has provided a second, direct-acting, servo to power the steam multi-valve assembly. In a sense, the Woodward servo serves as a pilot to the inlet servo motor. It is entirely probable that, in some other applications, the Woodward servo would have adequate power to turn the linkage of a less demanding engine. Proper linkage design can impart either rectilinear or rotary motion, as requirements dictate.

Here, in brief, is how the hydraulic governor system operates:

The gear oil pump draws oil from the sump and initially fills the system, pressurizing the accumulators until the bypass port is exposed allowing excess pressure to be relieved back to the sump. Thus, the accumulators act not only as pressure reservoir but as relief valves as well. Pressure, through an unimpeded passage, is exerted on the smaller side of the power piston, pushing the steam valves into the fail safe (closed) position. If the control land of the pilot assembly moves down, oil flows to the larger area of the servo, forcing the piston of the inlet servo motor down and, thus, opening the steam valves.

For the pilot assembly to have moved down one of two things must have happened:

1. A load was added to the engine (or turbine), slowing it down without the speed changer being reset. The force on the flyweights decreased and allowed the plunger to drop.
2. There was an increase in load and engine speed had to be maintained. The speeder spring, by resetting, forced the flyweights in, pulling the plunger down.

The opposite is true and the pilot plunger lifts itself, if the action terms are reversed.

The inlet servo motor provided by the turbine manufacturer to actuate the steam valve has a differential pressure piston which has its own pilot but lacks compensation. Both sides of the piston relieve pressure to the turbine oil console.

The *compensatory system* depends on a negative feedback signal to act on the compensating land (disk) to bias the speed signal. In other words, the negative feedback tends to slow down the response of the control, thus stabilizing the operation. The bias is exerted by means of a differential pressure between the upper and lower faces of the compensation land. The pressure difference is created through the buffer piston and regulating needle valve and produces a force that aids the speeder spring or flyweights to reposition the pilot plunger. This force is quickly dissipated, depending on the position of the needle valve.

Speed setting can be applied manually at the governor or remotely by an air pressure signal emanating from a pneumatic controller or remotely and manually by means of a gradual switch at the control board varying the air pressure to the speeder spring. Another means of resetting the speeder spring is by using a split field universal motor, which drives the speed adjusting linkage. The motor is equipped with a friction clutch to protect itself when the linkage is forced against speed limit stops.

6.13.8 *Hand Valves*

Hand valves, inexpensive to install, should be specified for steam turbines if automatic, sectionalized admission valves are not provided. Hand valves require manual adjustment as the load changes. Without hand valves, there will be enormous flows at low loads. Of course, if the operator is unwilling to operate the valves, it will be very costly. A comparison of curves 2-3 and 7-8 of Fig. 6-32 for the same load with all valves open and two valves closed, shows how important the difference can be.

The sectionalized valves have an overlap and thus show as curves rather than as clean ladder steps. The averaged sectionalized valve curve is more economical than the averaged hand valve curve in Fig. 6-32 and, on that basis, the sectionalized valve is more desirable. However, automatic sectionalized admission valves are incredibly expensive and the decreased steam rate must be weighed against their cost. In general, it is easier to justify their specification for large turbines.

6.13.9 *Governors and Parallel Operation*

The advantages of electrical governors in parallel generator operation were pointed out in Section 6.13.5. Electrical governors handle transient conditions best but do not perform better than other types of governors under steady state conditions.

There are two methods of performing a parallel, load-sharing operation with mechanical-hydraulic governors. In the first, all engines are run in droop, an operation that depends a great deal on speeder spring tolerances. This method permits frequency to decrease as load increases. The second method involves running one unit isochronously, while all the others are in droop. This method

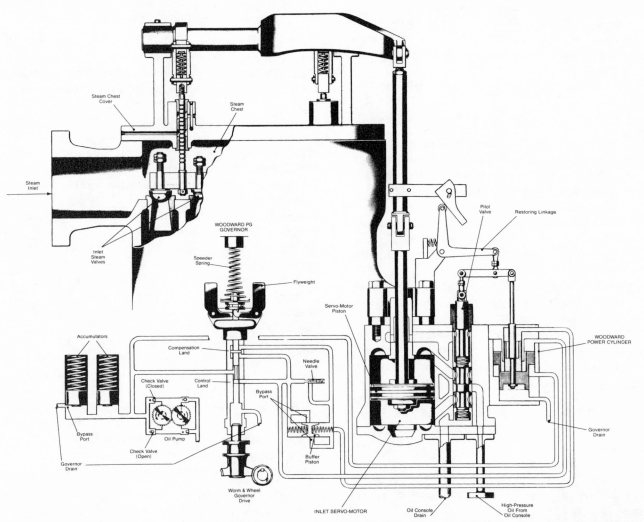

Courtesy of Elliott Co., Division of Carrier Corporation, subsidiary of United Technologies Corporation.

Figure 6-31. Combination of a hydraulic governor and a steam multi-valve assembly, with independent hydraulic system for the steam valve power cylinder.

maintains a steady frequency, but the single isochronous unit must accept all changes, risking overloading.

Synchronization is immeasurably easier with electric governors and it is directly dependent upon the use of the electronic synchronizer. Hydraulic governors must have speed-setting motors and additional load-sharing equipment to achieve synchronization.

Manual synchronization is effected by changing the speed-setting of the oncoming unit and closing the circuit breaker when the indicator shows that the voltages of the oncoming unit and the bus voltages are in phase.

Automatic synchronization is accomplished by various methods. The main control parameters are phase angle and frequency. One major manufacturer uses a paralleling phase switch and a digital synchronizer in a method which depends on monitoring of phase angle and frequency and closing the circuit breaker when frequency is within a bandwidth limitation of 0.5 Hz and a phase preselected angle tolerance of $15°$.

Peaking load control is another synchronization process, which involves connecting with a general network and assuming an assigned load. A digital synchronizer is used and the unit is brought to within $± 0.2$ Hz and $15°$ phase angle before the circuit breaker is closed. In order to unload, the speed-setting potentiometer must be reset.

Power transfer from a commercial to a local bus without frequency disturbances, providing a smooth operation which prevents the oncoming unit from taking on a large load suddenly, is another function of synchronization. It can be manual-automatic or completely automatic.

The *frequency trimmer* is another control which applies corrective signals to the governors of each unit to change each speed equally without disturbing the frequency or load-sharing ability of each unit.

In *time control*, the generator frequency is compared to the signal of a crystal oscillator. When transients occur, the time control mechanism increases speed to keep the average frequency accurate. This arrangement is useful when time clocks are connected to the controlled electrical system.

6.14 Automation and Operational Computers

Operational computers supervise, control and monitor the functioning of the entire plant or part of it, at all times or part of the time. The extent of automation is one of need, balanced against cost. In certain cases, as in nuclear power plants, there are forms of automation which are not allowed because they reduce the degree of safety.

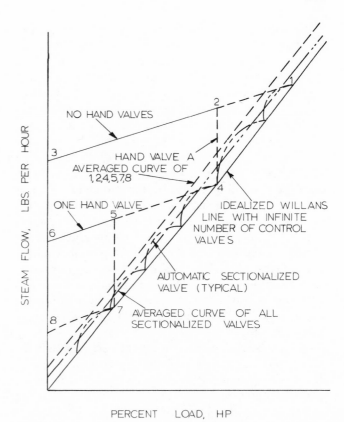

Figure 6-32. Performance of hand valves vs. automatic sectionalized valves.

Centralized automation and computerization is expensive and, if first cost is a primary objective, a careful examination must be conducted of all points that are proposed to be connected to a central system, eliminating any that are not organically functional. Central automation, in addition to cost considerations, merits comparison to other automation modes and calls for an examination of all pertinent factors which justify not only first cost but the required degree of sophistication as well.

The utility power plant, the industrial plant and the commercial central plant employ different criteria to justify first costs, type of installation and sophistication level, since the objectives of automation are distinctly different in each case. This is true despite the fact that some criteria fit all installations.

6.14.1 *The Rationale for Automation*

Central plant automation relies on computers, which make sophisticated systems possible. Although there can be automation without computers, the inclusion of computers defines a system as automated. With this distinction, we can use the term automation without reference to computers each time.

The utility power plant justifies high first cost in terms of safety. The operator is deprived of the opportunity to exercise any options by assigning these options to the computer as a first choice, with a manual intervention available when automatic operation fails for some reason. A fail safe philosophy predominates: in the event of malfunction, the machinery is shut down and a back-up unit may be activated to carry on the function.

In many cases, it is most difficult to justify automation of industrial plants from an economic point of view because the plant to be automated is not ordinarily part of the process complex and most often is not served by the process automation panels, which are reserved for process work only. Besides, safety codes and insurance policies dictate the separation of the power plant and require a licensed crew. A decision to automate fully requires complete backing by management and operating crew. Most often, partial automation is adopted, enough to provide adequate safety and some monitoring and alarming functions.

The commercial central plant is the best candidate for full automation because its design objective is to achieve the lowest operating costs, maximizing income potential of the buildings it serves by improving to the utmost the quality of the building's environment.

The basic reason for employing automation can never be attributed to only one factor. At times, it is a personal preference of a manager who has the power to decide. Quite often, it is a combination of the following reasons:

1. *Low operating costs*: The lowest operating costs, especially in commercial buildings where start-stop and monitoring operations are numerous and highly repetitive, often impell the engineer to consider automation as a manpower-saving device, although each automated function must be weighed against first cost. A cost-benefit study can result in a sophisticated, economical but not overinstrumented plant. It is estimated that there are savings of 3-4 operators per million square feet of office space.[10] Another way to estimate operator overhead is 0.14 to 0.02 manhours per point per day or anywhere between 50 and 350 points per operator per shift.[11]

2. *Differential savings*: In residential communities, hospitals, campuses and other large complexes, remote start-stop, control adjustments, monitoring and alarming are necessary with or without a central plant. The combination of central plant automation with that of the complex, located and operated at the central plant, reduces costs for automating the central plant since an investment must be made for the outlying structures of the complex anyhow.

3. *Energy management*: An important tool for reducing operating expenses, its basic function is energy conservation by implementing procedures which minimize electrical or fuel consumption but, nevertheless, meet all plant requirements. Computers are logical instruments for these complex operations.

4. *Shorter plant start-up time*: This is extremely desirable if generating equipment is involved. Fewer operators can attend the intricate complexities of starting prime equipment subject to destructive failure. In cases of completely automated start-ups, the operator acts in a standby capacity for administrative intervention in case of malfunction. A computer can schedule all major equipment and auxiliaries, thereby increasing plant capability to reach design loads at the fastest possible rate without exceeding equipment design limits. Automation in the start-up procedure is an area of compromise that takes into consideration manufacturers' recommendations, operators' experience and engineers' judgment but without affecting safety limits or equipment specifications.

5. *Elimination of operational errors*: A well-planned automation program is geared to eliminate lapses of judgment on the part of an operator faced with an instant crisis on the control board. Such a program will direct equipment or a process or both to a fail safe mode, preventing damage and avoiding a catastrophe. A computer codifies good judgment and places it on a sound technical foundation. Supervisory computer control can comprehend rapidly changing relationships among variables, can predict rapidly developing difficulties and can, with a pre-arranged logic, render them controllable before dangerous conditions dominate. Signal transmittal is almost instantaneous and reaction time is the shortest of which technology is capable.

6. *Availability of operators unnecessary*: Foolproof operation, *i.e.*, one that is immune to human error, saves manpower by allowing one or two operators to operate a huge plant from a central location under their direct control, an operation which otherwise would demand substantial manpower. In an emergency, an operator cannot be expected to react instantly. Some of his actions may be affected by his alarm or confusion, causing unnecessary tripping.

---

[10] Laszlo Bodak, "Savings and Improvements," *Actual Specifying Engineer* (now *Specifying Engineer*), September 1974, p. 97.

[11] ASHRAE, Systems Handbook, 1973, 34–29.

Some computers are capable of monitoring themselves by checking the calibration of control actions or of the instruments themselves. Under self-monitoring conditions, a trip is programmable and, therefore, occurs only under controlled conditions.

Programmable action can take place without the presence of an operator. However, large systems require the attendance of at least one operator either by law or because a process must proceed to the end on a manual basis in case of an automatic trip. In some complexes, the operator also acts as a security monitor or loudspeaker announcer.

7. *Maximum environmental conditions*: Such standards may be imposed by environmental agencies or may be part of a quality project. Systems in support of such standards are characterized by early detection, fast response or immediate corrective action by either highly intelligent operators or sophisticated computers.

8. *Uniform operation* can be the product of well-balanced plant automation. Irregular and upset conditions are liable to be less frequent and, if they occur, are less apt to disturb the entire system. Also, these disturbances can be localized or the system can be quickly restabilized. A stable and uniform operation contributes to efficiency and leads to reduced operating costs and equalization of running time.

9. *Reduced maintenance*. Precise control results in equipment operating with the least number of components, at the most efficient point. Equipment is forced to operate at regulated output, diminishing stresses and saving wear and tear.

   This area is most difficult to evaluate. Maintenance records of similar installations and repair cost records compiled by various manufacturers are helpful in determining the maginitude of maintenance savings versus capital costs of automatic controls.

10. *Better records* can be maintained with as much detail as desired. In the modern plant these data are essential for energy and load control and, at times, can serve accounting and tax purposes. Data collection can contribute to system performance analysis and can, in some circumstances, be useful in research and development by. providing experimental or scientific data.

6.14.2 *Automation Criteria*

Automation in the past used hard wiring from sensor locations to the main control board. All instruments had to be individually hard-wired. As a result, the main control board was easily crowded with recorders, alarm panels and switches to the point where boards became inconveniently long and difficult for one, or even more than one operator to observe with alertness. In other words, automation failed to provide the economic advantages, convenience and safety features for which it was intended. Instrument miniaturization resolved some of the control board inconveniences but not the high cost of hard-wiring.

Hard-wiring still persists in applications where ultimate safety and individual instrument display are requirements. However, in a fourth or fifth generation automation development, multiplexing came to the rescue. Multiplexing is a system of common wires connected to similar remote stations that use the principle of sharing time in the use of the wires. Multiplexing is an easy and rational solution and is reliable, flexible, durable and immune from interference from electrical transients. The cost has been reduced through mass production, not only of multiplexing equipment but also of electronic controls. These factors, in addition to two-wire interstation connections, have brought the cost of automation to accessible levels. Considering the high cost of hard-wiring, the cost advantages of multiplexing and the dramatic savings associated with it become easily demonstrable.

To translate functional design objectives into design requirements, the engineer needs to consider the following criteria:

1. The *ability to expand* is of prime importance. If the remote modules are not able to absorb new functions easily, the installation can choke itself. The capacity for expansion must be provided initially which, by current standards, can vary up to over 15,000 points. Initial design must consider scanning speeds. Slow speeds are cheaper but not safer or more reliable. Speeds can vary from 50,000 to 1,000,000 bits per second. The faster speeds allow more time for repeated interrogations and permit "readaround" time of 5000 points per second instead of a 700-800 mixture of analog and digital points. However, fast scanning speeds are not necessary in central energy plants.

2. *Noise and electrical interference* are significant aspects in selecting carrier frequency and modulation. Frequency shift and phase shift transmitters may be used for interrogation and response respectively to ensure that there is no error in interpretation.

3. *Immunity to electrical transients* in the entire system caused by variations in the case potential of each instrument is a system design consideration. Transformer couplings or other non electromechanical devices can offer protection against common mode failure caused by electrical transients.

4. *Good maintainability* involves prolonged operation without recalibrations, adjustments, automatic fault detection, isolation features for on-line tests or replacement of faulty components without disruption.

5. Desirable capabilities which increase reliability, independence and flexibility such as:

Built-in equipment to monitor and correct field unit addressing, check circuit integrity and indicate action that will restore full operation.

Redundancy, where required for safety.

Ability of the remote data collecting multiplexing panels (RDCMP) to accept a variety of signals and serve all conventional instruments. A guarantee that the system can accept most available input-output devices and can accommodate future or present computer use. The manufacturer must guarantee interface with computer hardware and must furnish assurance that software can be written easily to manage the interface.

System longevity, which, besides endurance, means addition, subtraction and change without wiring or component replacement.

6.14.3 *Multiplex Central Control Room*

The operation of a typical multiplexing system is shown schematically in Fig. 6-33.

The sensors measure the variables. The nature of the signal can be analog, digital, code or aural. The signal is "conditioned" by a transducer, converter, etc., to make it understandable to the remote data collecting multiplexing panel (RDCMP). On the other hand, output devices convert commands or instructions from the RDCMP into functional signals which, for instance, operate valves or dampers. All input-output (I/O) signals can be analog, digital, code and aural. I/O signals are received or dispatched by the RDCMP. These enclosures bear many names but their function is that of a terminal board for the transmission links and I/O devices. Some additional functions of the RDCMP are address storage, recognition command, multiplex logic, transmitting and receiving. Therefore, the RDCMP not only communicates with the central data processor (CDP) in two directions but recognizes commands and executes assigned functions.

Communication between the RDCMP and the CDP is accomplished in diverse ways. Among other media are included wires, telephone lines, microwaves or a combination thereof. Transmission can take place over hundreds of miles. The type of wiring used for transmission links is either coaxial, multiconductor or a single pair, the latter prevailing in most commercial installations. Transmission, outside transmission media, has other variables such as transmission mode, signal types and message format. Basic transmission configurations are shown in Fig. 6-34.

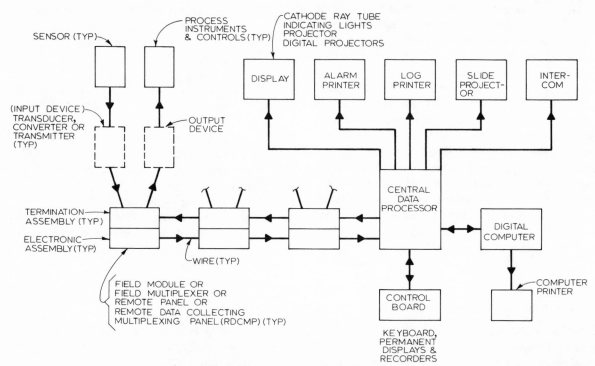

Figure 6-33. Architecture of a typical multiplexing automation system.

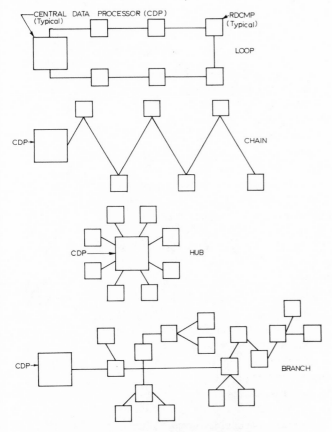

Figure 6-34. Transmission configurations.

The central data processor contains the management logic. It acquires and processes data continuously in an orderly fashion and controls timing and form of information to the system. The computer, with its memory, acts like a sophisticated component of CDP. There is peripheral equipment such as typewriters, printers, displays, memory files for the computer, etc.

6.14.4 *Automation and Computer Capabilities*

There are six basic functions for which automation and computers find their greatest use and widest application: (1) supervisory, (2) operational-functional, (3) energy management, (4) forecasting, (5) troubleshooting and (6) safety.

The first two categories involve staffing a project to make it operative. It goes without saying that automation lightens the manpower burden and relieves management of onerous tasks such as log keeping, start-and-stop and other routine operations. Savings in manpower must be weighed against an initial investment in automation equipment, continuing maintenance of automation components and equipment of a fairly sophisticated nature, and periodic replenishment and upgrading of systems and components.

Category 3, *Energy Management* (capital investment in components and systems such as runaround coils, variable speed pump drives, double-bundle condensers, etc.), involves two approaches to automation; one is purely instrumental, *i.e.*, closer fuel-air ratio, reduction in room temperature, night reset, later starting, etc. The other approach is one of promoting a more efficient operation, *i.e.*, sequencing compressors, boilers or turbines, resetting chilled water temperatures, using enthalpy controllers, etc.

Category 4, *Forecasting*, represents an area where the operator is offered economic options by having the computer force system electrical and mechanical components to behave in a predetermined manner to safeguard the benefits of this option. For instance, a computer and an associated control system can monitor the demand for electric power. The computer, by checking the trend, can predict if the demand will exceed the set limit and can shed electrical loads to reduce the demand below the established limit. In another instance, the computer can calculate whether to generate electricity or to purchase it by injecting information about fuel costs, purchased steam or inplant generated steam into the program.

Category 5, *Troubleshooting*, can be of a simple nature such as taking corrective action by resetting room temperature. Modern, high-priced computers, of the type employed in nuclear power plants, can automatically present the operator with a number of sequenced options upon receipt of an emergency signal, in a manner that does not call for the operator's participation. This type of operation is prohibitively expensive and has no place in central plants.

The following tabulations categorize numerous functions.

Supervisory

Remote start-stop
Audio system near motor (to hear operation)
Remote status monitoring
Recording
Logging
Alarms
Bearing monitoring (temperature or vibration)
Energy calculations for billing
Damper or valve position
Projector display
Alarm and logging printing
Maintenance schedule

Operational-Functional

Control of chilled water distribution
Remote lighting control
Monitoring of electrical distribution
Energy balance of cycle
Control point adjustment
Night setback
Remote changeover
Remote starting of boilers, generators and refrigeration machines
Calibration of controllers
Soot blower operation
Chemical treatment of boiler and cooling tower make-up

Energy Management

Automatic recording and billing
Minimum enthalpy change
Refrigeration plant efficiency
  a) Chiller efficiency
  b) Operating time reduced by thermal lag
Resetting of chilled water supply
Rate of change of monitored or calculated variables
Scheduled start and stop
Electric demand

Load shedding
Discharge air temperature reset
Supply water temperature reset
Instructions on running or calendar time
Summaries or profiles
Scheduled lighting control
Night setback
Energy and psychometric calculations

Forecasting (Optimization)

Chiller and chilled water pump control
Electric demand
Trend analysis
Economic evaluation of options
Optimum boiler loading program
Optimum chiller loading program
Optimum generator loading program
Optimized start and stop time for systems

Troubleshooting

Remote reset
Remote automatic changeover:
  Day - night, etc.
   Summer to winter or vice versa
Starting spare equipment
Signal diagnosis
Instrument calibration
Scheduling of equipment maintenance
Management information report
Energy demand control

Safety

Central programmed fire response
Central security functions (usually separate and in different locations)

BIBLIOGRAPHY

Computers and Automation

Anderson, R.H. "Programmable Controllers as Applied to Energy Management," *Instrument Society of America* (ISA), Paper 75-826, 1975.

ASHRAE Handbook and Product Directory, 1976 Systems, Chapter 34, "Automatic Control," *Centralized Control*, 34.25 to 34.30.

Bodak, Laszlo. "Savings and Improvement Through Automation Systems," *Actual Specifying Engineer* (now *Specifying Engineer*), (September, 1974), 97–104.

Chen, Steve Y. S. "Dissecting Computer Programs," *Heating, Piping, Air Conditioning*, (September, 1975), 41–47.

———. "The State of the Art," *Heating, Piping, Air Conditioning*, (October, 1975), 59–65.

———. "Existing Load and Energy Programs," *Heating, Piping, Air Conditioning*, (December, 1975), 35–39.

———. "Redirecting Load and Energy Program Priorities," *Heating, Piping, Air Conditioning*, (March, 1976), 72–77.

Cho, C.H. "Let's Examine the Role of the Computer in a Plant Energy Conservation Program," *Instrument Society of America* (ISA), Paper 74-524, 1974.

Cuesta, Armando E. "Computer Control of Natural Gas and Electric Power Demand," *Instrument Society of America* (ISA), Paper 74-425, 1974.

Davis, John C. "Small Chips but Big Stakes As Microprocessors Tackle Control," *Chemical Engineering*, (May 24, 1976), 75–77.

Douglas, Edwin S. "Electric Utilities Offer New Computerized Method for Consulting Engineers," *ASHRAE Journal*, (November, 1971), 41–43.

Dunn, George W. "Central Plant Design for Utilization of Computer Controls," *ASHRAE Journal*, (July, 1973), 55–59.

Elliott, Thomas C. "Understanding Microprocessors," *Power*, (May, 1977), 25–32.

Evers, William E. "E Cube Computerized Energy Analysis: Energy-Equipment-Economics," *ASHRAE Journal*, (September, 1971), 46–53.

Fling, John J. "Specifying Multiplex Data Systems," *Power Engineering*, (February, 1976), 50–53.

Hall, F.B. III and G.S. Jones. "Computer based Automation's Role in Good Building Management and a Look at Energy Conservation Through Power Demand Limiting," *Instrument Society of America* (ISA), Paper 75-530, 1975.

Hershfield, Walton, N. "How to Specify Building Central Control Systems," *Specifying Engineer*, (July, 1977), 82–89.

Hittle, Douglas C. *The Building Loads Analysis and System Thermodynamics Program*, Vol. I, NTIS Document AD-A048734 and *Users Manual*, Vol. II, NTIS Document AD-A048982. Springfield, Va.: NTIS, 1977.

Hordeski, Michael. "When Should You Use Pneumatics, When Electronics?" *Instruments & Control Systems*, (November, 1976), 51–55.

Kruger, Phillip. "Computer Simulation and Actual Results of an Energy Conservation Program," *ASHRAE Journal*, (October, 1974), 52–57.

Marciniak, T.J. *OASIS–A Computer Program for Simulation and Optimization of Central Plant Performance*. Argonne, Ill.: Argonne National Laboratory, 1977.

Meckler, Milton. "Flexible Approach to Computerized HVAC System Analysis," *Heating, Piping, Air Conditioning*, (March, 1975), 60–66.

Lokmanhekim, Metin, ed. *Procedure for Determining Heating and Cooling Loads for Computerized Energy Calculations*. New York: American Society of Heating, Refrigerating and Air Conditioning Engineers, 1971.

Lokmanhekim, Metin and Robert H. Henniger. "Computerized Energy Requirement Analysis and Heating/Cooling Load Calculations of Buildings," *ASHRAE Journal*, (April, 1972), 25–33.

Mikol, W.W. and Y.J. Yaworsky. "Complete Automation for Combined-Cycle Operation," *Proceedings of the American Power Conference*, Vol. 35, 1973, 638.

Shih, James Y. "Energy Conservation and Building Automation," *Transactions of the American Society of Heating, Refrigerating and Air Conditioning Engineers*, Vol. 81, Part 2, 1975, 419.

Stallings, R.D., W.C. Rochelle and S.L. Ferden. *Energy Systems Optimization Program (ESOP), Users Guide, Update 4 Vol. I & II; Economic Base Line Data, Vol. III*. Lockheed Publication LEC 5041. Houston, Tex.: Aerospace Div., Lockheed Electronics Co., 1974 (Vols. I & II) and 1975 (Vol. III).

Swanson, James R. "New Generation in Automation," *Heating, Piping, Air Conditioning*, (August, 1976), 53–58.

Refrigeration Equipment Controls

Anderson, R.E. "New Approach to Chilled Water Control," *Heating, Piping, Air Conditioning*, (October, 1967), 130–132.

Buehler, Leon, Jr. "Capacity Control," *ASHRAE Journal*, (November, 1968), 39–44.

Hazeltine, J.D. and E.B. Ovale. "Comparison of Power Efficiency of Constant Speed Compressors Using Three Different Capacity Reduction Methods," *Transactions of the American Society of Heating, Refrigerating and Air Conditioning Engineers*, Vol. 77, Part 1, 1971, 159–162.

Kirkman, C.G. "Automatic Hot Gas Bypass," *Heating, Piping, Air Conditioning*, (January, 1968), 64–68.

Lorentz, John W. "Hot Gas Bypass Boosts A/C Performance," *Power*, (June, 1971), 120–122.

Magliozzi, Thomas. "Control System Prevents Surging in Centrifugal Flows," *Chemical Engineering*, (May 8, 1967), 139–42.

May, D.L. *et al*. "Optimizing Plant Refrigeration Costs," Instrument Society of America (ISA), Paper 78–565, 1978.

Romita, Enzo and James Ross. "A Direct Digital Control for Refrigeration Optimization," *Transactions of the American Society of Heating, Refrigerating and Air Conditioning Engineers*, 1977, Vol. 83, Part 1, 384–92.

Trane Engineering Bulletin EB-CIV-16. "The Control of a Typical Single Centravac Installation," 1967.

Trane Engineering Bulletin B-6. "The Control of a Typical Two-Centravac Installation," 1967.

CHAPTER 7

Pumping Systems

An ideal pumping system serving a structure or a process delivers just the right amount of heating and/or cooling energy at the required moment to each conditioned space or process unit on a continuous basis. Actual operating systems fall far short of this goal, because of transient conditions that unbalance the system, unusual piping configurations, hydraulic problems and malfunctioning controls, as well as designs that were ill-conceived in the first place.

How many approaches to designing systems are there that promise satisfactory results? There are two main approaches, the primary and the secondary system designs. The primary can exist by itself, whereas the secondary system can exist only in conjunction with a primary system. A primary system encompasses both the source of energy (*e.g.*, a boiler, a chiller) and the energy user (*e.g.*, a radiator, a cooling coil). On the other hand, the primary side of a primary-secondary system incorporates the source while the secondary side includes the user. Primary-secondary systems are based on concepts of energy level and hydraulic separation. Variations in application of these two basic categories have their own advantages and disadvantages, which we shall discuss in this chapter.

7.1 Primary-Secondary Circuit Relationships

After a system is installed and operating, it is important for startup engineers and operating personnel to obtain design water flows both in the primary and secondary systems. Excess flow, *i.e.*, flow in excess of design flow, and the consequences of such flow will be examined. The assurances that design flow provides are: (a) performance according to design intent and delivery of energy as needed, (b) horsepower within design limits, (c) lower probability of hydraulic imbalance, (d) equipment performance within specification limits, and (e) water temperatures within design allowances.

The behavior of primary and secondary systems at design and non-design conditions is discussed in Section 7.1, while Figs. 7-1 through 7-6 point out the performance of these systems at a number of conditions. Equations 7-1 through 7-8 form the mathematical basis for the examination of system behavior.

In the heating mode heat is transferred from heat-producing equipment by the primary circuit, passed to the secondary circuit, then given up by the secondary circuit to heat-using equipment (heaters). In the cooling mode heat absorbed by the secondary circuit is transferred to the primary circuit to be given up by the primary circuit to heat-absorbing equipment. If we neglect losses in piping and equipment, then the law of conservation of energy dictates that heat absorbed by the primary circuit equals heat given up by the secondary circuit or vice versa.

In mathematical terms, the heat balance is

$$W_p c_p \Delta t_p = W_s c_s \Delta t_s$$

Where

W = Water flow, lb/hr
c = Specific heat of circulating water, Btu/deg F-lb
Δt = Temperature difference of circulating water, degrees F, within circuit
p, s = Subscripts designating primary and secondary circuits

In medium and high temperature water systems due to the difference in water temperature between the (hot) primary circuit and secondary circuit, the difference between c_p and c_s may be of some significance. In the range of temperatures encompassed by low temperature heating and chilled water cooling systems, however, the difference is negligible, so that

$$W_p \Delta t_p = W_s \Delta t_s \qquad (7\text{-}1)$$

At any given heating or cooling load 1, 2, 3 ... n, Equation (7-1) can be written as

$$(W_p \Delta t_p)_{1,2,3\,...\,n} = (W_s \Delta t_s)_{1,2,3\,...\,n} = K \qquad (7\text{-}1a)$$

Primarily, three factors affect mass flow in a primary water circuit: (a) supply temperature of primary water, (b) flow in the water circuit and (c) secondary circuit load and related water flow, which may be either uncontrolled and therefore constant at all loads, or modulated in direct proportion to the load.

To analyze the effect of primary water supply temperature on primary flow, we suppose that primary water return temperature is kept constant, while that of the supply is varied. Also the load is kept constant as the primary Δt fluctuates.

Let us assume that, at load condition (1), temperatures of primary supply (ps) and return (pr) are

$$(t_{ps})_1 = 38\,\mathrm{F}(3.3\,\mathrm{C}) \; ; (t_{pr})_1 = 58\,\mathrm{F}(14.4\,\mathrm{C})$$

Hence $(\Delta t_p)_1 = 20\,\mathrm{F}(-6.7\,\mathrm{C})$, and that flow, $(W_p)_1$, is unity, or 100%.
Thus, $(W_p \Delta t_p)_1 = K = 100\% \times 20\,\mathrm{F}$

$$(W_p)_1 = \frac{K}{(\Delta t_p)_1} = \frac{2000}{(58 - t_{ps})_1}$$

And, at *any* load condition *(n)*, given the same 58 F return water,

$$(W_p)_n = \frac{K}{(\Delta t_p)_n} = \frac{2000}{(58 - t_{ps})_n} \qquad (7\text{-}2)$$

Load (1) of Fig. 7-1 is a plot of Equation 7-2 and shows percentage of primary water mass flow as a function of supply temperature for various loads.

Secondary circuit water flow is either controlled or uncontrolled. Controlled flow implies that flow is modulated in response to load demand. We will discuss this later (see Section 7.5).

Uncontrolled flow, illustrated in Fig. 7-2, is unrestricted flow, usually for smaller installations of multi-zone, face-and-bypass and similar equipment. A 3-way valve in the circuit can change the proportion of flow among the lines it connects, but not total flow. Hence, uncontrolled flow is, essentially, constant flow.

If the 3-way valve possesses a modified parabolic characteristic it can provide fine throttling action at low flows

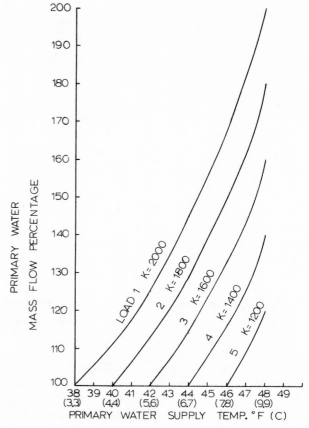

Figure 7-1. Effect of primary water supply temperature on primary water return temperature of 58 F (14.5 C) for different cooling loads. Return temperatures other than 58 F will produce different sets of curves.

but exhibits an approximately linear characteristic at the upper valve travel. This type of valve action is satisfactory for the bypass function of the 3-way valve. However, the 3-way valve was always intended for the small system, where pumping costs are not an important consideration and steady flow removes a degree of uncertainty associated with variable flow. Uncontrolled flow would not ordinarily be used in larger central plant applications but could be a sizable part of a large project. Therefore, uncontrolled flow merits examination since it is particularly vulnerable to the crippling effect of excessive flow in the secondary circuit.

As previously stated, excess flow is flow in excess of design flow in the secondary circuit. Excess flow is a measure of how unbalanced the secondary system is. The excess factor (*EF*), an indicator of this unbalance, has a direct effect on primary water demand. At $EF = 1$, secondary water flow is at design. At $EF = 2$, secondary water flow is twice the design quantity, and so on. Diversity factor (*DF*) is an indication of instantaneous load and

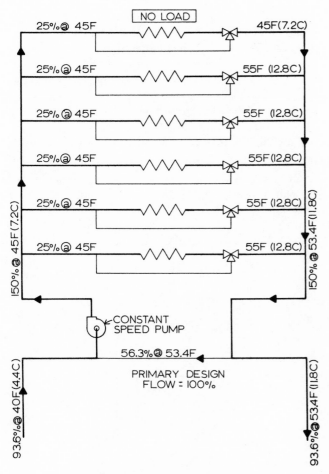

Figure 7-2. Uncontrolled water flow.

By definition, secondary flow, $W_s{}'$, at a particular excess factor

$$W_s{}' = EF \times W_s$$

It follows that, at constant load, the larger the flow, the smaller the Δt, or

$$\Delta t_s{}' = \frac{\Delta t_s}{EF}$$

By examination of Fig. 7-2, one can see that the following are true and are not functions of EF:

$$t_{ps}{}' = t_{ps}$$
$$t_{ss}{}' = t_{ss}$$
$$t_{pr}{}' = t_{sr}{}'$$

For any given excess factor, we can employ the following equation:

$$W_p{}' \times \Delta t_p{}' = W_s{}' \times \Delta t_s{}' = K$$

$$W_p{}' = \frac{W_s{}' \times \Delta t_s{}'}{\Delta t_p{}'} = \frac{[W_s(EF)](\Delta t_s/EF)}{t_{pr} - t_{ps}}$$

Where

$W_p{}'$ = Primary water flow, lb/hr at a given excess factor (EF) in the secondary circuit

$t_{pr}{}'$ = Temperature of return water in the primary circuit, deg F or C at given excess factor (EF) in the secondary circuit

$t_{ps}{}'$ = Temperature of supply water in the primary circuit, deg F or C at given excess factor (EF) in the secondary circuit

Adding and subtracting t_{ss} and making appropriate substitutions in the denominator, we obtain

$$W_p{}' = \frac{W_s \times \Delta t_s}{(t_{pr}{}' - t_{ss}) + t_{ss} - t_{ps}} = \frac{W_s \times \Delta t_s}{(t_{pr}{}' - t_{ss}{}') + t_{ss} - t_{ps}}$$

$$= \frac{W_s \times \Delta t_s}{\Delta t_s{}' + t_{ss} - t_{ps}}$$

$$W_p{}' = \frac{W_s \times \Delta t_s}{t_{ss} + \dfrac{\Delta t_s}{EF} - t_{ps}}$$

$$\text{(7-3)}$$

is defined as the ratio of instantaneous load or flow divided by the sum of all connected loads or flows. It is possible for a secondary system to have an *EF* or *DF* or both. A system with an *EF* in excess of unity means that the system pumps more than its design permits and that it has constant flow. A secondary system with a diversity factor of less than unity is pumping less than its design pumping capacity, not as a result of unbalanced conditions but because some of the water users have automatic 2-way valves which reduce flow at partial load. Therefore *DF* is an indication of controlled flow. If both *EF* and *DF* apply to a system, it implies that the system is not hydraulically balanced to circulate design flow at design conditions but also can reduce its flow when partial load conditions demand it if automatic 2-way valves are placed at the point of utilization. *EF* is equal to *DF* only when both are unity, indicating that the system is correctly balanced and that it is operating at full load under controlled flow conditions. *DF* is always unity when flow is uncontrolled.

Where

t_{ps} = Temperature of supply water in the primary circuit, deg F or C at excess factor (*EF*) equal to unity in secondary circuit

t_{ss} = Temperature of supply water in the secondary circuit, deg F or C at excess factor (*EF*) equal to unity in secondary circuit

Rewriting Equation (7-3) for any percentage load condition *n* at excess factor *EF*,

$$(W_p')_n = \frac{K}{(t_{ss})_n + \dfrac{(\Delta t_s)_n}{EF} - (t_{ps})_n} \qquad (7\text{-}4)$$

Equation 7-3 expresses the demand on the primary water system at any time and for any excess factor. Equation 7-4 is somewhat more convenient for plotting Fig. 7-3 or any other chosen load condition.

The following example is illustrative of the method used to plot Fig. 7-3. To establish *K* for any primary load factor, it is necessary to define all the design conditions at the chosen load factor.

Equation 7-4 may be used now to plot Fig. 7-3 by assuming t_{ps} of 44, 42, 40 and 38 F (6.7, 5.6, 4.4 and 3.3 C) and any excess factor. The same technique may be used to plot diversity factor by using Equation 7-5 for *K* = 2000.

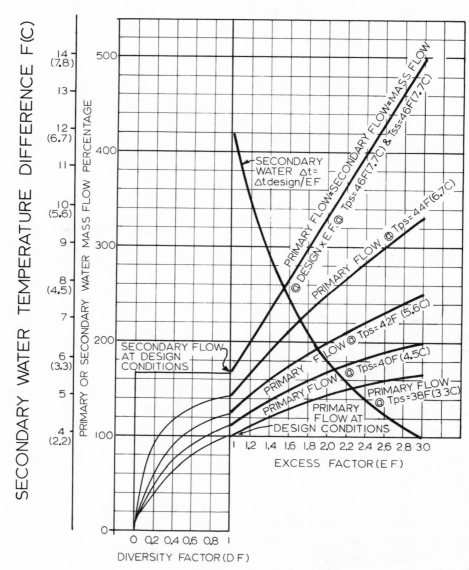

Figure 7-3. The effect of excess and diversity factors at 100% primary load is shown on both primary and secondary circuits for various primary water temperatures.

Example 1.

Given $t_{ps} = 38\,F$ (3.3 C), $t_{pr} = 58\,F$ (14.4 C), $\therefore \Delta t_p$
$\simeq 20\,F$ (11.1 C), $t_{sr} = 58\,F$ (14.4 C)
$t_{ss} = 45\,F$ (7.8 C), $\therefore \Delta t_s = 12\,F$ (6.7 C)
$W_p = 100\%$

Substituting in Equation 7-1 at $EF = 1$,

$$W_s = \frac{(100)(20)}{72} = 166.66\%$$

$$K = W_s \times \Delta t_s = 166.66 \times 12 = 2000$$
$$K = W_p \times t_p = 1000 \times 20 = 2000$$

What is W_p' and W_s' at $EF = 1.6$?
Using Equation 7-4

$$W_p' = \frac{2000}{46 + \dfrac{12}{1.6} - 38} = 129.03\%$$

$$W_s' = W_s \times EF = 166.66\% \times 1.6 = 266.66\%$$

check

$$\Delta t_s \text{ at } EF = 1.6 = \frac{12\,F}{1.6} = 7.5\,F \ (4.2\ C)$$

$$t_{pr} = t_{sr} = 46 + 7.5 \quad = 53.5\,F \ (11.9\ C)$$
$$\Delta t_p = 53.5 - 38 \quad\quad = 15.5\,F \ (8.6\ C)$$
$$W_p \times \Delta t_p = 15.5 \times 129.03 = 2000$$
O.K.

Example 2.

Assume all conditions being the same as those of Example 1, except that $t_{ps} = 40\,F$ (4.4 C). What is W_p at $EF = 1$ and at $EF = 1.6$?

$$W_p = \frac{K}{\Delta t_p} = \frac{2000}{58 - 40} = 111.11\% \text{ at } EF = 1$$

check: $$W_p' = \frac{2000}{46 + \dfrac{12}{1} - 40} = 111.11\%$$

$$W_p' = \frac{2000}{46 + \dfrac{12}{1.6} - 40} = 148.14\% \text{ at } EF = 1.6$$

check: $W_p \times \Delta t_p = 144.14 \times 13.5 = 2000$
O.K.

Diversity factor, as it applies to uncontrolled secondary water flow indicated in Fig. 7-1, affects only the secondary water temperature difference.

Hence

$$\Delta t_s' = DF \times \Delta t_s$$
$$W_s' = W_s$$

The effect of diversity factor on primary water flow is given by Equation 7-5, derived in a manner similar to that of Equation 7-3:

$$W_p' = \frac{W_s \times \Delta t_s \times DF}{t_{ss} + \Delta t_s(DF) - t_{ps}} = \frac{K\,(DF)}{t_{ss} + \Delta t_s(DF) - t_{ps}}$$

(7-5)

Note that, when $t_{ps} = t_{ss}$, secondary flow is equal to primary and the need for the primary bridge (see Figs. 7-14 and 7-15) is eliminated.

For uncontrolled secondary water flow, the following expression gives the combined effect of primary water supply temperature, excess factor and diversity factor:

$$W_p' = \frac{K\,(DF)}{t_{ss} + \dfrac{\Delta t_s\,(DF)}{(EF)} - t_{ps}} \qquad (7\text{-}6)$$

In practice, an excess factor of 2 indicates a secondary system with twice the design flow, afflicted with severe balancing problems such as noise, high water velocity, high pump head and unable to maintain design conditions in main equipment such as boilers, chillers, pumps, etc. However, there may be unusual systems with highly transient loads that cause wide fluctuations in flows and loads.

The second mode of operating the secondary water system is controlled flow whereby secondary water flow is reduced or increased in a direct, linear proportion to refrigeration load.

In Fig. 7-4, flow is controlled in six steps of 25% increments by closing each 2-way valve in succession. However, because the fact that the 2-way valves modulate, it is possible for each valve to respond individually to its own load. The sum of the instantaneous branch flows is in direct relationship to system diversity factor. Theoretically, at least, Δt_s is constant at all flow conditions. In actuality, this is not the case because the heat transfer characteristics of the various coils or heat exchangers connected to the secondary water system do not bear a direct linear relationship with flow. The designer of a system must evaluate heat transfer in relation to flow at all intermediate load points. Coordinating all heat exchangers in this manner for an instantaneous diversity factor must take into account zoning as well as internal and external load variations.

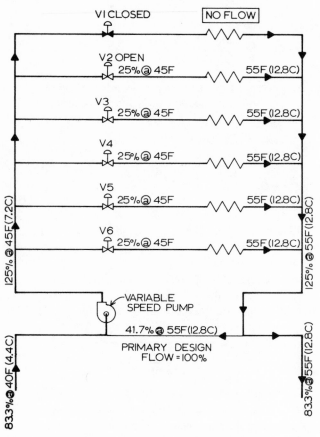

Figure 7-4. Controlled secondary water flow.

For controlled flow in the secondary circuit, the following relationships pertain:

$$W_s' = DF \times W_s$$
$$\Delta t_s' = \Delta t_s \qquad (7\text{-}7)$$

$$W_p' = \frac{W_s \,\Delta t_s \,(DF)}{\Delta t_p'}$$

To account for both excess flow and diversity in the secondary circuit, the following expression may be used:

$$W_p' = \frac{W_s \,\Delta t_s \,(EF)\,(DF)}{\Delta t_p'} \qquad (7\text{-}8)$$

Equation 7-5 is used to plot Fig. 7-5 and Equation 7-7 to plot Fig. 7-6. Figures 7-5 and 7-6 compare the performance of Fig. 7-2 and Fig. 7-4, i.e., uncontrolled and controlled flow in the secondary circuit. It will be noticed

that both secondary and primary water flows are significantly reduced under controlled flow conditions, resulting in substantial pump operating savings.

7.2 Compatibility Between Primary and Secondary Systems

The relationship between primary and secondary systems is apparent in the areas of flow, temperature, and differential pressure between them. It is a relationship of cause and effect with the effect being more apparent than the cause or any intervening events. Although quite a bit remains to be learned about large scale water distribution, enough is known to evaluate systems already in existence and to examine some of the principles that govern large distribution systems.

The key questions are: Which of the three factors, primary water supply temperature, secondary circuit excess flow or controlled versus uncontrolled flow in the secondary circuit, is a prerequisite for a controllable and stable operation? Under what circumstances? To what extent? The ultimate objective is to match secondary

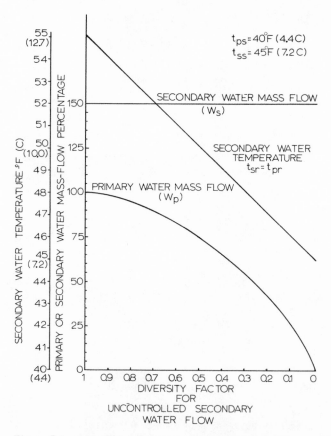

Figure 7-5. Diversity factor for uncontrolled secondary water flow at fixed primary and secondary supply temperatures.

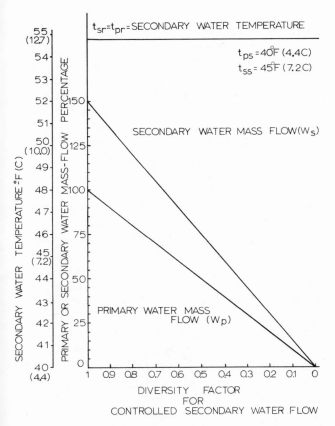

Figure 7-6. Diversity factor for controlled secondary water flow at fixed primary and secondary supply temperatures.

capacity as closely as possible to primary and chiller capacity with the least expenditure of energy and in the shortest response time.

Proper selection of components is not limited to the usual selection of the most efficient units for a particular operation at peak load. It also entails maintaining high efficiency at part load conditions, keeping a close, responsive and mutually profitable association between primary and secondary systems throughout the *entire* range of operation, since practically all systems operate most of the time at part load.

A complete grasp of the factors for successful central plant engineering is of fundamental value in steering a project in the right direction during its inception stage.

Flow and temperature are interrelated by Equation 7-1. However, pressure relationships between primary and secondary are affected not only by flow but also by such items as direct or reverse return, complex grid distribution patterns, proximity or remoteness of a load to a pumping station, constant or variable flow, 3-way or 2-way control valves, secondary pumps acting as boosters to the primary pumps or acting independently within the

secondary circuit, types and location of pressure controls. In some instances, a number of these factors may be acting jointly, complicating the analysis of a system.

An additional measure of complexity is added when a new central plant is called upon to absorb some existing building loads with their own idiosyncrasies. Depending on the magnitude of the central installation, the engineer might consider a constant flow, variable temperature rise system to avoid being subjected to the transients caused by existing systems or, at least, to minimize their impact

Primary flow may be (a) greater than, (b) equal to, or (c) less than, secondary flow. Case (a) occurs most frequently in the primary-secondary bridge connections due to unbalanced conditions in the primary circuit. Excess primary flow as a general design condition is of no particular interest or value except for the fact that it is energy wasting. Case (b) has few advocates in large scale projects, basically because both primary and secondary systems involve small temperature differences between supply and return water. Such arrangements frequently lead to the secondary pumps acting as booster pumps.

This leaves Case (c) as the only method suitable for extensive central plant installations. Secondary flows larger than primary flows are complex but also diverse and flexible. These systems, properly utilized, are economical in design and operation.

Customarily, the secondary chilled water system is designed for conventional temperature differentials—10-12 F (−12.2 to −11.1C)—but in many cases it reaches 14-16F (−10 to −8.9C). Primary differentials are usually larger—15-20F(−9.4 to −6.7C)—but sometimes are as high as 24F (−4.4C).

In systems where the secondary chilled water flow equals the primary, one temperature differential is used throughout, usually in the range of 14-16F (−10 to −8.9C). In general, the objective of large temperature differentials in the primary, whether for chilled water or high temperature water, is reduced flow, economical pipe sizes and substantial savings in pump motor horsepower.

The disproportion between primary and secondary flow imposes certain thermodynamic rules. Consequently, the primary chilled water supply temperature must be lower than the secondary supply temperature. On the other hand, the secondary return temperature must be equal to the primary return temperature. When dealing with chilled water systems, direct mixing of primary and secondary water systems is customarily accomplished through direct piping connections. The use of heat exchangers between primary and secondary systems is rarely used because the temperature difference between primary and secondary is not large enough to tolerate placing a heat exchanger between them. However, in some instances involving high-rise buildings, a heat exchanger is interposed in order not to transmit pressure from the upper stories to the chillers, which are normally sized for 150 psig (1034 kPa) water pressure.

In the case of high temperature water, the primary water temperature must be higher than the secondary. The piping systems explored in Section 7.5 are for the most part applicable to medium temperature water systems of up to 250F (121 C). Process loads are directly connected to high temperature water systems and flow is controlled by 2-way valves. For office spaces, heat exchangers are interposed between primary and secondary systems, which are at a lower thermal level. Occasionally, however, direct mixing is practiced using 3-way valves.

Arrangement VF-1 (see Section 7.5.1, "Variable Flow System VF-1" later in this chapter) indicates 3-way mixing valve control around the building pump and 3-way mixing or diverting valve control around the coils. This type of valve arrangement is not of thermodynamic importance since the end result is to introduce chilled water into the coil at the required quantity, bypassing the remainder. The mixing valve around the pump introduces secondary return water to provide water to the building at a constant supply temperature by blending the return water with an appropriate amount of primary supply water at a lower temperature than the secondary supply design temperature.

The ultimate objective is to match secondary capacity to primary capacity, so that only the minimum number of chillers, primary pumps, cooling tower cells and condenser water pumps are operating.

There are two ways to reduce secondary side refrigeration load. The first is by a sudden and sizable load withdrawal, as when an entire building or process block load comes off the line. Examples of this type may occur when an office building or an industrial process shuts down at the end of a shift. This type of load reduction is direct, in the sense that both primary and secondary flows and capacities are reduced proportionately. The second type of load reduction occurs when loads are reduced gradually due to internal or external factors. During this type of load change, the secondary water flow remains constant but the primary supply and return bleed lines introduce less and less primary water into the secondary system at an increasingly smaller secondary temperature difference.

Reduced secondary temperature differentials during part load conditions can be handled in two ways; first, by keeping the return temperature constant and allowing the secondary supply temperature to float and, second, by

providing constant secondary supply temperature and letting the return temperature find its own level. Secondary systems employing constant return temperature are rare because the apparatus dew point of the cooling coil is constantly raised as t_{ss} floats upwards with decreasing load, lessening the ability of the coil to remove moisture from return air. As a result, space relative humidity exceeding design conditions may be expected. In addition, the heat transfer characteristics of the coil change at higher t_{ss}, increasing flow requirements and, therefore, additional heat transfer surface. Under the circumstances, it is entirely possible that part load conditions may assume greater importance in coil selection.

Many trials may be necessary to pinpoint the design conditions. How to schedule the resetting of the supply temperature is a difficult problem because, aside from defining the measured variable, there may be several zones within the piping system serviced by one secondary pump. One supply temperature obviously cannot satisfy all zone requirements.

All these difficulties are avoided when we resort to a constant secondary water supply temperature and allow return temperatures to be determined by coil performance. By employing Equation 7-1 and assuming constant $t_{ps} = 38F$ (3.3 C) and $t_{ss} = 46F$ (7.8 C), and a plausible t_{sr} and t_{pr} schedule at part load, the following relationships may be developed:

% Load	t_{sr} & t_{ps} deg F (C)	t_{ps} deg F (C)	t_{ss} deg F (C)	% W_s	% W_p
100	58(14.1)	38(3.3)	46(7.8)	100	60
75	55(12.8)	38	46	100	52
50	52(11.1)	38	46	100	42
25	49(9.4)	38	46	100	27
0	46(7.8)	–	46	100	0

Similar relationships can be created for other primary and secondary temperatures.

7.3 Variable Speed Pumping

Energy conservation criteria dictate that, ideally, pumping capacity should match pumping requirements for maximum operating savings. A pumping efficiency index has been proposed[1,2] that would permit a system to be

[1] Herbert P. Becker, "Energy Conservation Analysis of Pumping Systems." *ASHRAE Journal*, April, 1975, pp. 43–51.

[2] Bunnelle, Philip R. "Reducing Pumping Power in Central Chilled Water Systems," *Transactions of the American Society of Heating, Refrigerating and Air Conditioning Engineers (ASHRAE)*. Vol. 81, Part I, 1975, 652.

evaluated on a pumping requirement basis at the design stage. A convenient index can be gallon-feet or kilogram-meters or any other combination of units. One of the authors[3] rates various common pumping arrangements at full load and partial load. This approach could be further developed into a computerized program that would not only give pumping system efficiency index rating but could also summarize annual pumping energy consumption and make economic comparisons between various pumping systems. A computer approach could account for such items as pump efficiency drive costs, load profiles, physical arrangement, pump-chiller arrangement, valve response, project size and other factors. However, the index as presented is useful for preliminary evaluation.

Although the superiority of variable speed pumping for energy-saving is apparent when it is viewed as an independent item with few pumping subsystems and a minimum number of pumps and chillers, it appears less so when viewed in connection with the complex demands of large central chilled or hot water installations.

Variable speed pumping is receiving serious consideration for possible application to central plants. However, a central plant with, let us say, six chilled water pumps already can vary its flow in six discrete steps. Whether variable speed within each of the six steps is an advantage is a question of technical compatibility and economic balance.

In general, variable speed pumping is more suitable for applications where a broad range of flows is expected, in contrast to constant speed pumping, which cannot tolerate wide variations in flow without instability and energy waste. Instability in constant speed systems can be corrected by means of elaborate controls but the energy waste is an inherent part of a constant speed system at partial loads.

In weighing the pros and cons of constant and variable speed pump drives, the engineer must not only examine the intrinsic value of each drive but must also resolve the hydraulic problems of a piping arrangement at a particular site and must harmonize the project load profile with all selected equipment. An overall view of system performance is necessary.

The following advantages of variable speed pumping must be noted in connection with all the points just discussed:

1. *Reduced pipe sizes* are possible, especially in chilled and hot water mains, where diversity plays an important role, and particularly in the main plant itself, where

[3]Becker, *op. cit.*, p. 50.

large pipe sizes predominate. Higher water velocities can be tolerated without fear of excessive noise for long periods, during the limited period of peak design conditions.

2. *Reduced pumping energy requirements* may be realized because of lower flows at lower load factors.

3. *Lower maintenance charges* result from lower speeds, flows and heads. Valve stuffing boxes are subjected to lower pressures and system leakage is reduced. Pumps exhibit smaller shaft deflections and reduced lateral forces on seals and bearings. Automatic valves experience smaller pressure drops with less wear and tear.

4. *Adjustments to actual operating conditions* are made very easily, compensating for miscalculations, safety factors, commercial tolerances, pipe age or winter-summer flow differences, excessive or deficient system flow.

5. *Incorrect initial selection is not as critical* as it is with constant speed pumps. A pump may be selected for both a future and present condition at nearly peak efficiency and without need to change impellers.

6. *System balancing* is easier to accomplish.

7. *Direct return piping is permissible* (see Section 7.5.3, "Variable Flow System VF-3"), if terrminal pumps are used, as each pump adjusts its head to serve the needs of its own subsystem. Theoretically, if there are many subsystems, each with its own terminal pump, no one system will have any excess head to influence any other subsystem, since they all are ideally balanced. On the other hand, if there are intermediate pumps introducing a specific head into the intermediate loop, the terminal pumps are in a position to take advantage of this head by adjusting their speeds to deliver their required subsystem heads. Thus, further pumping power savings can be effected.

8. *Complicated piping is eliminated* by not having to install bypasses, 3-way or 2-way control valves, crossover piping and balancing valves as one has to in most constant speed systems.

9. *Computer operations* are more suitable as systems expand in size, demanding greater operating flexibility in adjusting flow and pressure in the various loops.

Variable speed drives must be highly efficient and must not be of the type that wastes energy, as do eddy couplings or wound rotor motors.

Other factors that must be considered are greater control complexity and relatively high cost. Perhaps, for smaller systems, control complexity may be reduced by specifying pumping packages, factory-mounted and entirely prewired. When one deals with a number of variable speed pumps, it is a matter of some complexity as to what is the best way to control system flow and head, since one more variable, that of speed must be controlled.

Variable speed drives are at present very costly. Their justification must be based not only on convenience and better balanced pumping systems but, also, materials and energy savings. It is conceivable that many control valve stations may be eliminated by using a variable speed drive, so that first cost may not be a prime consideration because there are compensating cost reductions. Operating savings are an inherent part of variable speed pumping and should be easy to demonstrate.

7.4 Types of Distribution Systems

Distribution systems fall into two general categories, direct secondary and indirect secondary systems.

In the *direct* secondary systems, as the term implies, the secondary system pumps are in series with the primary pumps and each group of pumps affects the other hydraulically. For large central plants employing a number of large chillers, it is safe to assume that the number of chillers and chilled water pumps on line are related to the size of refrigeration load on the system. Variation in chilled water flow, usually in incremental steps, depending on the number of pumps on line, is the rule. Variable primary flow is thus defined within this context.

Some systems, in addition to primary and secondary loops, have intermediate pumped distribution loops in order to stabilize flow through the chillers (see Sections 7.5.2 and 7.6.1).

Secondary flow can be constant or variable, depending on pumping arrangement and use of 3-way or 2-way valves. Some of the most common systems are examined in this section from the standpoint of economy of operation and stability. Systems *VF-1* through *VF-6* (see Sections 7.5.1 through 7.5.6 and Section 7.6.1) belong to the direct secondary systems.

In the *indirect* secondary systems, the primary and secondary systems are connected through a very limited section of pipe common to both systems. The two systems are hydraulically separated, so that one can perform without the assistance of the other and, more importantly, without affecting the performance of the other. They can also be operated together or separately since their performance is not interdependent. The crossover bridge serves not only the purpose of maintaining hydraulic separation; it also helps to introduce primary water into the secondary system.

In all instances, primary water is introduced into the secondary system gradually, in response to system demand, whether or not the secondary system is constant

or variable flow. It will be noticed that there are more constant than variable flow systems. Three-way valves at the cooling coils and constant speed secondary pumps indicate a constant flow system, as shown in cases 4, 5 and 6 (Section 7.7.4). On the other hand, cases, 7, 8 and 9 (Section 7.7.5) exemplify variable primary and secondary flow. In an energy-conscious world, variable flow has an inherent attraction because of reduced pumping costs.

Direct secondary systems tend to utilize excess primary pumping head so, in that sense, they can conserve energy by utilizing it more efficiently. This is the principal rationale in their defense, a fact that must be balanced against the destabilizing effect of interacting pumping heads. In very large systems, employing grid distribution patterns and having hundreds of subloops, it is difficult from a practical point of view to justify the use of direct secondary systems, not only because it is heretical in defying the well-established principle of separating pumping heads but also because the large number of subloops can contribute towards a mammoth balancing headache.

Direct secondary systems often lend themselves to loop distribution, where balancing problems can be kept under control. On the other hand, indirect secondary systems are expensive, due to bridge piping and extra control valves. Also, experience in existing plants has established that they incur high pumping costs.[4] However, they are apt to perform more stably. Direct systems, if highly unbalanced, can cause not only concern but waste of energy as well. All these factors must be carefully weighed.

Abbreviations used in control diagrams and text that follow in this chapter:

PC	Pressure Controller
DPC	Differential Pressure Controller
TC	Temperature Controller
S	Switch
TT	Temperature Transmitter
FT	Flow Transmitter
FE	Flow Element
FC	Flow Controller
P or p	Pressure, psi

7.5 Direct Secondary Systems–Variable Flow

7.5.1 *Variable Flow System VF-1*

The main characteristics of this system are: On the primary side, variable flow and variable temperature rise;

---

[4] Maurice J. Wilson, "Pumping Techniques for Campus Chilling Systems," *Heating, Piping, Air Conditioning*, February, 1974, p. 50.

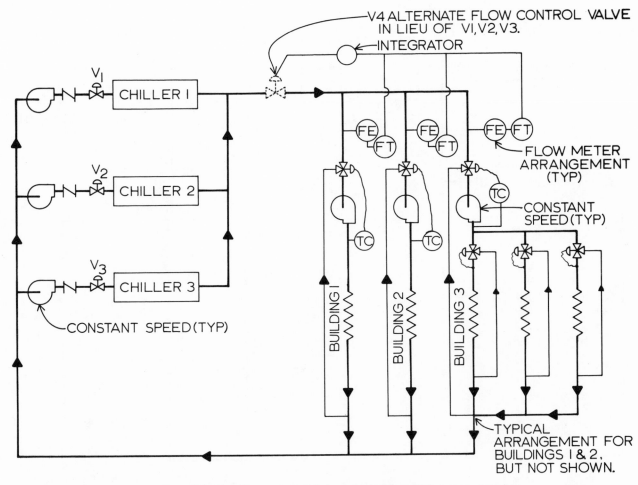

Figure 7-7. Variable or constant flow systems with variable temperature rises.

on the secondary side, constant flow and variable temperature rise.

V_1, V_2, and V_3 in Fig. 7-7 act as pressure regulating valves to maintain a fixed discharge pressure. V_4, as an alternate pressure regulating valve, serves the same purpose except that it subjects the chillers to the full pump pressure, is more economical and can be set to maintain a downstream fixed pressure. Also, it can be regulated by flow meters in the secondary branches, if the number of meters can be limited. These valves compensate for reduced friction in the circuit at less than design flows and they help to keep the pump functioning near its design operating point. Picking a centrifugal primary pump with flat curve characteristics aids in achieving a smooth operation at part load.

As one might expect, the branches nearest the pump receive excessive flow and higher pressure differentials while the most distant ones are starved. Some engineers, in order to offset this tendency, provide additional 2-way pressure regulating valves in front of the 3-way valves.

One of the undesirable features of this system is that the secondary pump acts like a booster pump at full primary flow through the secondary system.[5] At less than full primary flow, the imposition of the head of the primary pump on the secondary system is gradually less pronounced. In the first place, the flow characteristics of the secondary pump are affected by the impression of the primary head, and in the second place, the 3-way valve is called upon to dissipate the varying head between the primary supply and return headers, a task at which it is not very successful. Pressure variations within the primary system have repercussions within the secondary system and vice versa. As load varies, the relationship between primary and secondary systems constantly changes, creating a series of balancing problems.

---

[5] Erwin G. Hansen, "Central Chilled Water Distribution System Design," *ASHRAE Journal*, November, 1974, pp. 49–50.

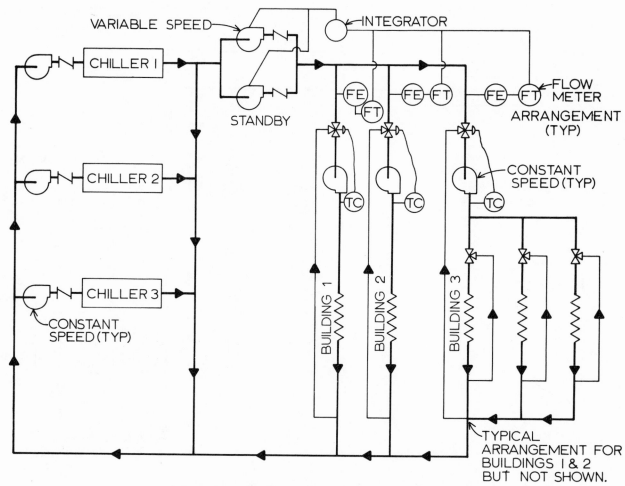

Figure 7-8. Typical three-loop primary system, *VF-2*.

The capacity of the primary system can be changed in incremental steps, depending on how many pump-chillers are on the line. However, the secondary system can absorb primary water in gradually increasing quantities. In order to resolve this apparent antithesis, many engineers install a bypass at the far end of the system to handle any difference between supply and demand, regulating the flow in the bypass by means of an automatic control valve set to maintain a fixed pressure.

7.5.2 *Variable Flow System VF-2*

This pumping system consists of three loops; on the primary and intermediate primary sides, variable flow and variable temperature rise; on the secondary side, constant flow and variable rise.

The primary (constant speed) pumps (see Fig. 7-8) serve a low head loop. Most of the pressure drop due to friction occurs across the chillers and it is expected that total dy-

namic head of the entire primary loop will not vary substantially as the number of chillers in operation change. The relationship between primary and intermediate is that of a primary-secondary arrangement and little influence may be expected on the primary by variations in the intermediate, especially if an attempt is made to keep the bridge piping as short as possible.

The intermediate loop's variable speed pumps adjust to secondary flow demand on sensing either discharge pressure, secondary loop differential pressure or flow into the secondary branches by means of flow meters. The secondary circuit, with its 3-way valve control, although fairly stable, is still subject to some pressure interactions from the intermediate system. However, the variable speed pumps tend to smooth out the extremes. This system is said to be able to operate at low overall horsepower requirements and low pumping costs.[6]

[6]Maurice J. Wilson, "Pumping Techniques for Campus Chilling Systems," *op. cit.*, p. 50.

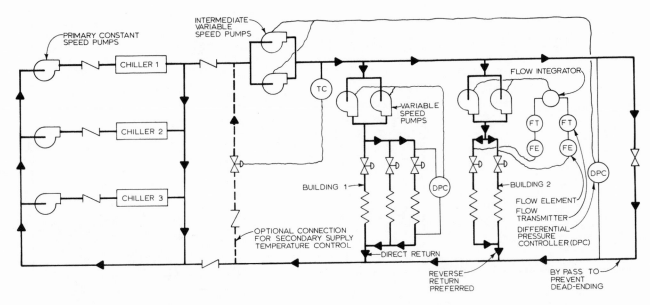

Figure 7-9. System *VF-3*.

7.5.3 *Variable Flow System VF-3*

True variable flow prevails throughout this system; in the primary, by incremental steps; in the intermediate, by means of variable speed pumps, down to a minimum flow level dictated by the differential pressure control in the bypass; and in the secondary, by means of variable speed pumps and 2-way valves. See Fig. 7-9.

System characteristics are on the primary, intermediate and secondary sides: variable flow, fairly constant temperature rise.

The differential pressure controller across the bypass maintains minimum pressure across the intermediate system, while the one across each building controls the head of its own pumps, thus compensating for any overpressure imposed by the intermediate pumps and taking advantage of it, thus conserving energy.

The flow transmitter indicates another method of accomplishing secondary system flow control. Each individual 2-way valve is controlled by its own space or duct temperature controller. This system appears to have very low pumping costs.[7] One would expect these costs to be lower than those assigned to system *VF-2*.

Unlike systems *VF-1* and *VF-2*, *VF-3* cannot furnish secondary water at a temperature different than that of the primary system. As it stands, this system cannot blend return water into the primary. This can be corrected by introducing another set of blending pumps con-

trolled by t_{ss}. Naturally, this would increase pumping costs. It is possible also to introduce return water into the suction of the intermediate pumps by means of a controlled return water connection. The control valve would be actuated by a temperature controller on the discharge of the intermediate pumps.

7.5.4 *Variable Flow System VF-4*

The operation of this system depends on $t_{ss} = t_{ps}$ and $W_s = W_p$. Therefore, secondary water mass flow is equal to primary water mass flow (see Fig. 7-10). According to the originator[8] of this system, the advantage in this is that no more water is pumped than is absolutely necessary. The diversity factor of the secondary system is that of the primary as well. In essence, this type of system may be limited to smaller installations because it is uneconomical to impose the severe design condition of $t_{ss} = t_{ps}$ on a large central system, where the pumping economies that accrue from large primary water temperature differentials cannot be realized. When $t_{ss} = t_{ps}$, the differential temperature of the secondary is the same as that of the primary, therefore, restricting the primary system to a small differential and a high pumping rate.

The controller is set for a fixed flow through the pilot line. Any pressure difference between supply and return is the effect of change in energy demand. The flow in the pilot line increases or decreases and the controller adjusts

[7]James B. Rishel, "Selecting Pumps for Secondary Chilled Water Systems," *Heating, Piping, Air Conditioning*, April, 1974, p. 61.

[8]G.F. Mannion and J. R. Mannion, "Diversification of Large Central Plant Pumping: a Solution," *Heating, Piping, Air Conditioning*, December, 1972, pp. 56–60.

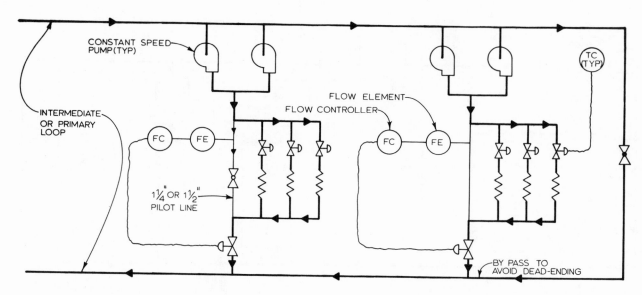

Figure 7-10. System *VF-4*.

the valve position to maintain its set point. Consequently, secondary flow is equal to primary flow. The constant speed pump must be selected so that the operating point can shift up and down the operating curve without difficulty. For a similar arrangement see Section 7.7.3, Modulating 3-way Valves, Case 6: 2-way Valve Located in Bridge Circuit with Pilot Line.

7.5.5 *Variable Flow System VF-5*

The only pumping power in this system is provided by the primary pumps. Intermediate or secondary pumps are entirely omitted. See Fig. 7-11. The designer of this system claims success, low cost and high system efficiency.[9] The subloops near the pump must dissipate the high pump head through a 2-way valve in the loop inlet branch controlled by a differential pressure controller set to maintain a fixed pressure drop across a distribution subloop. A blending pump, which injects adequate return water into the supply to maintain a fixed supply temperature, is selected for a fixed mass flow. Any excess is pumped back to the pump inlet. The blending and bypass valves are controlled by a temperature controller in the supply pipe. A flow controller in the subloop return line is set for minimum acceptable return flow. When this flow is reached, the flow controller resets the temperature controller higher so that flow through the coil is increased to prevent a laminar flow condition, which impairs coil performance.

---

[9] Hansen, *op. cit.*, pp. 51–52.

It is assumed that resetting the chilled water temperature to a higher level coincides with a low space load and does not affect space humidity adversely. One must remember that the subloops represent entire buildings or a number of buildings. Under the circumstances, the only system pumps, the primary chilled water pumps, must be sized to handle the mechanical space piping, yard piping and building distribution network. This must be done for a number of yard and building networks to pinpoint the worst condition. Theoretically, this amounts to one valve in some remote building. All other branch valves must throttle at all times to maintain a fixed pressure differential. The pumping head must be established for the worst condition and all excess pressure must be dissipated. This feature should be analyzed sufficiently to establish the superiority of a once-through system for the given circumstances of a particular project.

An additional design problem of this system is that a minimum head must be established for the chilled water pumps so that, when the inlet branch throttling valve is in the open position, this minimum head is sufficiently large to operate the subloops satisfactorily. This minimum head is required when only one chiller is active and, we assume, no other means are available to maintain a minimum head except the chilled water pumps. Low flow conditions must be investigated for adequate overall system performance.

7.5.6 *Variable Flow System VF-6*

A system of this nature, in spite of severe limitations, can be successful under certain circumstances. (See Fig. 7-12.)

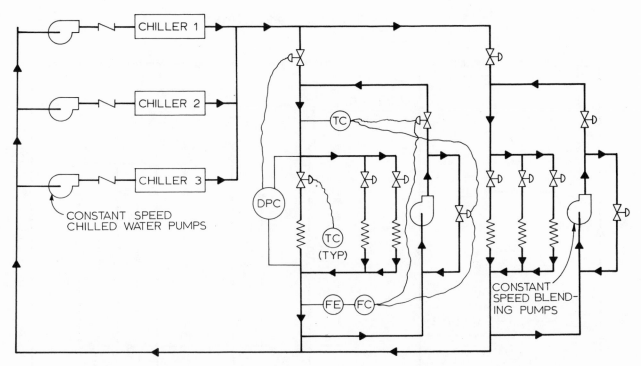

Figure 7-11. System *VF-5*.

There cannot be more than two chillers. The load can be split between them equally. This system has a greater chance of success if the load is split 30% and 70% and if the 30% is the first to be disengaged, leaving 70% of the flow on the line. The system friction change is not of very serious proportions and its ability to maintain a minimum pressure differential at the far end of the system is enhanced. The selected pump curve must be flat so that the throttling valve is not overburdened. Any water not needed by the system is bypassed back to the chillers. The throttling and bypass valves work in opposite directions. Any minor imbalances will be smoothed out by the coil valves. This arrangement works well with chillers in series as the flow is constant.

The performance of this system is better, if the piping runs are short and the coil valves are clustered. Some operators prefer manual operation of the throttling and bypass valves with adequate flow recording points to monitor performance. All this monitored information can be available at the central control board.

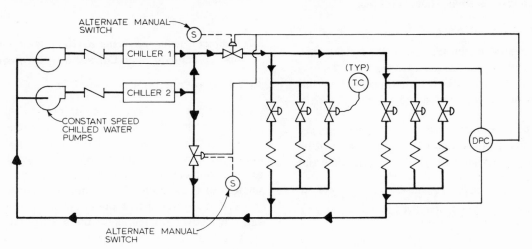

Figure 7-12. System *VF-6*.

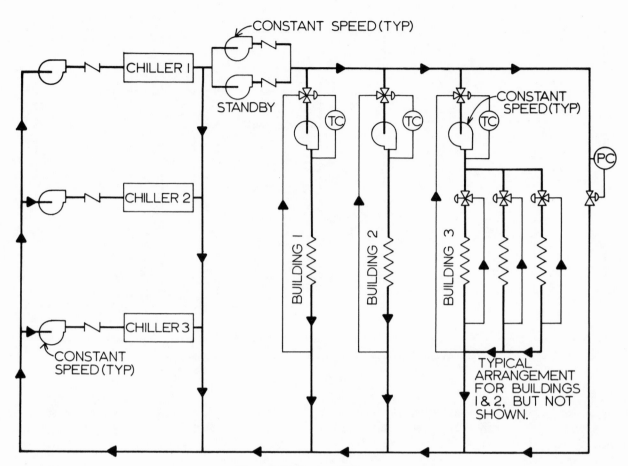

Figure 7-13. System *CF-1*.

7.6 Direct Secondary Systems—Constant Flow

7.6.1 *Constant Flow System CF-1*

In this system some of the features of the variable flow system, *VF-2* (Section 7.5.2) are modified to achieve a greater measure of stability in the intermediate loop but at the expense of increased initial and operating costs.[10] See Fig. 7-13.

The salient features of the system are: on the primary side, variable flow and variable temperature rise, and on the intermediate and secondary sides, constant flow and variable temperature rise.

The primary loop is as stable as that of system *VF-2*. In the intermediate loop, flow is stabilized by means of a by-

pass at the far end of every subsystem. Each is equipped with a pressure regulating valve scheduled to maintain a fixed pressure at the farthest point in the subsystem.

7.7 Indirect Secondary Systems: Basic Methods for Primary-Secondary Flow Control with Crossover Bridge

7.7.1 *Start-Stop Control*

Interruptible operation of on-off pump control is not suitable for large installations. Resulting cyclical temperature effects make it preferable to keep piping circuits to minimum runs and so limit space temperature variations to an acceptable level. A system of this type can be considered for large residential communities served by a central system. Intermittent pump operation is economical and will satisfy residential requirements. On-off pump control systems, as shown in Fig. 7-14, are frequently employed in residential hydraulic installations.

---
[10]Wilson, "Pumping Techniques for Campus Chilling Systems," *op. cit.*, p. 50.

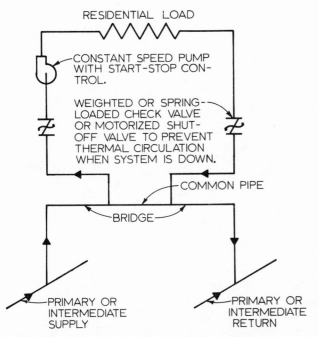

Figure 7-14. Start-stop controlled system serving residential load from central plant source.

7.7.2 *Variable Speed Ejection Pump*

The ejection pump withdraws return water from the secondary system. See Fig. 7-15. An equal amount of primary water is introduced into the secondary system through an injection riser. Ejection pump speed is reset in a constant temperature secondary supply by a space or outdoor temperature controller.

Ejection or injection riser size depends on maximum secondary system demand. Ejection pump control, if actuated by an industrial quality temperature controller, should give excellent results.

This method of controlling secondary water system operation is positive because it does not depend on negative pressure in the primary return. However, added controls and variable speed drive are involved at added cost. Because a variable speed pump is more difficult to service, this may be the pivotal consideration.

7.7.3 *Modulating 3-Way Valves*

There are three principal ways of using 3-way valves to control operation of secondary water systems, which depend on the valve's location:

Case 1 (Fig. 7.16): 3-way Valve Located in the Secondary Loop on the Injection Riser

System stability, which is the main advantage of continuous pump operation, is achieved at the expense of higher pumping costs, since full volume pumping is required at all times. Three-way valves cannot be used very successfully as pressure regulating valves. It is therefore advantageous to have a 3-way mixing valve controlled by temperature criteria only, either by an outdoor-indoor or secondary supply temperature controller.

An important characteristic of this system is that, with complete recirculation of the secondary system, it is possible that the primary pump head be excluded from the secondary system. Then, if the 3-way valve is too restrictive, under particular circumstances of low elevation head, secondary pump cavitation may occur. Hence, net positive suction head is a design concern. Sizing the 3-way valve

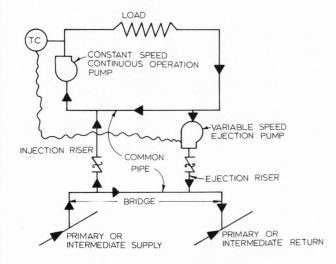

Figure 7-15. Variable speed ejection pump controls secondary flow.

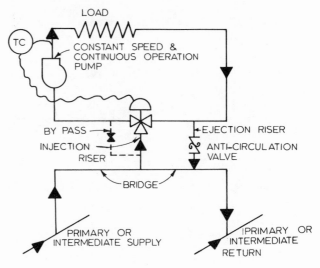

Figure 7-16. Control by 3-way valve on injection riser.

for the primary influx or the secondary recirculation flow is a problem, because the two flows are so disanalogous. If sized for the secondary flow, the dominant one, the valve is oversized for the primary flow, entailing poor control.

To prevent the complete absence of the primary head on the secondary system, it is entirely possible to install a small bypass around the 3-way valve, as indicated in Fig. 7-16.

Case 2 (Fig. 7-17): 3-way Valve Located in the Secondary Loop on the Ejection Riser

Most of the shortcomings of Case 1 are absent in this arrangement. The 3-way valve can be placed farther away from the pump. Primary head pressure then acts on the secondary through the injection riser, either by omitting the anti-circulation valve or installing a bypass around it or by using a lift check with a small orifice through the flapper.

Valve sizing remains a problem because the flow in this arrangement continues to be unequal through the two outlet ports. In general, if a 3-way valve must be used, this arrangement appears preferable to Case 1.

Case 3 (Fig. 7-18): 3-way Valve Located in Bridge Circuit

Note that, in this arrangement, the secondary water system is fairly free of large pressure drops and that the pressure drops that do occur are rather stable. Hence, stable secondary system operation may be expected. The 3-way valve is sized for the primary bridge flow and, therefore, it is easier to balance bridge flow by proper selection of 3-way valves.

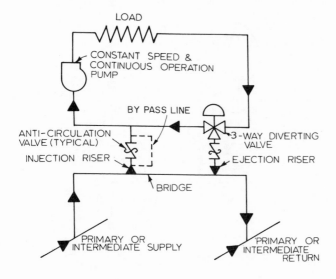

Figure 7-17. Control by 3-way valve on ejection riser.

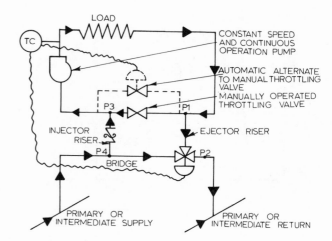

Figure 7-18. Control by 3-way valve in bridge circuit.

The hydraulic gradients of both primary and secondary systems must be well understood because the entire operation depends on the differences in pressure between them. The results can be distressing if the systems are not properly engineered. The restriction of the manually-operated throttling valve is necessary in order to make the 3-way valve operate. An easier, but more expensive, alternate is the automatic valve. The same leaving water temperature controller can operate both valves. Conditions for satisfactory operation are that $P_2 < P_1$ and $P_3 < P_4$.

7.7.4 Modulating 2-Way Valves (Constant Flow, Variable Temperature Rise in Secondary System)

Case 4 (Fig. 7-19): 2-Way Valve Located in Bridge Circuit

This often-used arrangement can limit the amount of primary water to what is actually needed. The 2-way valve is actuated by a temperature controller on the discharge side of the pump. This type of system can respond precisely to system diversity factor and still maintain different temperature levels in the primary and secondary systems.

The alternate valve position shown is less desirable because the primary head is not fully exerted on the secondary loop and leaves the secondary pump unshielded from the suction pressure of the primary. However, ejection is somewhat easier with the valve in the alternate position. Since the 2-way valve is controlled by temperature only, it is advisable to have an additional balancing valve in the bridge circuit.

Case 5 (Fig. 7-20): 2-way Valve Located in Ejection Riser

This system is the equivalent of *Case 3*, where a 3-way valve is used in the ejection riser. However, the 2-way

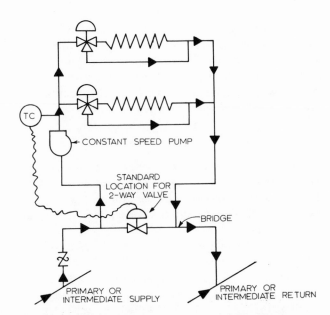

Figure 7-19. Control by 2-way valve in bridge circuit.

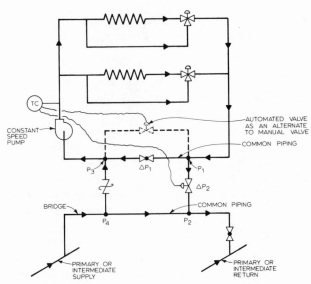

Figure 7-20. Control by 2-way valve in ejection riser.

valve lacks the positive feature of governing flow to the secondary and, through the bridge, flow to the primary return. This arrangement depends a great deal on the hydraulic relationship between ΔP_1 and ΔP_2. Note that there are two common lines and that the hydraulic balance is more complex than when there is only one.

In general, the following relationships must exist: $\Delta P_1 = \Delta P_2$ in the valve's open position and $P_2 < P_1$ and $P_3 < P_4$. An engineering investigation must establish these conditions throughout the system. "Improper application, however, can lead to serious operating problems ... 2-way valves should not be applied to primary-secondary arrangements unless there is complete understanding of the fact that a flow variation in the secondary circuit of substantial proportion can occur because of control operation of the valve and due to improper flow and pressure drop ratios ... Their proper application leads to superior performance."[11]

Case 6 (Fig. 7-21): 2-way Valve Located in Bridge Circuit with Pilot Line

A pilot line for measuring pressure differences between secondary water supply and return is used in system *VF-4* (Section 7.5.4). Unlike the present case, individual building circuits in *VF-4* are controlled by 2-way valves, not 3-way valves.

The arrangement of Fig. 7-21 is a constant mass flow, variable temperature system. The pilot line is instrumental

in maintaining a fixed pressure difference between supply and return. Whereas in system *VF-4* the function of the 2-way valve is to draw from the primary into the secondary on a one-to-one basis, in this arrangement, the 2-way valve, due to its combination with the 3-way valve, cannot fulfill this function and reverts to one of pressure regulation.

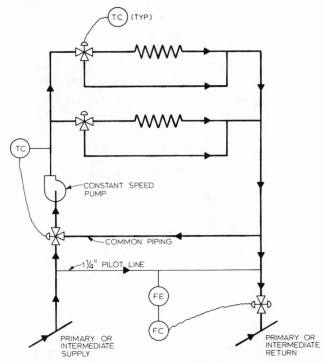

Figure 7-21. Control by 2-way valve in bridge circuit with pilot line.

[11]*Primary Secondary Pumping Applications Manual*, ITT Bell and Gossett, Morton Grove, Ill., pp. 46–47.

7.7.5 *Modulating 2-Way Valves (Variable Flow, Fairly Constant Temperature Rise in Secondary System)*

Case 7 (Fig. 7-22): 2-Way Valve Located in Bridge Circuit

Because of reduced secondary pumping costs, all variable flow systems become attractive for large central installations having many thousands of 2-way valves. Increased pumping savings can accrue from reduced flow demand in the primary system; operational economies combined with precise engineering can lead to a superior installation.

One of the main problems of the system in **Fig. 7-22** is the selection of a pump corresponding to the maximum simultaneous load. This is an indication that flows at other loads will not vary radically and that any variation can be accommodated within the operating curve of the pump, allowing the operating point to shift accordingly. This approach, however, places the burden of controlling pump flow on each and every 2-way valve in the system. Theoretically, repositioning of any valve causes a readjustment of all valves. In some cases, a large number of valves belong to a single zone and there can be mass repositioning and a direct, measurable effect on the remaining system valves. Some of the valves may not be able to dissipate the excess pump head, causing erratic malfunctioning by allowing more than design mass flow to pass through.

The attraction of this system is its simplicity and lack of complex controls. Careful engineering and follow-through during the construction period is essential to ensure that design features are incorporated in the installed system. The design should include as few zones as possible with as little fluctuation requirement between full and partial loads as possible.

Some engineers, in order to overcome the excess pump head problem, include a bypass around the pump operated by a differential pressure controller. The system is thus converted to one of constant flow, losing all of the advantages of variable flow. Hydraulically, it differs very little from a 3-way valve application.

Case 8 (Fig. 7-23): 2-way Valve in Bridge and on Pump Discharge

A solution for the over-pressure problems created in the system of Case 7 is to put an additional 2-way valve on the discharge of the secondary pump and assign to it some pressure reduction functions reserved for all individual valves in Case 7.

The valve located in the bridge is still regulated by a temperature controller while the discharge valve is regulated by a differential pressure controller sensing pressure differences between supply and return. In essence, the result is that, at low flows, the discharge valve dissipates enough head to relieve the coils' 2-way valves of the burden of dissipating the head that is now handled by the discharge valve. The result is to stabilize the system at low flows. The sensing points of the differential pressure controller must be so selected as to benefit all coil valves equally. A multiplicity of points may be necessary with an averaging relay to satisfy this design condition.

This system may be considered an improvement on Case 7 but suffers the expense of control simplicity and higher

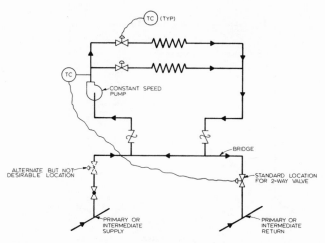

Figure 7-22. Control by modulating 2-way valve in bridge circuit.

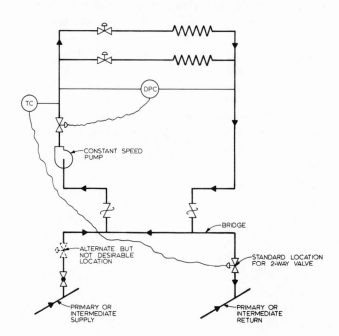

Figure 7-23. Control by 2-way valve in bridge and on pump discharge.

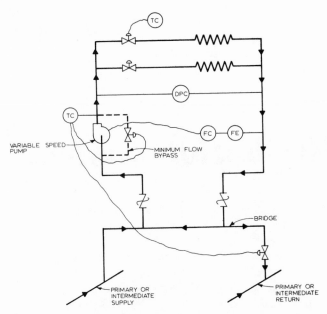

Figure 7-24. Control by 2-way valve in combination with a
variable speed pump.

The temperature controller regulates the bridge valve in order to introduce only the needed mass of primary water. A minimum pressure difference is maintained by the pressure differential controller. At this minimum condition, and only if the design requires it, the pressure controller resets the bypass valve to bypass any excess flow back to the pump. If necessary, the DPC will reset the temperature controller upwards to increase flow through the secondary loop. The flow controller resets pump speed.

first cost by incorporating additional valves and controllers.

Case 9 (Fig. 7-24): 2-way Valve in Combination with a Variable Speed Pump

This system: (1) resolves the question of overpressure at low flows, (2) regulates the problem of low flow by keeping a minimum flow without forcing any unwanted flow through cooling coils, (3) makes possible secondary supply temperatures different than those of the primary supply, (4) can provide operating savings with variable speed pumps and (5) controls bridge flow.

This variable flow system has a fairly constant temperature rise and combines a number of functions with the help of only one major control valve, that of the bridge circuit. It requires a number of controls which, nevertheless, have clear and distinct functions.

BIBLIOGRAPHY

Hydraulic Analysis

Becker, Herbert P. "An Analysis of Variable Speed Pumping for Hydronic Systems," *ASHRAE Journal*, (May, 1970), 71–77.

Carlson, G.F. "Hydronic Systems: Analysis and Evaluation—Parts I–VI," *ASHRAE Journal*, (October, 1968-March, 1969).

_____. "Central Plant Chilled Water Systems—Pumping and Flow Balance—Parts I-III," *ASHRAE Journal*, (February–April, 1972).

Rishel, James B. "Variable Pressure-Volume Pumping for Hot Water and Chilled Water Systems," *Transactions, The American Society of Heating, Refrigerating and Air-Conditioning Engineers*, Part I, Vol. 81, 1975, 646.

Wilson, Maurice J. "Campus Chilled Water Plants—District vs. Decentralized," *Heating, Piping, Air Conditioning*, (November, 1966), 106–109.

_____. "Four Plans for Expanding District Chilled Water Systems," *Heating, Piping, Air Conditioning*, (January, 1967), 148–154.

_____ "How Secondary Systems Affect Primary Flow, Chiller Arrangement," *Heating, Piping, Air Conditioning*, (May, 1968), 109–114.

_____. "Sequencing Chiller Operation, Using Distribution Storage Potential for Operating Economy," *Heating, Piping, Air Conditioning*, (July, 1968), 112–114.

CHAPTER 8

Distribution Systems

This chapter is about the delivery of energy in the form of chilled water, hot water, and steam, from a central plant to any number of buildings through a network of pipes.

The art of designing successful piping networks is a mixture of engineering principles and a great deal of empirical data accumulated over the decades. Hence, we attempt here to bring these practices into focus, to classify them, to take into account thermal and hydraulic considerations in developing criteria for design of systems.

We will also try to clarify areas which have been problematical and to delineate system limitations. Many distribution systems are hydraulically complex because they are the connecting link between disparate central plant and building systems, reflecting the problems of both.

Computerized hydraulic analysis can ease some of the problems of large scale distribution systems. Available computer analytical techniques for city water networks can be adapted to chilled water distribution systems. The University of Illinois has done some work in this area. The annual Purdue University Chilled Water Conferences are imparting new insights into distribution systems, setting up the theoretical foundation for further evolution in this field.

Before examining distribution systems, it may be worthwhile to survey the type of plants to which these systems are connected and to review briefly the status of coal fired plants.

8.1 Types of Central Plants

In addition to the fuel it burns, there are other factors that help to classify a plant; for instance, *pressure range* and *use* provide the following categories:

Low pressure heating plants are applicable where low pressure turbine exhaust steam or low pressure boilers are available.

High pressure heating plants distribute high pressure steam at 250 psig or somewhat less and supply campuses, hospital groups, business districts and industrial areas.

Electric utility steam systems are medium or high pressure and usually handle steam from turbine exhaust or turbine extracting points. Ordinarily the sale of steam is regulated by the state or city.

Turbogenerator exhaust from locally installed turbogenerators is used mostly for industrial or institutional loads. Pressures depend on project requirements.

High and medium temperature water plants distribute 250–300F (121–149C) (MTW) or 350–400F (177–204C) (HTW) water to service residential communities, campuses, hospital complexes, military bases and other installations.

High pressure steam and HTW plants. If electric generation and/or central plant refrigeration are involved, a combination of energy forms must be employed to have an efficient plant. Important auxiliaries and centrifugal compressors are very often driven by steam turbines, preferably at pressures of 600 psig (4134 kPa) or higher. Absorption machines may require 12 psig (83 kPa) steam and the heating system may be HTW, produced by shell and tube or cascade heaters.

8.2 Coal-fired Steam Plants

Most of the old steam plants had field-erected, coal-fired boilers in the 125–300 psig range. This practice continued into the early 1950's when package boilers and HTW boilers began to make serious inroads. HTW coal-fired boilers were still used in some plants but, like the coal-fired steam boilers, they gave way to oil-firing. Very few companies still possess the skill to manufacture coal-fired boilers for central plant applications. Considering the limitations in the supply of oil, it is probable that coal-fired boilers will regain dominance.

The following accounts for the decline of coal-fired steam plants: (1) Coal produces air pollution and dirt in the plant, the required complement of coal auxiliaries occupies considerable space, and coal-handling is replete with problems. (2) Coal-handling requires more labor than oil supply. Increasing labor costs discourage new coal plants. (3) The utilities had been promoting gas and oil to increase profits. (4) Increased labor costs continue to discourage field erection of boilers. (5) There has been a phenomenal increase in available package boilers with corresponding reduction in purchasing and installation costs. (6) There are a number of smaller problems, including corrosion in the condensate lines and steam pressure reducing stations. (7) The relatively greater supervision required for the overall operation tended to increase the financial burden imposed by coal-burning steam boilers.

Many of these problems have since been resolved through technological progress but some continue, while environmental standards impose more stringent criteria. Significant studies are being made in the design of packaged coal-fired boilers, such as fluidized-bed combustion, which burns both cleanly and efficiently.

8.3 Underground Piping Criteria

Underground piping systems must currently conform to numerous and more severe criteria and regulations, due to degrading environmental conditions, than an above-ground system. Also, the underground system operates unattended for its entire lifetime. Some of the most important criteria are:

1. The system must be low enough in cost, have long life, be reliable and have good thermal efficiency and short installation time.
2. The system must be able to resist the corrosive effects of the soil and the intrusion of moisture into its insulation. On the other hand, if moisture can enter the system, insulation must be driable and its piping, drainable.
3. Underground system design must allow for lateral and longitudinal expansion due to thermal cycles and must accommodate pipe supports, guides or anchors.

4. Buried systems are required to withstand superimposed structural loads caused by footings, moving vehicles, equipment loads, earth settlement and compaction. In some instances, appropriate bridging over the conduits may be required.
5. Insulation must stand up to humid conditions without substantially losing its insulating value on drying. It is expected that some parts of all underground systems will become wet some time. The overall thermal efficiency of the system depends on the frequency and duration of such occurrences and the ability of the system to recover. Insulation and encasement envelopes must be selected after considering the drainability and insulation value of the soil surrounding the pipes.
6. Since replacing the system is very inconvenient and expensive, one must expect it to perform well during its life without serious deterioration or need for replacement.
7. Part of the versatility of a system is its ability to keep operating despite a pipe failure. This can be accomplished by feeding a building from two directions and by judicious placing of isolation valves.
8. An important consideration is hydraulic balance at all loads, which is significant for water systems not only in terms of predictable pumping performance but, also, in determining minimum flow under low-flow conditions to avoid laminar stagnation.
9. In the case of non-welded underground pipes, a guaranteed low leakage factor must be met. Pipe support is of greater importance for non-welded pipe, since ground settlement or shifting can more easily result in a pipe break.

8.4 Types of Systems

There are three basic methods of installing distribution piping for central heating and cooling plants in order to deliver hot water, chilled water or steam and to return hot water, chilled water or condensate:

(1) Installation above ground, or
(2) in tunnels, or
(3) direct burial.

Above-ground piping systems become traffic barriers when crossing streets and are aesthetically objectionable. Ordinarily, pipes are placed on a trestle or in a covered walk. The insulation finish must be weatherproof. This type of installation is most often encountered at industrial sites.

Walk-through tunnels are readily accessible for maintenance, additions and alterations. Anchoring is much easier than in covered walkways. Tunnel construction, with its excavation, sheathing, dewatering, backfilling, etc., is very expensive. In many installations it is combined with required passageways between buildings and is therefore, more economical.

Direct burial is most often used for large central piping systems. Problems appear even in the most perfectly constructed installations, and sometimes continue to exist for some time after the last construction worker has left the site since they cannot be detected or easily repaired.

The latter were developed in response to the ever increasing labor costs entailed in field erection. Factory control of quality and use of non-corrosive materials have made the new systems attractive to owners and engineers. Steel pipe is still required for steam, condensate and HTW piping. Insulated, clad steel is now invading this last province of the pipe fitter: field-erected steel pipe.

8.5 Pressurized Casing System

This system consists of factory-fabricated 20 and 39 ft (6.1–11.8 meter) sections assembled or welded in the field. The sections contain a carrier pipe or pipes, a casing and space between the pipes and the casing. If there is one pipe, the space is annular. The casing is of helical corrugated steel and lap or butt-welded smooth joining strips, most often galvanized and coated with asphalt, coal tar, epoxy or fiberglass, depending on soil type, temperatures, dielectric resistance or mechanical strength requirements. (See Fig. 8-1.)

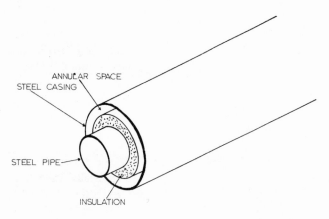

Courtesy Johns-Manville Corp.

Figure 8-1. Casing and carrier pipe assembly.

The annular space between carrier pipe and casing must be adequate to contain an insulation envelope and pipe supports. Steam or hot water pipe material is usually steel, A–53 grade B or A–106 grade B; for chilled water, it is A–53 and for steam condensate extra strong steel, Yoloy, type K copper, IPS copper and brass are most often encountered. Condensate return pipe is frequently put in a separate casing. Because of the corrosivity of condensate return, pipe life is shorter and must be replaced more often than pipes serving steam or water.

The casing is designed to be air testable, drainable and dryable. The interior of the casing is constructed so as not to create structural dams. The designer must be careful to pitch the casing to drain or pumping stations so as to avoid flooding the casing if leaks develop. If, on the other hand, flooding occurs, the casing must be anchored sufficiently to offset the buoying force of the rising water table. Following interior flooding, the space around the carrier pipe can be used as free passage for hot, drying air to remove moisture driven out of the insulation by the hot pipe(s).[1] The space around the carrier pipe, besides furnishing additional resistance to heat flow, also provides the means to leak test the casing by pressurizing it to 7.5–15 psig (51–103 kPa). The testing medium is air from an air compressor or, infrequently and to a limited extent, nitrogen. Very often pressurization is maintained as a system operating condition. The need for permanent pressurization and level of pressurization arises from system design requirements to resist groundwater infiltration. Typically, casing pressure is maintained at 5 to 8 psig (34–55 kPa). The criteria for permanent pressurization are frequency of flooding conditions, casing leak tightness, length of line, nature of soil, presence or absence of cathodic protection and desired system life. Monitoring pressure alarms can be installed that govern compressor operation and signal the loss of pressure when leaks develop. If leaks are small, the compressor can maintain the required pressure to keep groundwater out. The Academy of Sciences defines four groundwater condition classifications for government installations: severe, bad, moderate and mild. [2] The first two are tested at 15 psig

---

[1] For guidelines in drying out wetted underground insulation, see National Academy of Sciences, Building Research Advisory Board (BRAB)-Federal Construction Council (FCC), Technical Report 39-64, *Evaluation of Components for Underground Heat Distribution Systems*, paragraphs B.5.2 and C.4.3.

[2] *Criteria for Underground Heat Distribution Systems*, Academy of Sciences, BRAB-FCC, Technical Report No. 66, Washington, D.C., 1975, p. 15.

(103 kPa), the latter two at 7.5 psig (51 kPa).[3] Government underground piping systems are subjected to the following tests, in addition to the casing pressure test: impact load, thermal stress, simulated failure and surface load.[4]

Piping insulation, for the most part, is glass fiber or calcium silicate; some urethanes and foam glass are also acceptable insulations, which can be dried after being water-soaked. Besides coatings or in conjunction with them, sacrificial anodes may be used for external corrosion control of buried casings. The resistivity of the soil must be known before cathodic protection is undertaken. Protection of the casing interior surface and of the carrier pipe is more difficult due to short circuits and because water flows into the casing from flooded manholes or faulty end seals. Short circuits occur because of metallic contacts between casing and carrier pipe, such as pipe supports and anchors and also because of the presence of electrolytes (water) at the bottom of the casing. Dielectrically insulated metallic pipe supports pass current from the casing to the carrier pipe, depriving considerable lengths of pipe of cathodic protection. Current discharging from the inner surface of the casing through the electrolyte can cause corrosion on the inner surface. To reduce the probability for corrosion, coating defects on the casing and carrier pipe and metallic contacts must be kept to a minimum. Under adverse conditions, cathodic protection of the inner casing surface and carrier pipe in addition to that of the outer casing surface may be necessary subject to exhaustive analysis. Corrosion of the inner casing surface is accelerated because air inside the casing is usually warm and humid. Condensation takes place on the cool interior casing walls.

Casing systems, whether pressurized or nonpressurized, involve considerable field handling and weight lifting, extensive field welding labor costs, costs for cathodic protection when required and difficult replacement procedures. These factors contribute to higher owning and operating costs. Their use must be preceded by a thorough investigation of soil conditions, economic considerations and project requirements.

Casing systems require more space than other systems defined in Sections 8.7–8.10, because the conduits are much larger than the carrier pipes and allow less cross-over space for adjacent pipes. It is also prudent to

---

[3] *Ibid.*

[4] *Underground Thermal Distribution Systems* (35F to 250F), Academy of Sciences, BRAB, Washington, D.C., 1970, Appendix A, pp. 29-35.

place only one pipe inside each conduit, although more than one is possible. Expansion U loops are provided at regular intervals, infringing on adjacent space and making it unavailable for other pipe runs. Manholes are required at specified intervals for isolation valves and expansion joints, adding to the cost.

Both the carrier pipe and casing are field-welded in sequence. The pipe is welded and tested first and subsequently insulated. The testing of the casing follows. Adjacent pipes are welded in time sequence so as to allow adequate space for the field crews. Due to the unusual amount of field labor and steep labor costs, these systems are becoming increasingly expensive. They were often employed in military complexes and elsewhere for steam and high temperature water systems, for which they are highly qualified. Their need for air pressurization and vulnerability to corrosion outweigh the advantage of long pipe lengths (hence, fewer field joints) that they possess over other systems. In time, air pressurizing of the casing becomes a maintenance headache as advancing corrosion pierces it.

8.6 Nonpressurized Conduit

The casing itself can be tile, a concrete trench which is poured in place, of the single or double type, or concrete or asbestos cement pipe. (See Figs. 8-2, 8-3, 8-4 and 8-5. These systems are field-erected.)

Some seepage of groundwater into the conduit can be tolerated and is within design limits. The bottom of the conduit is sloped to allow for complete drainage, usually to a pump station.

In the case of sectional tile or concrete pipe, the joints must be waterproofed. The elimination of joints in the case of the poured-in-place concrete trench, whether single or double, makes the conduit almost immune to water penetration. Where high groundwater levels reach above the bottom of the conduit, a trench may be the only solution. Unlike the glazed tile or concrete pipe conduit, concrete trenches can be constructed to any size to contain a number of pipes for different services. The double trench is particularly effective in wet or damp areas; the space between the double conduit is used for primary drainage and is ordinarily filled with crushed stone. High labor costs reduce the practicality and effectiveness of concrete trenches.

8.7 Asbestos-Cement Casing with Internal Asbestos-Cement Pipe

This piping system is factory-fabricated and accommodates water between 35-200F (2-93C) at 150 psig

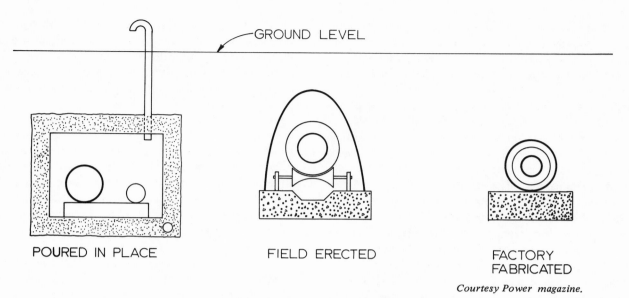

POURED IN PLACE   FIELD ERECTED   FACTORY FABRICATED

Courtesy Power magazine.

Figure 8-2. Glazed tile conduit.

(1034 kPa) pressure. It used push-on joints which consist of pipe joints and couplings as shown in Fig. 8-6 or bell and spigot ends that interlock by pushing one pipe into another. Introduced in the mid-1960s, it has achieved a certain degree of popularity. The epoxy-lined asbestos cement carrier pipe is very smooth, with a Williams-Hazen-roughness coefficient, C of 150. Insulation between pipe and asbestos-cement casing is close-celled urethane, foamed in place at the factory. The entire assembly is produced to the manufacturer's own standards but under rigid and exacting quality assurance procedures. Short pipe lengths (only 13 ft long) are joined by an asbestos-cement coupling with rubber gasket seals, as shown in Fig. 8-6. The coupling accommodates thermal expansion, gasket adjusting without sliding during expansion. Pipe sizes range from 1½″ to 30″ (38–762mm), so that the system is suitable for many installations.

This system is said to eliminate some of the thorniest problems of underground piping networks by providing: (1) high corrosion resistance, in that the asbestos-cement carrier and casing are immune to corrosion; (2) complete encasement of insulation, keeping it dry. Caps at both ends of each section with water and heat resistant end seals ensure impermeability and thermal efficiency; (3) elimination of expansion joints or loops because the rubber rings contain linear expansion by adjusting to it; (4) little friction and no chemical attacks due to the epoxy-lined inner surface of the carrier pipe.

The conduit can be placed directly in an earth trench, without a concrete foundation, and the trench back-filled. It has the greatest unit weight (pounds per linear foot) of all compared materials described in Sections 8.5, 8.8, 8.9 and 8.10 and, despite its easy assembly, the installed unit cost is relatively high. Fittings are not of the same material or construction but of cast iron with either ring or mechanical ends. Permissible leakage rates and longevity of rubber seals are questions to which the engineer must address himself. There is some field experience which must be evaluated before a final decision is made. Available pipe sizes are 3″ to 30″ (76–762mm).

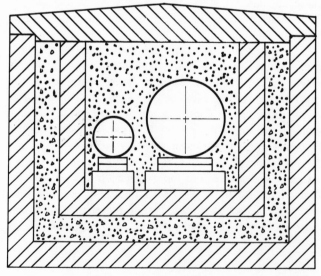

Courtesy Air Conditioning, Heating and Ventilating.

Figure 8-3. Double concrete trench.

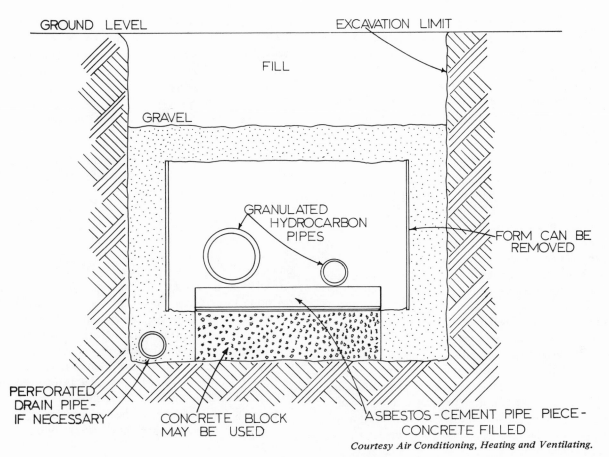

Figure 8-4. Single concrete trench.

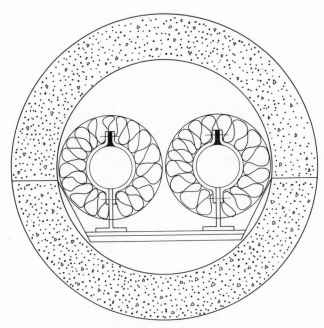

Courtesy Air Conditioning, Heating and Ventilating.

Figure 8-5. Insulated pipes in concrete casing.

8.8 Metallic Pipe with Integral Asbestos-Cement Casing

The basic difference between this piping system and the asbestos-cement pipe with integral conduit is that the carrier is steel pipe and, instead of rubber rings, there is a

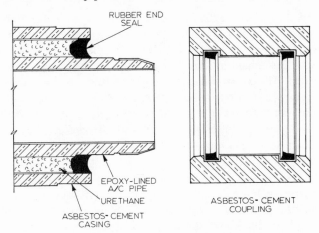

Courtesy Johns-Manville Corp.

Figure 8-6. Pipe ends are compatible with coupling and are machined to seal joint.

Teflon-covered stainless steel ring backed by a secondary O-ring seal. The core is a layer of calcium silicate and a second layer of polyurethane. The outer casing is asbestos cement pipe.

Introduction of this piping system could be a major development in the prefabricated, push-on joint piping assemblies, as the system is designed to carry HTW up to 450 F (232 C) and 500 psig (3445 kPa) operating pressure.

Groundwater that seeps into the end seal cavity flashes into steam and escapes to atmosphere through the casing. Fittings are of the same material and construction as the pipe.

Encasement of insulation, elimination of corrosion and absence of expansion loops are major advantages, according to the system vendors. Their contention that field labor is reduced has not been borne out; the difference in cost between conduit pipe systems and this system is not substantial. Actually, unit cost is fairly high and, when the system is selected, it will not be for economy.

Problems accentuated by high temperatures involved are increased linear expansion and resulting thrust forces, calling for thrust block construction, high pressures and the possibility of increased leakages or even rupture and deterioration of materials due to severe underground operating conditions. Since the product is fairly new, field experience and data are limited and no firm conclusions can be drawn.

Available pipe sizes are 3″ to 12″ (76-305mm), with a maximum heating load capacity in the vicinity of $200 \times 10^6 - 250 \times 10^6$ Btu/hr ($58.6 \times 10^3 - 73.3 \times 10^3$ kW).

8.9 Plastic Pipe with Integral Plastic Conduit (Push-on Joints)

Another less expensive version of the prefabricated integral conduit, push-on joint, piping system consists of a polyvinyl chloride (PVC) or fiberglass carrier pipe, PVC casing and integral, foamed-in-place urethane insulation. The assembled conduit units adhere to the manufacturer's own quality control standards and the materials ordinarily follow the ASTM specifications. The carrier pipe's smooth interior has a Williams-Hazen coefficient C, of 150. The manufacturers rating of 160 psig (1102 kPa) at 73 F (22.8 C) qualifies it for chilled and potable water services. The system is completely immune to corrosion but has limited capacity for withstanding substantial earth loads, the size of which depend on the depth of bury and the nature of the backfill material. Laying lengths are either 10′ or 20′ (3.1 + or 6.1 + m), depending on the manufacturer. Diameters range from 1½″ to 14″ (38-356mm). Unit weight is the lightest of

all comparable systems, which reduces installation costs.

Apparently this system is suitable for smaller installations, branch mains, and branches of larger installations. End seals are less efficient than those associated with asbestos-cement casings.

8.10 Field-Joined Pipes with Integral Conduits

This designation applies to different combinations of piping and casing materials, rather than to a specific combination. It is more a manufacturing technique than a specific assembly of materials.

The carrier pipe, whether of steel, iron, plastic, aluminum or asbestos cement, may be in 20′ or 40′ (6.1 or 12.2 m) lengths. The only limitation on pipe diameter is that imposed by the piping material itself. Depending on material, they vary from 1″ to 48″ (25-1219mm).

Insulation is bonded to the pipe, forming an anticorrosion layer. Over this, there is an intermediate layer of closed-cell urethane and, finally, a third layer of high density urethane. However, urethane properties vary, depending on the manufacturing process. Approximately 6 in. of uninsulated carrier pipe is left at the ends of each section for the field joint, be it welded or coupled, as in the cases of plastic pipes. Urethane insulation is foamed in place in the field and finished in the same manner as the factory-applied finish through special kits that all manufacturers provide.

The outer layer, if prefabricated PVC pipe, limits pipe diameters to 12″ (304mm). Another method, for pipes up to 48″ dia., involves the outer wrapping of the assembly with polyvinyl chloride-butyl rubber laminate tape or other appropriate materials.

Depending on pipe material, liquids up to 275 F (135 C) with jacket temperatures of up to 185 F (85 C) can be accommodated.

There is much to recommend these systems, especially those with PVC jackets. They are simple to install and immune to corrosion if the carrier pipe is non-ferrous, even if moisture penetrates the jacket. These are direct burial systems, occupying minimum trench space, and are comparatively lightweight, reducing the work in the field.

More field data are necessary for an evaluation of the long term behavior of factory wrappings under adverse conditions. Quality control of insulation and jackets is critical, so vendors must be closely questioned concerning factory facilities and past performance records. In addition, strict field quality control is required for adherence to specifications. Field joints are subject to field labor failures. Only strict control can ensure acceptable installations. One additional difficulty is encountered in

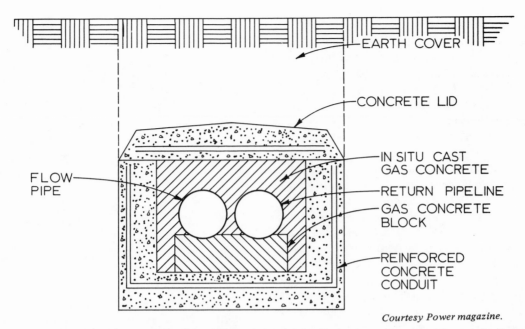

Figure 8-7. Typical cross section of pipe trench. If form is left in place, allowance must be made at bottom for drainage.

anchoring pipes and, to some extent, in guiding them. Plastic pipes and, to a lesser extent, metal pipes, experience considerable expansion, which tends to overstretch insulation and wrapping at elbows, creating the possibility of water penetration.

These piping systems hold promise because of their flexibility, ease of handling, economy and immunity to corrosion. They can be applied to chilled water and low to medium temperature water systems. Installed in shallow trenches, they can be replaced readily and can be used in areas where high water tables and ground moisture are prevalent.

8.11 Field-Joined Pipes with Poured-in-Place Insulation

Among the totally field-fabricated and field-erected systems are those that use a poured-in-place insulation, which also helps guide and support pipes, keeps moisture penetration to a minimum and offers cathodic isolation. The four most used insulation envelopes are granulated hydrocarbons, light concretes, urethane foam, and protexulate powder.

8.11.1 *Granulated Hydrocarbons*
The asphaltic hydrocarbons are load bearing, act as insulation and surround the carrier pipes, as in Fig. 8-7. It used to be necessary to cure the hydrocarbons around the pipe by applying heat through the pipe surface. Now, improved asphaltic hydrocarbons are supplied in a mixture of granule sizes such that the smaller granules fill

the voids left by the bigger ones, preventing water from reaching the pipe. Seeping rainwater can easily be kept out but groundwater requires a subdrainage tile that normally is pumped out at a convenient location, such as a manhole.

The material, which is found in nature, is available in various grades covering a temperature range from 35 to 520F (1.6–271C). If a chilled water line is to be located adjacent to a hot pipe in the same trench, two different grades must be used.

If manholes are not used, gravel vents must be installed every hundred feet or so.[5] Gravel vents are breathing pipes placed in the upper gravel layer of Fig. 8-7 and terminating in an elbow above ground. The vents allow moisture driven out by hot pipes to escape to atmosphere. If compacted to 49 lb/cu ft, the hydrocarbon will undergo no further compaction from its earth cover. Nevertheless, if the soil backfill above is less than 24 in. thick when compacted and is expected to carry vehicular or railroad loads, a reinforced concrete slab is absolutely essential as pictured in Fig. 8-7.[6]

Problem areas are created around elbows due to expansion, which leaves voids that collect rainwater. Chilled water applications are sensitive in that pipe surfaces can attract ground moisture through the porous granular mass.

[5]Steve Elonka, "Underground Piping Systems," *Power,* April, 1965, p. 222.

[6]William O'Keefe, "Hydrocarbon Granules for Pipe Insulation," *Power,* April, 1969, p. 87.

8.11.2 *Light Insulating Concretes*

All concrete casings, except the economy type, are surrounded by a PVC envelope that keeps water out and prevents cathodic attack. Insulating concrete has a high resistance to water migration, only 0.002 gal/sq ft/hr for average hydrostatic heads.[7] Vent tubes carry off any small quantities of water that are present, usually into manholes. Any vent leakage is a telltale indication of water entering the concrete casing. Pipe anchors are required at calculated intervals and provision for pipe expansion is made either through U-loops or expansion joints in manholes. Figure 8-8 illustrates pan construction that allows U-loop elbow movement that can be easily drained. Although this detail applies to concrete construction, it applies equally to asphaltic hydrocarbons. Smaller pipes can be supported by a 4 in. concrete sub-slab. Larger pipes, or a number of pipes, may need more substantial support structures. In order to replace a section of destroyed pipe, it is necessary to remove a section of concrete equal to one pipe length and then proceed to withdraw the destroyed section of pipe and reintroduce new pipe. The concrete envelope is subsequently repaired.

There are four basic types of light concretes.[8]

1. *Concrete Conduit for Government Service*

The specification and details of Fig. 8-9 are standards proposed by the Federal Construction Council for Government services that adopt this code, mostly for military installations.[9] The outer membrane consists of PVC tested for aging and reactant exposure. The membrane joints are made with heat resistant adhesives. The casing rests on a 4 in. thick concrete base. The insulating concrete is pumped into the PVC envelope and consists of Portland cement, water resistant vermiculite and waterproofing admixtures. Corrugated cardboard or other parting media, when attached to the outer surface of the pipes, allow free thermal expansion. Conservative government design provides four drain and vent pipes. The upper two are more for venting than draining purposes.

2. *Industrial Conduit*

Industrial casing design (Fig. 8-10) is a derivative of the government service application and is almost identical except for the two top vents which are omitted and

[7]George E. Ziegler, "Insulating Concrete Thermal Conduits," *ASHRAE Journal,* December, 1964, p. 29.

[8]*Ibid.,* p. 28.

[9]Academy of Sciences, Federal Construction Council Technical Report R30, Publication 660, 1959, National Research Council.

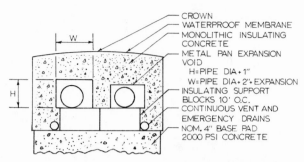

Courtesy ASHRAE Journal.

Figure 8-8. Section showing metal pan construction. Pipe can expand withing metal pan space.

the addition of a special waterproofing admixture of Portland cement and vermiculite to develop resistance to water migration.

3. *Economy Conduit*

The mass around the pipe can be self-supporting by adding a special water resistant vermiculite and maximum weight of Portland and liquid cement waterproofing admixture. This construction (see Fig. 8-11) offers the greatest load carrying capacity and resistance to water

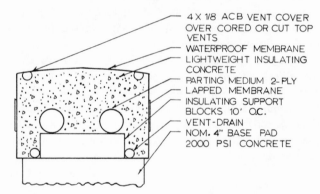

Courtesy ASHRAE Journal.

Figure 8-9. Section showing waterproof membrane and top and bottom vents.

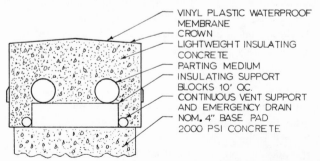

Courtesy ASHRAE Journal.

Figure 8-10. Section of industrial conduit.

penetration, eliminating the PVC envelope. The engineer has the option of providing a concrete support pad or a plastic rain shed at the top.

4. *Aggregate-asphalt Conduit*

The lightweight asphaltic concrete is a special high melt-point asphalt instead of Portland cement. It is applied to the aggregate hot. Brine temperature can be as low as 0F (-18C) and hot water up to 250F (121C). This insulating concrete is particularly suitable for both chilled and hot water systems. (See Fig. 8-12.)

8.11.3 *Urethane Foam*

A newer technique for underground piping is to pour or spray in place rigid urethane foam, in densities of 2-3 lb/cu ft (32-48 kg/cu m), to form a monolithic insulating envelope. It is best to apply in an above 60F (15.6C) ambient or pipe temperature, as temperature affects density, and that it be poured into a dry cavity. Pour-in-place machines mix the two components and drop the liquid to froth in place and expand to thirty times its original volume. Urethane sets in seconds, cures in minutes and is ready for immediate backfill, which must be free of rocks and must be placed carefully over the sides and not compacted to more than 25 psi (172 kPa) at the casing surface. A parting agent is placed around the pipe to provide for thermal expansion and the entire pour is wrapped in a plastic membrane. Internal vent drains are necessary for keeping the insulation dry. Thermal expansion at U-loop elbows, up to 2-3 inches (50–75mm), is allowed by wrapping pipe with cellular plastic material. For a limited size and number of pipes, the foam may be poured directly into a carefully excavated cavity lined with a plastic membrane. For pipes larger than 4 in. (102 mm) it is advisable to provide a 4 in., concrete sub-slab of 2000 psi (13 780 kPa) minimum compressive strength. See Fig. 8-13.

Urethane is the most efficient insulator of all commercially available materials, as shown below[10]:

Urethane foam (1.8 pcf)	k = 0.17 Btu/hr ft² (F/inch) at 100F mean temperature	24.5 W/m² deg C
Calcium silicate (15 pcf)	k = 0.38	54.7
Cellular glass (8.5 pcf)	k = 0.42	60.5
Glass fiber (7.5 pcf)	k = 0.26	37.4
Polystyrene (1.8 pcf)	k = 0.27	38.9
Insulating Concrete (35 pcf)	k = 0.75	108.0

[10] ASHRAE, *Handbook of Fundamentals*, 1977, Table 3B, p. 22.17.

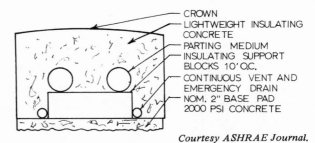

Courtesy ASHRAE Journal.

Figure 8-11. Section of economy conduit.

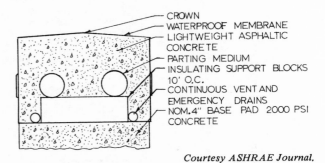

Courtesy ASHRAE Journal.

Figure 8-12. Section of aggregate asphalt conduit.

Effective temperature range is $-$ 300 to $+$ 250F ($-$ 184 to + 121C). An unbroken vapor barrier must be provided for chilled water applications as urethane loses its effectiveness on becoming moist.

8.11.4 *Protexulate Powder*

Calcium carbonate finely ground and coated with aliphalic acid becomes protexulate, a loose fill hydrophobic powder having a thermal conductivity equal to 0.78 Btu/hr ft² (F/inch) (0.112 W/m² deg C), a bulk density of 60 pcf, resistivity equal to 10^{14} ohms/cm/cm², a load bearing strength of 11,000 lb/ft² (53 680 kg/m²) and excellent dielectric properties.

The material is poured into the trench, most preferably in a PVC envelope, so it does not disperse, and on top of a 4 in. (102 mm) concrete base.

A National Bureau of Standards test[11] indicates that temperatures of up to 350F (177C) are safe, that at 400F (204C) there is smoke and discoloration and, at 450F (232C), smoke and a slight weight loss. The test report states that the material retained its insulation qualities in a wet underground environment for 200 days and in another instance when it was tested at 400F (204C) in a wet trench for 8 months, it retained its hydrophobic characteristics intact.

[11] T. Kusuda and W.M. Ellis, "Report Test to Tri-Services, Laboratory Investigation of Three Underground Systems," U.S. Department of Commerce, National Bureau of Standards, 1974.

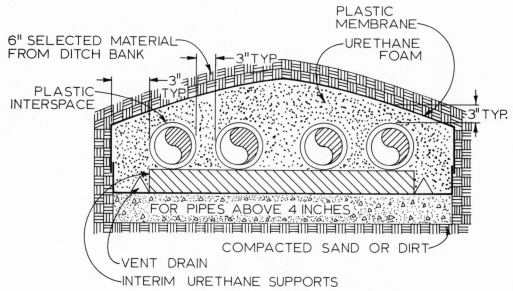

Figure 8-13. Typical section for chilled water.

A United States Air Force engineering report [12] on the material concludes that protexulate powder is an effective insulator and can be used for repairing underground heat distribution systems. Furthermore, it prevents corrosion and makes separate cathodic protection unnecessary. The report states that the insulating quality of the material can be improved by using preformed calcium silicate insulation over the hot pipes prior to the addition of protexulate to improve the thermal quality of the insulation system.

8.12 Underground Piping Heat Transfer

For aesthetic reasons and to avoid interference with traffic on the surface, most central plant distribution piping runs underground. For economy, direct burial is the most common system.

How much insulation, if any, to provide has a major effect on capital cost. The decision is based on a prediction of heat gain or loss to the earth over the life of the project. Although it may take years to reach steady-state heat exchange between piping and the earth, a point is reached where the surrounding earth has either cooled down or warmed up and further changes in temperature differential are insignificant. Equations 8-1 through 8-4 enable the engineer to predict the heat transfer with moderate effort.

[12] T. F. Lewicki, "Use of a Hydrophobic Loose Fill Powder to Insulate and Waterproof Underground Heat Distribution Pipe," Engineering Report AFCEC ER 74-9, Air Force Civil Engineering Center, Tyndall Air Force Base, United States Air Force, 1974, p. 5.

Figure 8-14 represents the cross section of an insulated cement asbestos carrier pipe with an outside cement asbestos casing.

These symbols are used in deriving the equations that follow:

α = Thermal diffusivity, soil $= \dfrac{K_s}{(\rho c)_s}$, ft^2/hr.

ρ = Density, lb/ft^3.

μ = Absolute viscosity, lb/ft-sec.

c = Specific heat, Btu/lb-deg F.

h_i = Inside film coefficient of carrier pipe, Btu/hr-ft^2-deg F.

K_M = Conductivity of any piping material, insulation, carrier pipe or casing, Btu/hr-ft^2(deg F/linear ft).

K_s = Soil conductivity, Btu/hr-ft^2(deg F/linear ft).

L = Linear length of pipe, ft.

Q = Heat transfer rate, Btu/hr-linear ft.

R_i = Inside film resistance $= 1/h_i$, hr-ft^2-deg F/Btu

r_i = Inside radius of carrier pipe, inches or feet.

r_2 = Outside radius of carrier pipe, inches or feet.

r_3 = Inside radius of casing, inches or feet.

r_o = Outside radius of casing, inches or feet.

R_c = Resistance of asbestos cement casing, hr-ft-deg F/Btu.

R_M = Sum of the carrier pipe, insulation and casing resistances, hr-ft-deg F/Btu.

R_p = Resistance of asbestos cement carrier pipe, hr-ft-deg F/Btu.

R_T = Total resistance = $R_i + R_M$, hr-ft-deg F/Btu.

T_i = Temperature at radius r_i, inside film temperature, deg F.

T_o = Temperature at radius r_o, outside casing surface temperature, deg F.

T_s = Soil temperature, deg F.

T_w = Hot water, chilled water or steam temperature in pipe, deg F.

U_p = Overall heat transfer coefficient of entire underground pipe per linear foot of pipe, Btu/hr-deg F-linear ft.

V = Linear velocity, ft/sec.

From a conservation of energy point of view, heat lost by the pipe equals heat picked up by the ground and vice versa. Heat travelling from the interior of the pipe to the ground must overcome the combined resistance (R_T) of the interior film, carrier pipe, insulation and casing.

$$R_T = R_i + R_T$$

Referring to Fig. 8-14, the resistance per linear foot of pipe of the inside film, carrier pipe, insulation and casing is as follows:

$$R_i = \frac{1}{2\pi r_i (h_i)}$$

$$R_p = \frac{T_2 - T_i}{\frac{\ln\frac{r_2}{r_i}}{2\pi K_p}}$$

$$R_{ins.} = \frac{T_3 - T_2}{\frac{\ln\frac{r_3}{r_2}}{2\pi K_{ins.}}}$$

$$R_c = \frac{T_o - T_3}{\frac{\ln\frac{r_o}{r_3}}{2\pi K_c}}$$

To find heat transfer due to conduction through inside film, carrier pipe, insulation and casing:

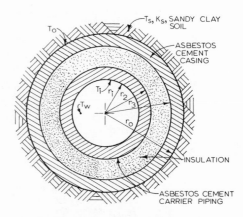

Figure 8-14. Cross-section of underground pipe of Example 1 and illustration of symbols used.

$$Q = \frac{T_i - T_o}{2\pi r_i h_i + \frac{\ln\frac{r_2}{r_i}}{2\pi K_p} + \frac{\ln\frac{r_3}{r_2}}{2\pi K_{ins.}} + \frac{\ln\frac{r_o}{r_3}}{2\pi K_c}} = U_p(T_w - T_o) \tag{8.1}$$

Example 1 will demonstrate that resistance through the inside film is insignificant and that resistance through the carrier pipe and casing amounts to approximately eleven per cent of total resistance, increasing heat transfer by that amount, if omitted from Equation 8.1. Hence, $T_w = T_i$.

Under these conditions Equation 8.1 may be rewritten as

$$Q = \frac{T_w - T_o}{\frac{\ln\frac{r_3}{r_2}}{2\pi K_{ins.}}} \tag{8.2}$$

To find heat transfer due to soil conduction[13]:

$$W = K_s (T_s - T_o) F(z) \tag{8.3}$$

And,[14]

$$z = \frac{\alpha t}{r_0^2}$$

[13] L.R. Ingersoll *et al.* "Theory of Earth Heat Exchangers for the Heat Pump," *Transactions of the American Society of Heating and Ventilating Engineers* (ASHVE now ASHRAE), Vol. 57, 1951, p. 172.

[14] *Ibid.*

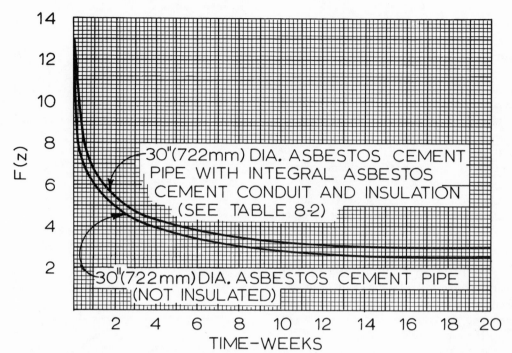

Figure 8-15. A graphic representation of Table 8-2, which is a worksheet for the asbestos cement pipe with integral casing and insulation. The 30 in. (762 mm) dia plain asbestos cement pipe is plotted only for purposes of comparison.

Equation 8.2 and Equation 8.1 which requires more work equals Equation 8.3, Hence:

$$\frac{T_w - T_o}{\dfrac{\ln \dfrac{r_3}{r_2}}{2\pi K_{\text{ins.}}}} = K_s(T_s - T_o)F(z) \qquad (8.4)$$

At this point a word of caution is necessary. Both sides of the equation must be tested out to determine which is numerically smaller. The smaller number governs. Usually, if the underground pipe is insulated the left side governs, since heat transfer is numerically small and can be readily absorbed by the earth. If the pipe is insulated and pipe outside surface temperature is high, the right side of the equation governs in most instances, since the earth cannot absorb more than its resistance will allow it to absorb under steady state conditions.

Values of $F(z)$ are given in Table 8-1. The heat transfer rate between ground and buried pipe is a function of z and it is a matter of plotting $F(z)$ against time to determine when steady state conditions are reached. A practical manner for doing so is to construct Table 8-2. Diffusivity values may be sought in heat transfer or civil engineering texts. Ordinarily K_s, ρ_s and c_s may be sought

Table 8-1

Values of $F(z)$ (See Eq. 8-3)

z	$F(z)$	z	$F(z)$	z	$F(z)$
0.01	38.508	0.70	6.926	100	2.172
0.02	28.096	0.80	6.634	120	2.114
0.03	23.474	1.00	6.180	150	2.042
0.04	20.716	1.2	5.844	200	1.956
0.05	18.829	1.5	5.467	250	1.893
0.06	17.433	2.0	5.029	300	1.846
0.07	16.351	2.5	4.726	400	1.775
0.08	15.472	3.0	4.500	500	1.721
0.09	14.744	4	4.174	600	1.683
0.10	14.130	5	3.947	800	1.622
0.12	13.135	6	3.776	1000	1.578
0.14	12.361	8	3.529	1200	1.543
0.16	11.737	10	3.354	1500	1.502
0.18	11.217	12	3.224	2000	1.454
0.20	10.777	15	3.074	2500	1.418
0.25	9.916	20	2.898	3000	1.393
0.30	9.277	25	2.774	4000	1.350
0.35	8.779	30	2.678	5000	1.319
0.40	8.373	40	2.539	7500	1.266
0.45	8.037	50	2.440	10000	1.233
0.50	7.751	60	2.363	15000	1.184
0.60	7.288	80	2.251	25000	1.131

separately and combined into the diffusivity formula given in the symbols of this section.[15] The project soils engineer may be in a position to furnish this information. Once the value of diffusivity is obtained, plot $z = \alpha t/r_0^2$ for values of t selected from Table 8-2 and complete Table 8-2. Figure 8-15, which is a graphic representation of Table 8-2, may now be drawn.

This heat transfer method does not account for surface effects, i.e., the distance of the buried pipe from the earth's surface and its effect on heat transfer or the effect of other buried pipes in the immediate vicinity. Some references explore certain aspects of surface effects and proximity of other pipes.[16,17,18,19,20]

In estimating the heat transfer rate, the problem amounts to defining T_o at various time intervals. The solution involves using Equation 8.4 or Equation 8.1 and solving for T_o at time $t, t+1, t+2 \ldots t_{n-1}, t_n$. Example 1 illustrates a step by step procedure for determining various characteristics of underground heat transfer.

Example 1: 30 inch (762 mm) dia. asbestos cement pipe with integral insulation and asbestos cement casing (see Fig. 8-16) carrying water flow that varies from 1500 gpm (95 l/s) to 15,000 gpm (950 l/s); water temperature at 40F (4.4C) and soil temperature at 60F (15.5C); soil is sandy clay at 15% moisture, $K_s = 0.53$ Btu/hr-ft²(deg F/linear ft) (0.076 W/m-deg C), $\alpha = 0.015$ ft²/hr (0.039 × 10⁻⁵ m²/s).
Find:
(a) The time at which the system arrives at steady state conditions.

(b) The heat transfer rate from ground to pipe at $t = 24$ hours.
(c) Minimum acceptable flow to prevent laminar flow.
(d) How does a pipe carrying 400F (204C) water compare in terms of heat transfer with both insulated and uninsulated chilled water pipes.

(a) In order to work out Table 8-2, first determine α, if not given, and then substitute various values of t in the expression for z.

Steady state conditions, with or without insulation, are substantially established in approximately eight weeks. The first three weeks after startup are the most crucial and the period during which most energy loss occurs.

(b) In order to find the heat transfer rate between ground and insulated pipe, it is necessary to evaluate the significance of the inside film resistance and the numerical value of the overall heat transfer coefficient, U_p, as shown in Equation 8.1. Using Equation 8.1 and Equation 8.4 we can find T_o and Q at various time intervals.

In order to use Equation 8.1 it is necessary to calculate the inside film resistance, carrier pipe resistance, insulation resistance and casing resistance. The first step in calculating the inside film resistance is to find the numerical value of the inside film coefficient versus flow rates. For water temperatures of 40F (4.4C) to 220F (104.4C) Equation 8-5 can be used or Fig. 8-17, its graphical representation.

[15] For values of conductivity and density of some soils see G.S. Smith and Thomas Yamauchi. "Thermal Conductivity of Soils for Design of Heat Pump Installations," *Transactions of the American Society of Heating and Ventilating Engineers* (ASHVE now ASHRAE), Vol. 56, 1950, 355-370.

[16] L.R. Ingersoll and H.J. Plass. "Theory of the Ground Pipe Heat Source for the Heat Pump," *Transactions of the American Society of Heating and Ventilating Engineers* (ASHVE now ASHRAE), Vol. 54, 1948, 339-348.

[17] E.W. Guernsey, P.L. Betz and N.H. Skau. "Earth as a Heat Source or Storage Medium for the Heat Pump," *Transactions of the American Society of Heating and Ventilating Engineers* (ASHVE now ASHRAE), Vol. 55, 1949, 321-335.

[18] W.A. Hadley. "Operating Characteristics of Heat Pump Ground Coils," *Edison Electric Institute Bulletin*, Vol. 17, No. 12, December 1949, 437-461

[19] Donald M. Vestal, Jr. and Billie J. Fluker. "The Design of a Heat Pump Buried Coil," Texas Engineering Experiment Station, College Station, Tex., Bulletin 139, 1957.

[20] *Thermal Tables for Underground Heating and Cooling Systems*. Johns-Manville Corp.: Denver, Colo, 1975.

Table 8-2
Time vs. Transient Heat Transfer Function, $F(z)$

t (Hours)	$z = \dfrac{\alpha t}{r_0^2}$	$F(z)$
24	0.12	13.135
48	0.24	10.088
72	0.36	8.698
168 (1 week)	0.85	6.525
336 (2 weeks)	1.70	5.292
504 (3 weeks)	2.55	4.703
672 (4 weeks)	3.40	4.370
1008 (6 weeks)	5.10	3.930
1344 (8 weeks)	6.80	3.678
1680 (10 weeks)	8.50	3.463
2520 (15 weeks)	12.75	3.189
3360 (20 weeks)	17.00	3.004

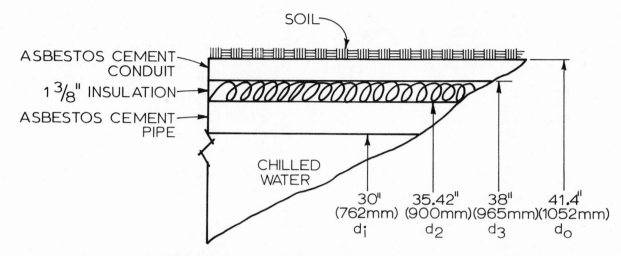

Figure 8-16. Detailed section of pipe shown in Figure 8-14 as it applies to Example 1.

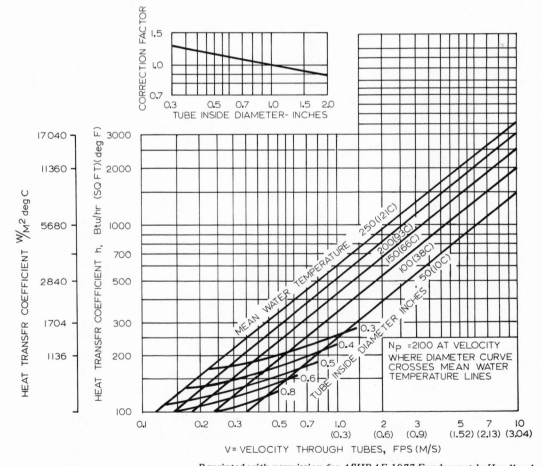

Reprinted with permission for ASHRAE 1977 Fundamentsla Handbook.

Figure 8-17. Heat transfer coefficient for turbulent flow inside pipes.[21]

[21] ASHRAE *Handbook of Fundamentals*, 1977, p. 2.16.

$$h_i = \frac{150(1 + 0.011T)(V)^{0.8}}{(d_i)^{0.2}} \qquad (8.5)[22]$$

The following tabulation indicates the variation of h_i versus flow rate.

Flow rate, gpm (l/s)	Velocity, fps (m/s)	Inside film coeff., h_i Btu/hr-ft²-deg F (W/m²-deg C)
1500 (95)	0.66 (0.202)	120 (682)
4500 (248)	1.99 (0.608)	280 (1590)
9000 (568)	3.99 (1.216)	490 (2783)
12000 (757)	5.32 (1.621)	630 (3578)
15000 (946)	6.65 (2.026)	770 (4377)

The resistances are as follows:

$$R_i = \frac{1}{2\pi r_i(h_i)} = \frac{1}{2\pi(15/12)\,120}$$

$$= 0.001 \frac{ft^2\text{-hr-deg F}}{Btu} \left(0.176 \frac{m^2 \deg C}{W}\right)$$

@ $h_i = 120$ (negligible)

$$R_i = \frac{1}{2\pi(15/12)\,770} = 0.0001 \text{ @ } h_i = 770 \text{ (negligible)}$$

$$R_{asb.cement pipe} = \frac{\ln(r_2/r_i)}{2\pi K_{asb.cem.}} = \frac{\ln(35.42/30)}{2\pi\,5.5/12} = 0.075 \quad (0.0132)$$

$$R_{insul.} = \frac{\ln(r_3/r_2)}{2\pi K_{ins.}} = \frac{\ln(38/35.2)}{2\pi\,0.178/12} = 0.828\,(0.145)$$

$$R_{asb.\ cement\ conduit} = \frac{\ln(r_0/r_3)}{2\pi K_{asb.cem.}}$$

$$= \frac{\ln(41.4/38)}{2\pi\,5.5/12} = 0.024\,(0.00422)$$

$F(z)$ at 8 weeks = 3.678 (Table 8-2)

$$R_{soil} \text{ at 8 weeks} = \frac{1}{K_s \times F(z)} = \frac{1}{0.53 \times 3.678} = 0.513$$

$$R_{total} = 0.001 + 0.075 + 0.828 + 0.024 = 0.928$$

$$U = \frac{1}{R} = 1.08 \text{ Btu/ft}^2\text{-hr-deg F (6.13 W/m}^2 \deg C)$$

By comparing resistances, it becomes apparent that only soil and insulation resistances are of any significance. It

[22] William H. McAdams, *Heat Transmission,* 3rd ed. (New York: McGraw-Hill Book Company, 1954), p. 228.

should be noted that h_i in Equation 8.5 is very insignicant, and regardless of the fact that its value varies between 120 and 770, heat transfer to the ground remains unaffected.

Equation 8.4 may be rewritten as follows:

$$Q = U(T_o - T_w) = K_s F(z)(T_s - T_o)$$
$$T_s = 70F (21.1C)$$
$$T_w = 40F (4.5C)$$
$$K_s = 0.53 (0.76)$$
$$F(z) = 13.135, \text{Table 8-2}$$

Substituting, we obtain

$$1.08(T_o - 40) = 0.53(13.135)(70 - T_o)$$
$$T_o = 65.96F (18.86C)$$

Solving for Q, we obtain $Q = U(T_o - T_w) = 1.08 \times (65.96 - 40) = 28.03$ Btu/hr-linear ft (27 W/linear meter).

Figure 8-19 illustrates not only a composite pipe but, also, bare steel and asbestos cement pipes for comparison purposes, plotted for a period of 20 weeks.

(c) To find what is minimum acceptable flow, we can test for 1500 gpm (95 l/s), which is the minimum design flow, by investigating if Reynolds number is greater than 2300, the limit for laminar flow.

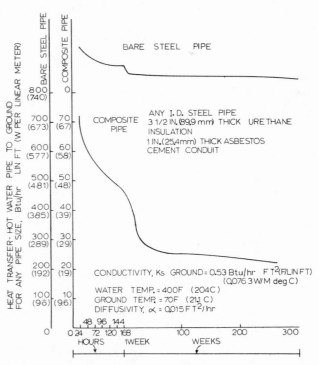

Figure 8-18. Heat transfer to ground from 12 in. (305 mm) dia high temperature water line.

Cross sectional area of 30 in. dia pipe = 4.9 ft^2

Although minimum N_{RE} = 3000 for turbulent flow, the area between 3000-10,000 is still one of uncertainty; it is best that minimum N_{RE} = 10,000.

$$N_{RE} = 10000 = \frac{\rho VD}{\mu} = \frac{62.4 \text{ lbs/ft}^3 \times V \times 2.5 \text{ ft}}{1.04 \times 10^{-3}}$$

Dynamic viscosity μ = 1.04 lb mass/ft sec at 40 F
V at N_{RE} = 10000 = 0.066 feet per second
Q minimum flow in gallons/minute

$$= \frac{0.066 \dfrac{\text{ft}}{\text{sec}} \times 4.9 \text{ ft}^2 \times 60 \dfrac{\text{sec}}{\text{min}} \times 62.4 \dfrac{\text{lb}}{\text{ft}^3}}{8.33 \dfrac{\text{lb}}{\text{gal}}}$$

Q_{min} = 145.35 gpm (9.2 l/s)

A minimum flow of 145.35 gpm (9.2 l/s) is in the turbulent range and is acceptable. Laminar flow has the potential disadvantage of temperature stratification and uneven chilled water supply temperature to various branches.

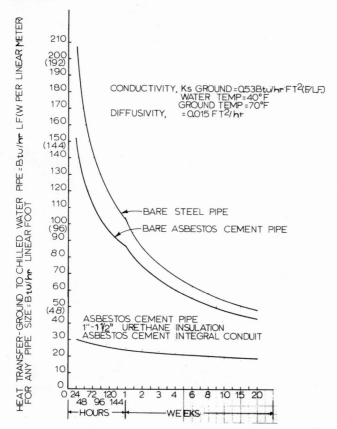

Figure 8-19. Heat transfer from ground to 30 in (762 mm) dia chilled water lines.

(d) Heat transfer rates for a hot water pipe buried in the ground are illustrated in Fig. 8-18 for both a bare steel pipe as well as for a composite pipe. The basis for the calculations are Formulas 8.1 through 8.8 used for a 12 in. (305 mm) dia. pipe.

It will be noted that heat conduction through the pipe wall and insulation depends only on the ratio of the radii and is independent of cylinder diameter. The right hand side of Equation 8.5 indicates that heat transfer depends on time, conductivity and a constant temperature difference. Function $F(z)$ indirectly depends to some extent on the inverse of r_o^2. The earth has a large thermal inertia. Regardless of the efficiency of a piping assembly, heat transfer is almost the same for quite a range of pipe sizes. Larger pipes simply transfer less heat per unit area. Figure 8-19 holds for pipes of 24, 20 and 18 inch dai. (607, 508 and 457 mm) which the author calculated. Undoubtedly the same holds true for Fig. 8-18 although no calculations were performed, except for 12 (305 mm) inch dia. It goes without saying that smaller buried pipes, when used as a heat source, are more efficient for heat pump applications.

For the example above, the temperature of the earth surrounding the pipe was assumed to be 70 F (21.1 C). Ordinarily, the engineer must define ground temperatures and duration of each temperature bin. L.R. Ingersoll and other sources[23],[24] treat this subject as well as temperature gradients caused by a heat source or influenced by earth surface conditions. This subject assumes importance because the depth at which pipes are buried affects excavation and backfilling costs and heat dissipation at the earth's surface, where grass or other vegetation growth can be hampered by excessive temperatures. Heat transfer is distinctly influenced by the type of soil surrounding the pipe and moisture conditions. Underground water movement in excess of 0.01 ft per hr (8.46 × 10^{-4} mm/s) has unpredictable results on heat transfer.[25] A wet environment calls for enclosing of pipes in concrete casings or tunnels for complete protection. Unfortunately, no mechanical or electronic device exists for measuring underground water velocity. Civil engineers (soils specialists or hydrologists) calculate underground water velocity by using the Darcy Weisbach formula and the flow net method, knowing topography, difference in elevation be-

[23] L.R. Ingersoll, *op. cit.*

[24] T. Kusuda and P.R. Achenbach, "Earth Temperature and Thermal Diffusivity at Selected Stations in the United States," *Transactions of the American Society of Heating, Refrigerating and Air Conditioning Engineers* (ASHRAE), Vol. 71, 1965, 61.

[25] L.R. Ingersoll, *ibid*, p. 186.

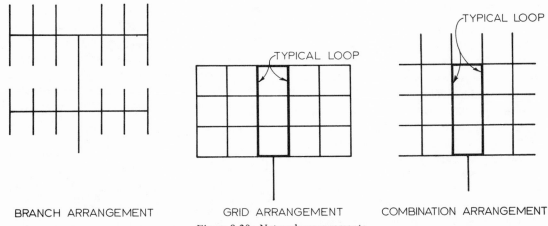

BRANCH ARRANGEMENT   GRID ARRANGEMENT   COMBINATION ARRANGEMENT

Figure 8-20. Network arrangements.

tween water source and the location being examined, the type of soil and its density and other relevant factors.

Development of heat transfer data is important in a cost analysis of direct burial of insulated or uninsulated pipes. The question revolves around the advisability of omitting insulation, the effects of this on long term pipe preservation, the economic penalty of losing or gaining heat to or from the ground and increase or decrease in the size of boilers or chillers. Many engineers omit insulation when burying chilled water pipes, although few do in the case of hot pipes. An economic study will prove which arrangement is most beneficial.[26] One must include backfilling, dewatering, sheathing, manholes, expansion loops and other pertinent items for a complete study. In heating applications, it is almost axiomatic that insulation is necessary. Figure 8-19 shows that heat losses from bare steel pipe are intolerable.

8.13 Distribution Networks

Distribution systems fall into three main configurations: branch, grid and combination arrangements illustrated in Fig. 8-20. The grid is preferred to a branch system because each point can be supplied from at least two directions. Ordinarily, sectioning valves are strategically located so that, in case of a line break or shutdown for repairs, services to any building can continue from at least one direction.

A branch arrangement is not this flexible since all branches terminate in dead ends. One more basic advantage of a grid system is that a main is not required to circulate the entire water volume required by the buildings or points connected to it but only a portion, since

any point can be fed from two directions. Substantially smaller pipes may be expected at considerable economy. Conversely, all points on a branch system are fed sequentially, so that flow is cumulative, requiring larger pipes. At this point, we must explain that the size of a branch line can be sizable since it has the capacity to feed several buildings or processes. Depending on arrangement and layout, it is difficult to predict which arrangement demands longer pipelines. The National District Heating Association has investigated twenty-four distribution patterns and has assigned relative piping costs to each.[27]

System configuration is determined by many factors such as physical arrangement of buildings, topography, height of buildings, location of central plant, degree of site development, construction staging and economics.

Design of large distribution systems follows some basic rules: on mountainous or hilly sites, several distribution zones may serve the purpose of separating high pressure from low pressure zones, to avoid imposition of needlessly high hydrostatic heads on the low lying districts. The same may hold for a district that includes both tall and low buildings. Separation barriers must be established. In general, pressures must be adequate to guarantee flow requirements. Excessive pressures increase the potential for leakage. The most economical sizing is based on the ability of the system to accommodate its simultaneous loadings. Any future system expansion must be accounted for in the original analysis. Due to the high capital outlay for large systems, design based on incomplete data is very risky.

[26]*Underground Thermal Distribution Systems (35 to 250F), op. cit.* pp. 1–62.

[27]*District Heating Handbook,* 3rd ed. (Pittsburgh, 1951), pp. 153-154.

8.14 Hydraulic Design of Primary and Intermediate Water Loops

Almost all primary water systems are designed for variable flow. Constant primary flow, although used occasionally on smaller systems, is very wasteful of energy and is, therefore, indefensible.

Correct system design consists, first, in establishing individual building heating or cooling peak loads. It is necessary to establish diversity for the entire project, which helps in selection of principal equipment and to establish flow requirements, pipe sizes, and pumping capacity for each section of the project. Additional diversity is created in each project not only by variations of outdoor or indoor conditions but by the particular loads imposed by different types of structures, whether residences, industrial or office buildings, business centers, etc., by peaking at different times, having diverse demand and occupying distinctly strategic locations on the site.

Design concepts and calculations must account for these variables. Unfortunately, field data are not available to confirm theoretical calculations as there are in town water distribution networks. This is more reason for exerting greater care in establishing parameters and concepts for the projects. Computer technology now makes it possible to run most of these calculations on specially designed programs with a sophistication that was unavailable some time ago.

There are two basic methods for pipe network analysis. The equivalent pipe method, which can be used by anyone dealing with simple piping systems, and the Hardy Cross method, which can be calculated for smaller systems or can be run on a computer for the more complicated ones. The Newton-Raphson and linear theory methods are primarily intended for large water works and are beyond the scope of this book.

8.14.1 *The Equivalent Pipe Method*

In this method, a system of different pipes is replaced by one equivalent pipe which has the same hydraulic characteristics as the system of pipes it replaces. *Example:* For the piping system illustrated in Fig. 8-21, (a) What is the equivalent length of 14 in. (356 mm) dia pipe for branch *ABC*? (b) What is the equivalent length of 24 in. (610 mm) dia pipe for the sum of branches *ABC* and *ADC*? (c) What is the equivalent length of 30 in. dia (762 mm) pipe for the entire system?

(a) Replace *ABC* with 14″ dia pipe. Assume a flow of 10 cfs (283 l/s). From the Hazen-Williams Nomograph, Fig. 8-22:

$$\text{Friction loss of } 18'' = 9.4'/1000'$$
$$12'' = 70.0'/1000'$$

$$\text{Friction } ABC = 9.4 \times \frac{300}{1000} + 70 \times \frac{700}{1000}$$
$$= 51.82' \ (155 \text{ kPa})$$

Friction loss of 14″ @ 10 cfs = 33.3′/1000′ (98.6 kPa/ 305m)

$$\text{Length of } 14'' = \frac{51.82}{33.3}(1000) = 1556' \ (474 \text{ m})$$

(b) Replace *ABC* and *ADC* with one 24″ (610 mm) dia pipe.

Assume a friction drip of 10 ft (29.9 kPa) between points *A* and *C*.

Using Fig. 8-21 a and Fig. 8-22:

Friction loss of 14″ (356 mm) branch *ABC*

$$= \frac{1000}{1556} \times 10 = 6.42'/1000' \ (19.1 \text{ kPa}/305 \text{ m})$$

Friction loss of 20″ (508 mm) branch *ADC*

$$= \frac{1000}{1200} \times 10 = 8.33'/1000' \ (248 \text{ kPa}/305 \text{ m})$$

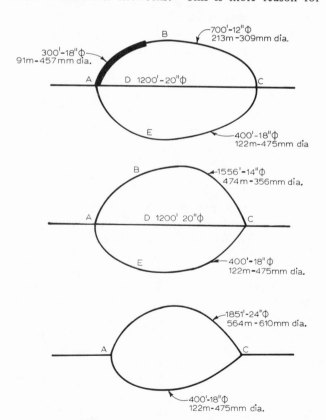

Figure 8-21. Middle drawing is equivalent pipe of top; bottom drawing is equivalent pipe of middle.

Friction loss of 18″ (4.57 mm) branch *AEC*

$$= \frac{1000}{400} \times 10 = 25.0'/1000' \ (74.5 \ kPa/305 \ m)$$

Flow in branch *ABC* = 4.3 cfs (122 l/s)
Flow in branch *ADC* = 12.4 cfs (351 l/s)
Flow in branch *AEC* = 17.3 cfs (490 l/s)
Total flow = 34.0 cfs (962 l/s)
Flow in *ABC+ADC* = 16.7 cfs (473 l/s)

Friction loss of 24″ dia @ 16 cfs = 5.4'/1000' (16.1 kPa/305 m)
Equivalent 24″ (610 mm) dia pipe to replace 14″ (356 mm) and 20″ (508 mm) dia pipes

$$= \frac{10}{5.4} \times 1000 = 1851' \ (570 \ m)$$

(c) Replace *ABC, ADC* and *AEC* with one equivalent 30″ (762 mm) pipe.
Friction loss of 30″ @ 34 cfs = 7.6'/1000' (215 kPa/305 m)

Equivalent length of 30″ pipe $= \frac{10}{7.6} \times 1000 = 1315'$ (401 m)

Figure 8-22 is based on *C* = 100, which is for fairly rough pipe. There are other nomographs and tables which can be used with other friction factors.[28,29,30]

8.14.2 *The Hardy Cross Method*[31]

The equivalent pipe method is limited in that common pipes and multiple outflow points are unsurpassable obstacles for this method but offer little difficulty for the Hardy Cross approach, which can analyze complex networks at different flow conditions.

The application of the Hardy Cross method to hydraulic circuits is analogous to Kirchhoff's laws for electric circuits. The Hardy Cross method assumes that (1) flow into a junction is equal to flow out of the junction, and (2) the pressure at a junction is the same for all pipes. Referring to Fig. 8-21a

$$Q_A = Q_{ABC} + Q_{ADC} + Q_{AEC} = Q_C$$

$$\therefore h_{f\,ABC} = h_{f\,ADC} = h_{f\,AEC}$$

[28] Fluid Flow, Louisville, Tube Turns, a Division of National Cylinder Gas Co., 1951, p. 13 and p. 21.
[29] Flow of Fluids, Technical Paper No. 409, Chicago, Crane Co., 1942, (Figs. 48–51) pp. 77–83 and (Figs. 42–45) pp. 65–71.
[30] *Hydraulic Institute, Pipe Friction Manual.* 3rd edition. Hydraulic Institute: New York (now in Cleveland, Ohio), 1961.
[31] Hardy Cross, "Analysis of Flow in Networks of Conduits or Conductors," University of Illinois, Bulletin 286 (November, 1936).

A flow quantity and direction is assumed for every pipe of the network. It is a relaxation method which, by repeated trials, applies corrections to the assumed values until all values stabilize. A considerable amount of labor and accuracy are required to obtain dependable results in a reasonable number of trials. It is very important that the initial assumed values and flow directions be as close to the final values as possible so as to reduce the number of trials. For very large systems, it is common practice to skeletonize the system or to reduce a number of pipes to an equivalent pipe in order to rationalize system complexity. It is important to recognize that the Hardy Cross technique is mostly applicable to large water works.

The Hazen-Williams formula applies to water at about 60F (15.6C) or other liquids having a kinematic viscosity of about 1.1 centistokes.

The Hazen-Williams formula is given by

$$h = \frac{3.023 \ V^{1.852}}{C^{1.85} \ d_i^{1.167}} \qquad (8.6)^{[32]}$$

V = Average flow velocity, ft/sec
d_i = Inside diameter of pipe, ft
h = Unit loss in head, ft of fluid per foot of length
C = Constant accounting for surface roughness, 60-150 range.

Equation 8.6 by manipulation can assume the form

$$H_f = KQ^n \qquad (8.7)$$

H_f = Friction head in a section of pipe, ft
K = Constant or a particular piece of pipe, of a given diameter; may be $K_1, K_2, K_3 \ldots$ etc.
n = Hazen-Williams constant, 1.85

The nomograph of Fig. 8-22 is based on a conservative Hazen-Williams factor of *C* = 100, which corresponds to very rough and old pipe. Current water treatment practice and pipe finishes indicate use of a *C* in the 140-150 range for large chilled water systems. The Hardy Cross method is suitable for chilled water systems. It has not been used extensively with high temperature water systems. The value of *n* may be different and other formulas than the Hazen-Williams must be used to arrive at realistic pressure drops. It must also be remembered that, in high temperature applications, return water is at a substantially lower temperature than the supply and not subject to the same pressure parameters, although return and supply piping are parts of one system. In chilled water applications, the return water system can be treated hydraulically as an extension of the supply piping.

[32] Tube Turns, *op. cit.*, p. 33.

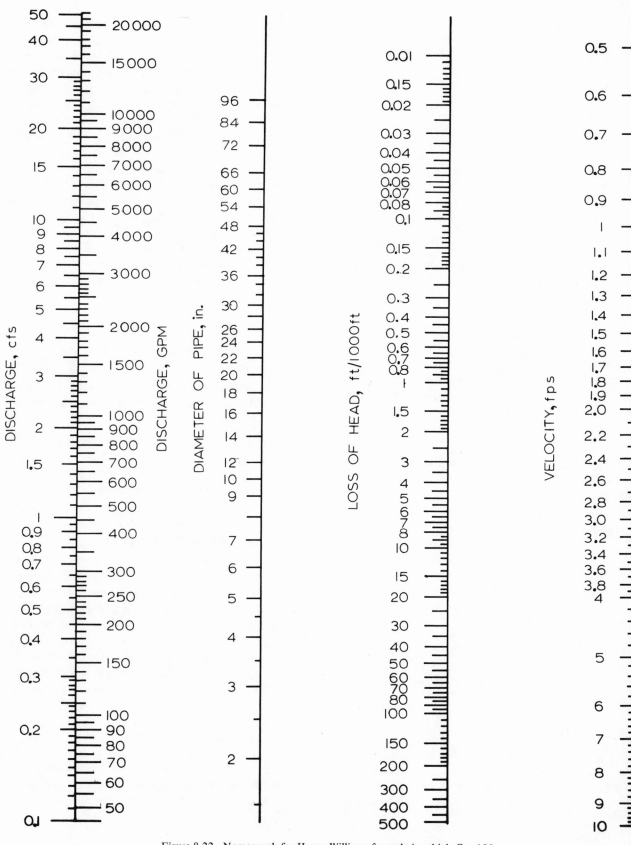

Figure 8-22. Nomograph for Hazen-Williams formula in which $C = 100$.

It can be shown that by expanding Equation 8.7 binomially, the following expressions can be derived:

$$\Delta Q = -\frac{\Sigma H}{n \Sigma \dfrac{H}{Q}} \qquad (8.8)$$

$$\Delta H = -\frac{n \Sigma Q}{\Sigma \dfrac{Q}{H}} \qquad (8.9)$$

In using Equation 8.8, one assumes flow quantities and direction and adopts a positive or negative flow sign convention. For example, the counterclockwise direction can be positive and the clockwise negative. $\Sigma H/Q$ has no sign and is the absolute sum of H's, regardless of sign. There are two ΔQ's applicable to a loop; first, the ΔQ developed from Equation 8.8, applied with the same sign to all loop pipes; secondly, an additional ΔQ applies to pipes that are common with another loop. The sign of this ΔQ is the reverse of the sign applicable to the adjacent loop. Table 8-3 illustrates this technique. In the example of Fig. 8-23, pipe *3* is common to loops I and II and pipe *4* is common to loop I and III. Therefore, pipes *3* and *4* are subject to separate corrections.

A similar procedure is used for Equation 8.9 by assuming values of H and then balancing the flows by correcting the assumed heads.

Example: Given the network of Fig. 8-23 with an inflow of 10,000 gpm (631 l/s), 10 distribution points of indicated outflow and pipe sizes and lengths as indicated, what is the water flow in each individual pipe and what is the friction drop in each loop?

The calculation method is given in Table 8-3, which is in the form of a worksheet. It is worth noting that the friction drop for loops II and III are approximately the same, indicating a balanced condition. The calculated flows are in error by a small percentage because of errors in reading values from Fig. 8-22. Computerization by a converging process can provide exact figures.

A technique used to study the effect of varying inputs and outflows is examined by McPherson,[33] who employs a proportional flow method to key points in the system on the assumption that each system demand fluctuates about its mean value in direct proportion to total system fluctuation. This technique prevents complete network analysis for each flow change, since the percentage flow in each pipe is constant irrespective of the magnitude of total demand.

Use of computers can give the engineer the flexibility of being able to change design flow conditions without painfully having to recalculate an entire network. Most of the computer work is dedicated to water works but is adaptable to chilled water systems.[34,35] More sources dealing with computerization and network analysis can be found at the end of the chapter.

8.15 Steam Distribution Systems

There was a surge of central steam plant and steam distribution system construction following World War I, mainly for downtown urban areas and for hospital and campus complexes. Pressures fell into three main categories: 7-40 psig (48-276 kPa) (low pressure systems), 125 psig (86 kPa) (medium pressure systems) and 250 psig (1722 kPa) (high pressure systems). [Although long practiced in Europe, high temperature water systems were not employed extensively in the United States until the 1940's, and then mainly in military installations and, during the 1950's, in institutional complexes.]

Steam distribution is a highly specialized engineering endeavor whose principal base is working experience. Currently, the main practitioners are utilities that sell steam as a byproduct of electric generation and who have in their hands extensive operational and design data which, unfortunately, are not public.

Cost is also affected, among other factors, by plant location, piping heat losses, feedwater treatment, percent of condensate return, leakage, groundwater conditions, ground conditions, number and type of reducing stations, number of manholes, sectioning valves and steam pressure level.

A field survey of existing conditions is indispensable for any underground distribution work. Proximity of other utility lines such as sewers, water mains, telephone and electric lines, etc., can increase costs.

Pressure level affects not only pipeline costs but the boiler room installation as well. The physical configuration of the service area bears on distribution design and, therefore, on cost. The energy requirements of the area

[33]M.B. McPherson, "Generalized Distribution Network Head Loss Characteristics," Proceedings of the American Society of Civil Engineers, *Journal of Hydraulics Division*, Vol. 86, No. HYI, January 1960, pp. 75–91.

[34]For an elaborated Fortran program see, John W. Clark, Warren Viessman and Mark J. Hammer, *Water Supply and Pollution Control* (Scranton: International Textbook Co., 1971), pp. 159–162.

[35]Ronald W. Jeppson. *Analysis of Flow in Pipe Networks*. Ann Arbor: Ann Arbor Science, 1976.

Table 8-3
Computation for Hardy Cross Method Example

1	2	3	4	Trial 1							Trial 2						
Loop No.	Pipe No.	Pipe Dia. in.	Length L, ft	Q, gpm (5)	S, ft/1000 (6)	Items 4×6, H, ft (7)	ΣH (8)	Items 7÷5, H/Q (9)	$n\Sigma \frac{H}{Q}$ (10)	$\Delta Q = \dfrac{H}{n\Sigma H/Q}$ (11)	Q (5)	S (6)	H (7)	ΣH (8)	H/Q (9)	$n\Sigma \frac{H}{Q}$ (10)	ΔQ (11)
I	1	18	4000	+6700	19.8	+79.2		0.0180		+108	+6808	20.5	+82.00		0.0120		−96
	2	18	1000	+4408	8.7	+8.7		0.0019		+108	+4516	9.5	+9.50		0.0020		−96
	3	10	1100	+1300	17.0	+18.7		0.0143		+108 + 117 = 225	+1525	23.5	+25.85		0.0162		−96 − 153 = −249
	4	8	1100	+633	12.9	+14.2		0.0224		+108 + 41 = 149	+782	19.0	+20.90		0.0267		−96 − 32 = −126
	5	12	2000	−3300	39.0	−78.0		0.0236		+108	−3192	36.0	−72.00		0.0227		−96
	6	12	2000	−2397	20.7	−41.4		0.0172		+108	−2289	19.1	−38.20		0.0166		−96
	7	12	2000	−1772	11.7	−23.4		0.0172		+108	−1664	10.8	−21.60		0.0131		−96
						−22.0	−22	0.1106	0.2046				+19.25	+19.25	0.1093	0.2022	
II	8	16	1200	+2900	7.4	+16.28		0.0056		−117	+2783	6.5	+7.80		0.0028		+153
	9	16	1000	+2830	7.2	+7.20		0.0025		−117	+2713	6.4	+6.40		0.0023		+153
	10	14	600	+2275	8.8	+5.28		0.0023		−117	+2158	7.8	+5.46		0.0025		+153
	11	8	1100	−250	2.3	−2.53		0.0101		−117 + 41 = −76	−326	3.7	−4.07		0.0124		+153 − 32 = 121
	3	10	1100	−1300	17.0	−18.7		0.0143		−117 − 108 = −225	−1525	23.5	−25.85		0.0162		+153 + 96 = 249
						+7.53	+7.53	0.0348	0.0643				−10.26	−10.26	0.0362	0.0669	
III	11	8	1100	+250	2.3	+2.53		0.0101		−41 +117 = 76	+326	3.7	+4.07		0.0124		+32 − 153 = −121
	12	14	2000	+2525	11.1	+22.20		0.0087		−41	+2484	10.5	+21.00		0.0084		+32
	13	14	700	+2248	8.8	+6.16		0.0027		−41	+2207	8.2	+5.74		0.0026		+32
	14	14	1700	−1711	5.3	−9.01		0.0052		−41	−1752	5.5	−9.35		0.0053		+32
	15	14	1000	−1433	3.8	−3.80		0.0026		−41	−1474	4.1	−4.10		0.0027		+32
	4	8	1100	−633	12.9	−14.19		0.0224		−41	−782	19.0	−20.90		0.0267		+32 + 96 = 128
						+3.89	+3.89	0.0517	0.0956				−3.54	−3.54	0.0581	0.1074	

Trial 3

1	2	3	4	Q, gpm (5)	S, ft/1000 (6)	Items 4×6, H, ft (7)	ΣH (8)	Items 7÷5, H/Q (9)	$n\Sigma \frac{H}{Q}$ (10)	$\Delta Q = \dfrac{H}{n\Sigma H/Q}$ (11)
I	1	18	4000	+6712	20.1	+80.4		0.0119		+100
	2	18	1000	+4412	8.8	+8.8		0.0019		+100
	3	10	1100	+1276	17.0	+18.7		0.0146		+100 + 24 = 124
	4	8	1100	−654	13.9	+15.3		0.02		+100 + 13 = 113
	5	12	2000	−3288	37.5	−75.0		0.0228		+100
	6	12	2000	−2385	21.0	−42.0		0.0176		+100
	7	12	2000	−1760	12.6	−25.2		0.0143		+100
						−19.01	19.01	0.1064	0.1968	
II	8	16	1200	+2936	7.4	+8.9		0.0030		−24
	9	16	1000	+2866	7.2	+7.2		0.0025		−24
	10	14	600	+2311	9.3	+5.58		0.0024		−24
	11	8	1100	−205	1.6	−1.75		0.0085		−24 + 13 = −11
	3	10	1100	−1276	17.0	−18.7		0.0146		−24 − 100 = −124
						+1.23	+1.23	0.0310	0.0573	
III	11	8	1100	+205	1.6	+1.75		0.0085		−13 + 24 = 11
	12	14	2000	+2508	10.9	+21.8		0.0086		−13
	13	14	700	+2231	8.6	+6.02		0.0026		−13
	14	14	1700	−1734	5.4	−9.18		0.0052		−13
	15	14	1000	−1450	3.9	−3.90		0.0026		−13
	4	8	1100	−654	13.9	−15.3		0.0233		−13 − 100 = −113
						+1.19	+1.19	0.0508	0.0939	

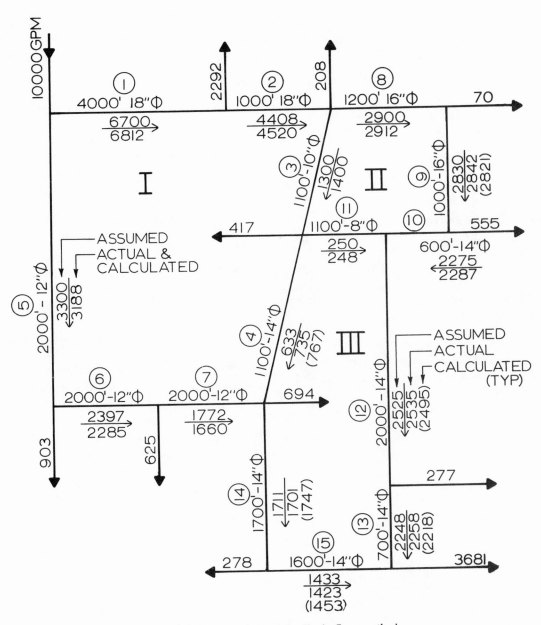

Figure 8-23. Network for Hardy Cross method.

and the manner in which they appear in terms of magnitude, physical separation and time concentration influence design to a large extent.

Pressure level also determines the extent of an area that can be served economically as well as the type of distribution and related design that will accomplish this purpose. There is, therefore, an interrelationship between area, pressure and steam volume.

In designing a steam system, the pressure at the most remote point in the system is set and the line size and pressure drop are calculated back to the plant, establishing along the way not only service loads but simultaneous loads for mains and transmission lines.

The following velocity and pressure guidelines are offered by International District Heating Association.[36]

[36]*District Heating Handbook*, 3rd edition; Pittsburgh, National (now International) District Heating Association, 1951, p. 151.

	Velocity range, fpm (m/s)	Pressure drop, psi/ 100' (kPa/30.5 m)
Low pressure system 7-40 psig (48-276 kPa)	11,000-14,000 (55.9-71.1)	0.3-1.0 (2.06- 6.89)
Medium pressure system 125 psig (861 kPa)	12,000-17,200 (61.0-87.4)	2.0-4.0 (13.8-27.6)
High pressure system 250 psig (1722 kPa)	12,700-19,200 (63.5-97.5)	4.0-8.0 (27.6-55.2)

Pressure selection must be guided by well-established criteria of safety, serviceability, economy and general plant objectives.

Once a path is marked, the feasibility of line installation must be confirmed and locations for trenching machines must be selected. It is well to place heating lines under walkways, thus providing a snow-melting facility.

Steam Piping Sizing

There are several formulas for sizing steam pipes. The most well-known are those by Unwin and Fritzsche. The Unwin formula is given as[37]

$$p = \frac{0.0001306 \, w^2 \, L \, (1 + \frac{3.6}{d})}{Y d^5}$$

Where

p = Pressure drop, psi
w = Steam flow, lb/min
L = Length of pipe, ft
Y = Average density of steam, lb/cu ft
d = Inside pipe diameter, in.

References in the bibliography carry detailed information on pipe sizing and steam flow.

8.16 High Temperature Water Pressurizers

The term "pressurizer" is used here rather than "expansion" or "compression tank," which are employed in low temperature or chilled water systems. High temperature water (HTW) systems are particularly subject to corrosion. Thus, air pressurization has been abandoned on account of its oxygen content. In high or medium temperature systems, steam and nitrogen are used almost exclusively. Nitrogen seems to be prevailing over steam

---
[37]*Ibid.*, p. 194.

for reasons that will become apparent in the following pages.

Pressurizers are examined here for several reasons:

1. They are an important part of the distribution system.
2. There is a great deal of confusion as to their use and usefulness.
3. Controversy abounds about the use of steam or nitrogen for cushioning pressurizers.
4. Pressurizers are an integral part of, and contribute to the stability, safety and controlability of medium and HTW systems.
5. The relationship among pressurizers, primary and secondary systems is not entirely appreciated.

Only the role of the pressurizer is examined here and, although any relationship between the pressurizer and the rest of the system merits attention, a design procedure for the rest of the system cannot be handled in the short space here devoted to the pressurizer.

The following points often emerge during the design of HTW systems. It is worthwhile to examine them before proceeding to other aspects of pressurizer philosophy.

1. A pressurizer would be outrageously large if it were to handle the expansion between the cold and hot states. The initial expansion, when filling up the system, is bled out the system. For reasonable pressurizer sizes, the pressurizer is normally dimensioned to accommodate the expansion between design supply and return water temperatures.
2. A self-contained pressurizer, i.e., with no auxiliary vessels and one that can handle expansion without discharging nitrogen or water, is the most desirable but also the largest in size. This type of pressurizer is recommended for rigorous process needs and whenever ultimate dependability is required.
3. Nitrogen and treated water are expensive. Their loss and spillage cannot be considered as a design basis except, perhaps, for relief purposes. Their costs should be part of system economic analysis.
4. Dual pressurizers, each capable of handling half of total expansion, are recommended for systems that cannot be shut down for pressurizer inspection and repairs. One tank can be valved off and emptied while the other continues to serve.
5. Insulation of all pressurizers is needed to prevent steam condensation and/or absorption of nitrogen in the water at lower temperatures.
6. The size of the pressurizer and accessories depends on a number of variables, all of which must be

satisfied for a successful operation. The factors that affect size are: type of pressurization, system water content, system average temperature level, temperature differential between supply and return, design pressure level of equipment and piping materials, and the magnitude of the system. Some of the system components depend on the rate of expansion and contraction. In some systems (hereinafter designated Method 3A — Nitrogen Pressurization), the number of reversals (expansion-contraction) per hour and their magnitude can affect the size of the pressurizer, condensate receiver and makeup pump. The same may be said for the size of the auxiliary vessels and compressor in Method 3B–Nitrogen Pressurization (Nitrogen Recovery). It can easily be seen why sizing a pressurizer involves every aspect of system safety, controls and operation. The engineer is responsible for resolving all these requirements in a unified approach.

7. The expansion line (pipe connecting system and pressurizer) must be sized for low velocity for the highest expected surge. However, the velocity must never be so low as to allow free convection currents during stable operating conditions. These demands are contradictory to some extent and the designer must resolve the contradiction when sizing the expansion line. Surge is a result of heat addition or subtraction from the system and can be calculated by considering the heat transfer characteristics of the boiler. Examples have occurred where this amounted to as much as 80 to 100 gpm (5.04 to 6.31 l/s) in the first short period following flame failure.[38] By the same token, volume expansion occurs if load is removed and boiler operation continues undiminished. As a precaution, adequate friction or flow impediments must be placed in the expansion line to prevent thermal currents between system and pressurizer.

8. Discharge piping from code safety valves and powered relief valves must be sized to prevent back pressure from exceeding 20% of valve setting. Manufacturers' catalogues can be consulted for permissible back pressures.

9. Pressurizer code safety valves must be placed on the vessel itself, in accordance with ASME Boiler and Pressure Vessel Code, Section VIII. On steam pressurizers, it is preferable to pipe steam safety valves to the roof. On nitrogen-cushioned pressurizers, it is customary to use water-relieving safety

[38] Owen S. Lieberg. "N$_2$ Pressurizing for HTW Systems," *Air Conditioning, Heating and Ventilating*, (August, 1962), p.45.

valves located below the water line, which prevents nitrogen from escaping. Occasionally, water-relieving safety valves are placed on top of horizontal nitrogen pressurizers but are connected to the water mass by means of dip tubes. Water discharge involves higher impact forces and discharge piping anchors and supports must be designed accordingly. Another method of mounting gas-relieving safety valves without risk of leaking is to mount them in the vapor space and connect them to a double U water trap. No insulation is provided for the trap, so as to have water vapor condensing and filling the double U trap. Relieving nitrogen directly to atmosphere is not recommended as seats deteriorate with time, wasting nitrogen.

10. Powered relief valves are an engineering option for additional safety. They serve as backup in case of failure of the safety valve and help to relieve pressure after the safety valve has an opportunity to act. The valve is actuated by the operator from the control room at his discretion. A third line of defense would be manual drain valves, which are normally provided for manual blowdown. Any blowdown receiver must be equipped with quenching facilities.

11. Safety measures include shutting down the burner or blocking the oil lines, or both, when the low pressure level is reached or when the system recirculation pump(s) fail. It appears that, in addition to these steps, attainment of the high pressure level should also shut down burners and block oil lines because this condition is indicative of heat being added to the system without equivalent heat removal. However, a high pressure level may also be the result of makeup pump malfunction. If precautions for a pipe break are to be taken to keep all parts of the system operating except the ruptured part, isolation valves may be strategically located to effect this goal.

12. Pressurizer integrity can be analyzed by considering filling the vessel with water and adding insulation weight, loadings exerted by piping connections and safety valve loads, if they are coincident. Similar load combinations apply to supports and anchors. In highly seismic areas, seismic acceleration forces may have to be included.

For nitrogen-cushioned pressurizers connected to the system by means of an expansion line, it is important to calculate accurately heat transfer to the environment and the number of surges of hot water into the tank. Pressurizers are heavily insulated. Over a period of time the water in the tank cools down and is at a lower temperature than the water circulating in the lines. Hence, nitrogen pressure is the dominant pressure in the pressurizer. Surges of hot water result in increased water vapor pressure in the vessel. Heavy insulation prevents the liquid in the tank from cooling down rapidly and restoring the previous balance between nitrogen and water vapor pressure. The liquid bulk temperature in the tank determines the partial water vapor pressure which, for temperatures above 400 F (204 C), can exceed the partial pressure of nitrogen by a substantial margin, converting the nitrogen pressurizer into a steam pressurizer. A cool pressurizer is desirable and an expansion line keeps the tank contents out of the hot water stream. That is one of the reasons that engineers avoid through flow in the pressurizer.

In years past, flow into or out of the pressurizer, viewed as a measure of system instantaneous load, was used to regulate the firing rate of the boiler. This method, used as a criterion of pressurizer appropriateness, has been by-passed by modern solid state instrumentation. Measurement of local loads may be transmitted instantaneously to a central location where a totalizer can regulate burner operation, avoiding the time lags and uncertainties of an anachronistic practice.

Another refinement can be effected in large systems in connection with the operation of the makeup pump which ordinarily operates between fixed levels of the pressurizer, regardless of thermal considerations. It is possible that the system could be instrumented to consider pressurizer level versus average system temperature in addition to the fixed levels, so that the pump would go on only if water leaked out of the system rather than on a drop in level caused by thermal shrinking.

Most systems tend to operate well under steady load conditions. However, the pressurizer is mainly for accommodating positive and negative surges caused by load transients and is instrumented to provide stability and control. The interface between liquid and vapor can be upset by the transients. Proper design procedure must, as a consequence, consider introduction and withdrawal of loads and transient conditions. The procedure must be clear and the function of the various components and instruments must be programmed to effect a safe transition. Procedures for the following events must be completely articulated:

1. Starting up a system and maneuvering it to the hot condition.
2. Adding load from zero to 100% in increments.
3. Removing load from 100% to zero in increments.

4. (a) Effecting a hot shutdown, i.e., having the system in the hot, standby condition, ready to take on load.
   (b) Effecting a cold shutdown, i.e., taking off all load and gradually deactivating the system.

The designer must investigate the type of conditions and the manner in which load may be shed; the effects of full and sudden volume reduction; the state of safety of the various types of equipment; the specific functions of various equipment and instruments during the shutdown process. Similar questions must be addressed in connection with the assumption of load, although the shutdown process seems to be more critical.

8.16.1 Pressurizers Versus Primary-Secondary Hot Water Systems

In most HTW systems, the building pump circuits are completely separated from the HTW mains by means of intervening heat exchangers. This vessel provides complete hydraulic separation between the pumping circuits as well as thermal isolation (no bleeding of high temperature water).

Sometimes the hydraulic separation is part of a logical design technique. For instance, if high rise buildings are to be part of a complex, connecting them directly to the primary circuit would impose the entire elevation head of the building on the primary circuit, over-pressuring equipment and piping alike.

Hydraulic separation is usually used to protect the secondary system from the high pressure rating of the primary systems, which are rated at 300 psig for a supply temperature of 400 F (204 C). If interconnected by a primary-secondary bridge, the secondary system must also be rated at 300 psig (2067 kPa), adding substantially to construction costs. Some engineers object to 300 psig (2067 kPa) secondary systems as jeopardizing building safety. This does not seem plausible to the author since secondary water temperatures are in the low temperature range. In industrial applications and, selectively, in commercial applications based on economic and safety criteria, the heat exchanger can be eliminated and a free, primary-secondary bridge connection can be applied.

Under a hydraulic separation regime, each secondary system must have its own expansion tank to accommodate its expansion. When primary-secondary systems are directly interconnected, the central pressurizer can accommodate the expansion of the primary as well as all the secondary systems. Deaeration of the secondary systems can be effected by using venting bottles at the high points. Under free interconnection conditions, providing individual expansion tanks in each building and a central pressurizer can actually be counterproductive, because system reaction to all the transients caused by the various expansion tanks is uncertain. Elevation head from high rise buildings on tanks at lower elevations may cause operating difficulties and flooding of the lower tanks.

Thus, one way to avoid expansion tanks in the secondary systems is to raise their pressure rating. An alternate way is to lower the pressure rating of the primary, say by lowering supply to 330 F (165 C) and return to 180 F (82 C). The temperature difference, 150 F (65 C), is similar to those used in high temperature systems. The saturation pressure at 330 F is 103 psia (710 kPa); therefore, 125-psi (861 kPa) fittings can be used. This concept is combined with a "push-pull" pumping system.[39,40] Both the supply and return loops are equipped with a pump and there are control valves on either side of load points to stabilize pressures in both loops, by actually attempting to mirror supply system pressures negatively, but well above vaporization pressures. A great deal of engineering attention is required to avoid vaporization in the return loop. However, the 150 F (65 C) differential cannot be applied everywhere and cannot be used for reverse return. According to Owen S. Lieberg[41] "MTW systems are best suited for installations having heat loads from one million Btu/hr to 15 million Btu/hr (293 kW-4395 kW) or having a water content of the system ranging from 500 to 15,000 gallons (1895 l to 56 850 l)." However, these figures might have shifted somewhat over the years. Increasing labor and material costs have tended to narrow this scope further for systems in the 250-275 F (120-135 C) range, MTW and HTW seem to be merging into the 300-350 F (155-175 C) range in West European district heating practice. However, it is possible that American economics may dictate a higher range.

In interconnected primary-secondary systems, free access from the secondary to the primary system must be provided at all times to transfer volume variations in the secondary to the pressurizer.

In chilled water systems, the only restriction that prevails is that of tall buildings imposing their elevation head on the primary. Otherwise, direct interconnection is indicated. Temperatures and pressure ratings are low in both primary and secondary systems and only one central pressurizer is required. Air-cushioning is quite customary in chilled water systems.

8.16.2 *Pressurizer Criteria*

All the previous discussion concerned general functions and important pressurizer design features that contribute towards safer and more controllable systems. What criteria should be used for judging the overall validity of any system?

Two engineers, who have written extensively on the subject, believe that the following eight points are the main criteria for HTW systems:[42]

1. *Minimize fluctuation in system temperature, pressure, flow rates and heat content.*
   As a result the following may be expected:

   (a) Improvement of combustion efficiency.
   (b) Smaller pressurizers.
   (c) Control stability.

2. *Detect average heating loads and translate into firing rates.*
   Loads must be detected as they occur. As pointed out in Section 8.16, modern instrumentation has reduced the immensity of this task and, therefore, the pressurizer is no longer a primary criterion for load detection.

3. *Maintain system pressure to prevent flashing into steam regardless of temperature change.*
   Minimum pressure must be maintained above the flash point at the remotest, highest (elevation) or lowest pressure point experiencing the highest system temperature. Monitoring stations at strategic points are recommended to provide the operator at the control room the assurance that the system is operating safely, above the critical pressure at selected design points. The operator can decrease the firing rate to counteract an emergency at the monitoring stations.

4. *Simplicity* amounts to having as few pieces of equipment as possible, of durable construction, requiring uncomplicated controls and simple maintenance procedures, and capable of being serviced by average personnel.

5. *Minimum maintenance* signifies that equipment should be readily accessible and serviceable, long lasting, not subject to deterioration and rapid replacement.

[39] E.G. Hansen. "Push-pull Pumping Permits Use of MTW in Building Radiation," *Heating, Piping, Air Conditioning*, (May, 1966), p. 97.

[40] *ASHRAE 1976 Systems Handbook*, Chapter 17, pp. 17.6-17.7.

[41] Owen S. Lieberg. "N$_2$ Pressurizing for HTW Systems," *op. cit.*, p. 46.

[42] J.S. Blossom and P.H. Ziel. "Pressurizing High Temperature Water Systems," *ASHRAE Journal*, (November, 1959), pp. 47-48.

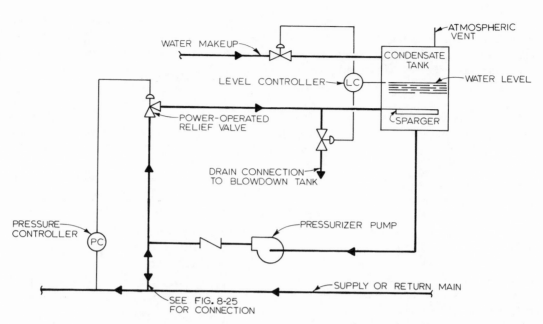

Figure 8-24. Basic schematic diagram for hydraulic pressurizer.

6. *Freedom from corrosion* dictates that oxygen must not be introduced into the system. Keeping makeup water to a minimum can help in this direction.

7. *Low cost* encompasses first as well as operating costs, both of which must be reasonable for the financial success of the project.

8. *Flexibility* is required in locating the pressurizer in a fashion that will achieve optimum system results.

8.16.3 *Methods of Pressurization*

Pressurization of HTW [and of water above 212F (100 C)] is required to prevent it from flashing into steam, followed by water hammer and boiler tube burnouts.

There are three basic pressurization methods:

1. Using a hydraulic pump. 2. Pressurizing the pressurizer vessel with steam. 3. Pressurizing the pressurizer vessel with nitrogen.

Method 1—Hydraulic Pump Pressurization

The operation of a hydraulic pump imposes a pump head as an external pressure on the entire system and does so on a continuing basis. A small capacity, high head pump is adequate to increase pressure throughout the system and is connected to the system as indicated in Fig. 8-24.

This method of pressurization dispenses with the pressurizer and is useful in two situations: (1) As a temporary

substitute for a pressurizer that is taken out of service for inspection or repairs (see Fig. 8-25),[43] (2) As a permanent arrangement, eliminating the need for a pressurizer.[44]

In both cases, water is withdrawn from the condensate tank and is injected into the system by the pressure pump. Excess pressure is relieved back to the condensate tank through the power-operated relief valve. The condensate tank can be at atmospheric pressure or can be nitrogen-cushioned to prevent air from entering the system.

The temporary operation in Fig. 8-25 works as follows: valves 1, 2, 3 and 4 are shut and valves 5, 6, 7 and 8 are opened in a prearranged, gradual manner, putting the pressurizer out of operation. On reactivation, bypass valve 1A is cracked open for a period and water is pumped into the system until a steady condition is reached. The process is repeated until the pressurizer is back in full service, at which time valves 1, 2, 3 and 4 are opened and valves 5, 6, 7 and 8 are gradually closed.

This system has been applied sporadically but never caught on. There is no cushion space and the system depends on relief valve(s) for its operational safety. In actual applications, double redundancy in relief valves and pumps (3 each) is recommended. Two pumps should operate at all times to guarantee continued operation in

---

[43] R. Lilly. "Pressure Pump System Permits Inspection of HTW Expansion Drum," *Power.* (May, 1969), pp. 76–79.

[44] D.A. Carafano. "District MTW System is Pump Pressurized, Has open Tank Storage," *Heating, Piping, Air Conditioning,* (July, 1970), pp. 68–70.

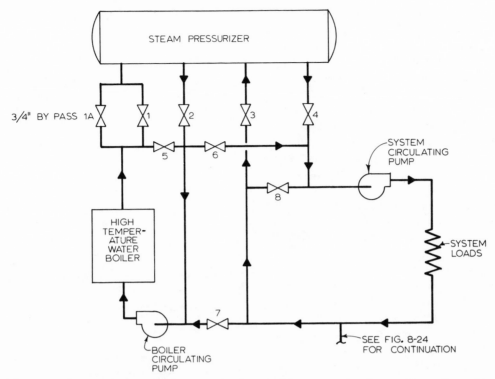

Figure 8-25. Basic schematic diagram showing pressurizer bypass piping arrangement for vessel inspection and repair.

case of the failure of one. Scheduled testing of relief valves and pumps is recommended to verify equipment integrity and operational readiness. The question of dissolved air in the makeup water could be resolved by passing it through a deaerating heater before injection into the system.

Unfortunately, not enough experience has been accumulated for a full evaluation. It appears that this system fails to satisfy criteria 4, 5, 6 and 7, above.

Method 2—Steam Pressurization

Boiler discharge water is pumped into the pressurizer, keeping it at the highest system temperature. Pump suction is taken from the pressurizer. This is normal for steam-cushioned systems because suction water is quenched by mixing pressurizer water with a portion of return water to assure adequate NPSH for the system pumps. Pumping into the pressurizer would upset the temperature-pressure relationship in the system. Piping and equipment arrangement is shown in Figs. 8-26 and 8-27. Ordinarily, enough return water is mixed into the suction to lower the mixture by 10F (5.55C) while the pressure remains the same. This temperature depression plus the elevation head on pump suction together with the system pump head provide adequate pressurization throughout the system. However, there are some qualifications to this statement.

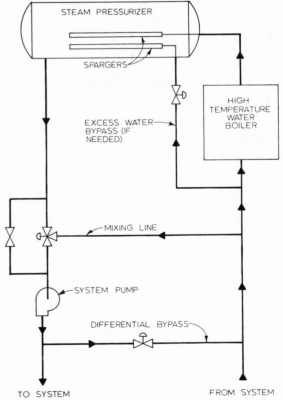

Figure 8-26. Steam pressurization with one pump.

It is difficult for single pump systems to maintain the 10F (5.55C) differential as load drops. As more water is diverted to the mixing line, less is available to circulate through the boiler, which needs full flow to avoid tube burnouts. A two-pump system corrects this deficiency and meets the individual head and flow requirements of the boiler and system circuits. Variable speed system pumping is acceptable in the two-pump arrangement.

In general terms, a one-pump system is more applicable to short piping runs and fairly steady loads. A two-pump arrangement is related to larger, more complicated, variable demand systems.

Sizing of steam-cushioned pressurizers is less critical than sizing nitrogen pressurizers; the former is more of a mechanical ritual and follows accepted rules, while the latter obeys formalized calculational methods.

Whereas nitrogen coexists with steam and exerts its own partial pressure, steam pressure in a steam pressurizer adjusts in time to the water temperature. Steam may experience some compression or expansion for a limited period of time but soon saturation equilibrium returns, regardless of the size of the cushioning space. The cushioning and submerged spaces, therefore, are arbitrarily chosen, according to some well-accepted ASHRAE rules:[45]

Total volume = *steam cushion + volume expansion + water reserve space*

Steam cushion = *20% (volume expansion + water reserve space)*

Water reserve space = *40% of volume expansion (covers spargers and submerged headers)*

The first step in sizing the steam pressurizer is to calculate total system volume expansion. One is cautioned to note that the specific volumes and specific weights of water vary significantly between supply and return headers due to the different temperatures in each. Dealing with volume expansion can lead to errors unless water temperature is accounted for.

Two shortcut methods are used to avoid the tedious labor involved in a complete piping and equipment take-off:

(a) Calculate the expansion between the lowest and highest system temperatures for *half* the system volume, since all supply lines are at the highest temperatures which do not substantially change during operation.

(b) Calculate the expansion between the average and lowest temperatures for the *entire* system volume.

_____
[45] ASHRAE 1976 Systems Handbook, Chapter 17, p. 17.4.

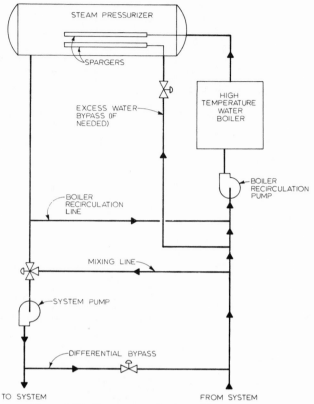

Figure 8-27. Steam pressurization with two pumps.

The average temperature is the sum of the highest and lowest temperatures divided by two.

These two shortcut methods do not account for differences between supply and return pipe size and routing and tend not to give strict accounting of the status of the various system vessels. If this approach is justified for steam pressurizers, where steam cushion space is not critical, it can be misleading in sizing nitrogen pressurizers, where an accurate water content and status inventory is essential, if system pressures are to be predicted with any accuracy. Hence, it is best in most instances to adhere to the following method:

Calculate water weight and temperature of every piping section and vessel between the full load and no load conditions. These data can be translated into volume increases which, if added, can give the total volume expansion. An alternate method would be to find the average water temperature of the system and apply it towards finding the total volume expansion in the manner of the second shortcut method above.

Some additional rules for sizing a steam pressurizer come from another source:[46]

_____
[46] Paul L. Geiringer. *High Temperature Water Heating. op. cit.,* pp. 148–151.

(a) Suggested drum diameters and lengths are:
6′ dia × 15-20′ long (1.82 m × 4.57-6.09 m) for medium plants. 8′ dia × not less than 40′ long (2.43 m × not less than 12.19 m) for large plants.

(b) Twenty percent of the vessel diameter above the highest water mark must be left as steam space.

(c) The highest water level above center line must be no closer than 12-15″ (0.30-0.38 m) to the vessel center (for horizontal pressurizers).

(d) Minimum center line for return water pipes must be no less than 12″ (0.30 m) from vessel bottom.

(e) Center line of boiler water sparger should be approximately 3′ (0.91 m) from vessel bottom.

(f) Pumps should not be less than 15′ (4.57 m) below drum.

(g) Best aspect (length to diameter) ratio is 3.5; water level at 3/5 of vertical height (horizontal tank) should amount to 10-15% of total system capacity.

The horizontal position for steam pressurizers is preferred because it offers the largest contact area between vapor and liquid, accommodates submerged headers and spargers and promotes more widespread and intimate mixing. The steam drum must be above the boiler and preferably above the entire system. Pipelines leading to the pressurizer must ascend to it without dipping. Figure 8-28 illustrates a typical steam pressurizer cross-section.

There are some important advantages to steam pressurizers, as follows:

1. Steam pressurizing appears to be highly applicable to very large systems. Past practice seems to point in this direction. However, an economic analysis ought to establish the facts.

2. The cost of replenishing nitrogen, at times high, is entirely eliminated.

3. The size of the steam-cushion is not critical.

4. The pressure rating of a steam vessel is markedly less than that of a nitrogen pressurizer.

These important advantages are counterbalanced by an impressive array of disadvantages, which ought to be weighed carefully before a final decision:

1. Associated items are numerous and undoubtedly add to cost, although some items are common to both steam and nitrogen systems. Consider the following: internal distributors, mixing devices, safety valves, blow-off connections, large number of reinforced nozzles, perforated horizontal inlet pipes, vertical outlet pipes, antivortex fittings, vertical drain pipes, thermometers, pressure gauges, test wells, relief valves, water columns, vents, level controls, overflow controls, feedwater controls, emergency feedwater controls, high and low alarms.

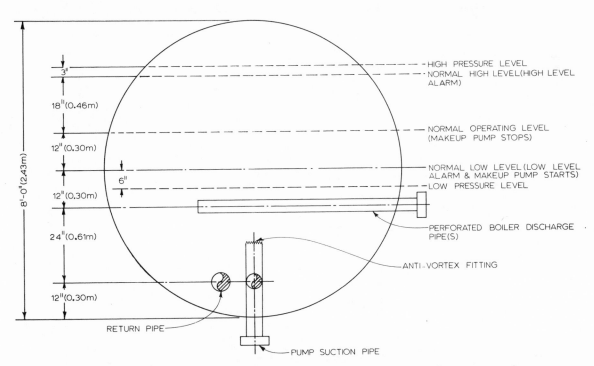

Figure 8-28. Functional and dimensional cross-section of typical 8 ft (2.53 m) dia steam pressurizer.

Anchors, supports, heavy insulation, trim and numerous connections render the tank expensive.

2. Flashing is difficult to prevent. When the system shuts down, a vacuum develops at the high points. Prevention of flashing should be one of the main design concerns. More operator attention and vigilance is needed to safeguard the system against flashing.

3. The pressurizer must be at the highest point in the system, calling for heavy steel framing to support a full vessel and conceivably increasing building volume.

4. Quenching of suction water is essential to maintain net positive suction head on the pump. Close control and operator alertness over the quenching operation is required.

5. Excess return water to the pressurizer can cause a temperature depression.

6. Rolling or sliding contacts to accommodate thermal expansion of the tank must be provided. Large tanks are anchored in the middle and are free to expand on both ends. Small tanks are anchored at one end.

Method 2A--Electric Steam Pressurization

An alternate method of pressurizing is by means of an electric heating element submerged in the steam pressurizer.

System pressure is maintained at a constant level independent of entering water temperature by activating an electric coil at the bottom of the pressurizer. General system arrangement is shown in Fig. 8-29.

The electric coil generates steam as required to maintain a set pressure in the steam cushion. Vapor locking anywhere in the system is precluded and the pump can be located on the cooler return main where more NPSH is available.

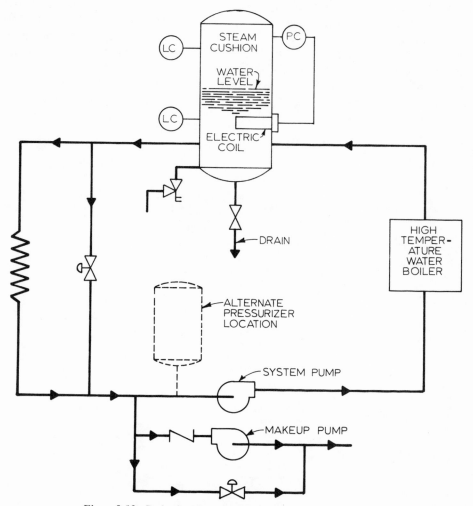

Figure 8-29. Basic electric steam pressurizer arrangement.

In the vertical tank arrangement, level is not of particular significance. Water level is in direct proportion to system volume expansion. A vertical tank in this instance is preferable because it reduces water surface radiation and at the same time encourages thermal convection.

This arrangement combines the noncriticality of the steam cushion and the constant pressure associated with nitrogen pressurization. However, there are two disadvantages which can easily be corrected. The first one is the through flow of the basic arrangement, which accentuates turbulence and thermal stresses. By using the alternate pressurizer location and arrangement (shown in Fig. 8-29), this disadvantage disappears. The second disadvantage is that, upon loss of the electric coil or its power, flashing would occur immediately in some parts of the system. A redundant electric coil connected to a small emergency generator can safeguard the system not only against loss of coil but loss of normal electric power as well. Coil size has to offset tank radiation and convection losses and is estimated not to exceed 10 kW.

Why not use electric boilers? They do not appear practical because tank size is large while heating capacity is very small. In addition, commercial electric boilers are not usually built to withstand the high pressures required in HTW applications.

There has been no significant experience with this type of system, which nevertheless is based on sound engineering principles. If an additional electric coil is provided, this system will give satisfactory service.

A similar approach is used in nuclear pressurized water reactors. The expansion line(s) emanate from the reactor vessel. The pressurizer, in addition to the electric heating coil, incorporates a water spray to reduce steam pressure transients in the steam cushion. The relief valves are connected to the steam cushion on top of uninsulated, double **U** seal traps (see Section 8.16, item 9).

Method 3–Nitrogen Pressurization

The use of air as a pressurizing medium has fallen into disuse because of its oxygen content, which encourages corrosion. Instead, nitrogen, one of the most commercially available inert gases, is used almost exclusively for gas pressurization.

Method 3, illustrated in Figs. 8-30 and 8-31, adopts a design of a self-adjusting and free-riding water level in the normal operating range. No loss of nitrogen takes place since it is bottled up at the top of the pressurizer. Water relief is rare and any discharge is contained within a small condensate receiver.

In order to contain the entire expansion between return and supply temperatures, a price is paid in the form of a

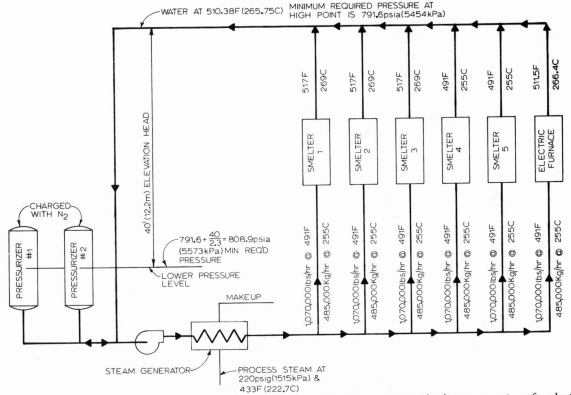

Figure 8-30. Hypothetical water system, recovering heat from industrial smelters and converting it to process steam for plant use.

large pressurizer. Large diameter ASME vessels are expensive and final vessel diameter ought to be chosen in consultation with the tank manufacturer to obtain the most economical size. For large systems or processes requiring tank redundancy, two tanks may provide a solution to the large expansion space requirement. Another solution may be to use the reduced tank size of "Method 3A–Nitrogen Pressurization (Water Recovery)" discussed later in this chapter. Another variation of nitrogen pressurization is that used in "Method 3B–Nitrogen Pressurization (Nitrogen Recovery)," which follows the 3A discussion.

A numerical example in this section indicates the complexity and number of criteria that must be counterbalanced to achieve a tank size that can accommodate all system design conditions.

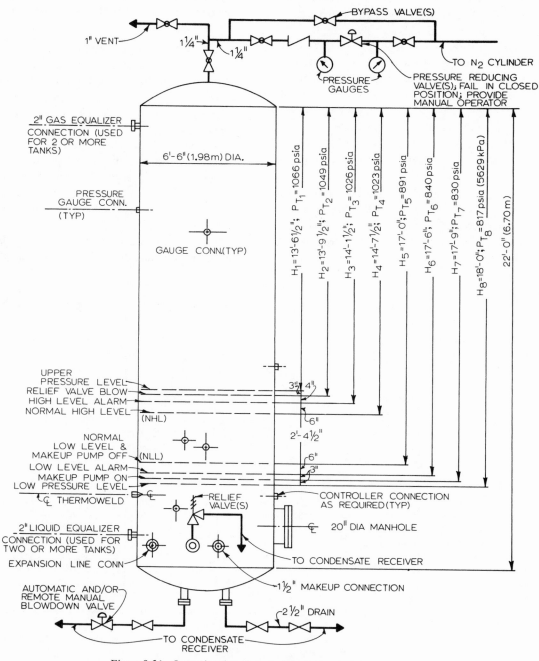

Figure 8-31. Operational model of nitrogen pressurizer.

Before proceeding with the numerical example, it is worthwhile to examine some of the advantages and disadvantages of nitrogen pressurization. Among the advantages, one may list the following:

1. The pressurizer may be installed anywhere, including the operating floor, since there is no fear of flashing. Elevation head, unlike the steam pressurizer, is not of significance. Vertical installation saves floor space and facilitates inspection and maintenance. Floor installation reduces boiler room height and provides potential construction savings.

2. The amount of water in the tank amounts to only 5-7% of total water capacity versus 10-15% for steam-cushioned tanks. The number of connections is reduced and spargers are no longer required.

3. Plant expansion can be handled by connecting additional pressurizer units without disruptions in plant operation or unbalancing of the system.

4. Flame failure of an oil-fired boiler is not as critical as it is for a steam-cushioned pressurizer. Both water and nitrogen can be added to maintain system pressure until the boiler cools off to some extent.

5. More trouble-free operation can be expected. NPSH on pump and the pressurization of the entire system does not have to be monitored as closely as it does with a steam-cushioned system.

However, there are some serious disadvantages, which must be weighed in the balance:

1. Loss of nitrogen, if unchecked can turn into a sustained economic drain.

2. Partial vapor pressure becomes significant above 400F (204C), to the point where the partial nitrogen pressure may be so small that the tank is essentially steam-cushioned. In such cases, pressurizer through flow is to be avoided as it keeps water temperature in the pressurizer at a high level. In general, through flow is used in isolated cases.

3. Gas pressure has no effect on pressurizer size.

4. Reduction in pressurizer size requires auxiliary pumps and receivers, adding complications, if not cost (see Method 3A).

5. Operating pressures tend to be higher in nitrogen pressurized than in steam-cushioned tanks.

6. Insulation is required to prevent water from cooling off and absorbing nitrogen.

The following numerical example pertains to industrial practice and is used to illustrate some of the principles discussed.

Example: Design nitrogen pressurizer(s), using Method 3, for the system illustrated in Fig. 8-30.

The pressurizers are connected to the system through expansion lines. This arrangement keeps the tanks at an assumed 300F (148.8C) (a temperature that would normally be calculated). In this process, the steam generator water outlet temperature is held constant, since the inlet to the smelters must be at 491F (255C). (In district heating applications, the opposite is true, *i.e.*, the outlet temperature at the heat addition station (boiler) is held constant.)

Normal system expansion and contraction takes place between normal low level (NLL) and normal high level (NHL). Adequate system pressure must exist when the water is at the low pressure level. The upper pressure level is used to calculate the pressure rating of the vessel and must be within 10% of the safety relief setting as per ASME rules.

Holding tank temperature below 300F (148.8C) has the advantage of keeping water vapor pressure low yet temperature high enough to prevent most of the nitrogen from being absorbed by the water. If tank temperature were kept at 510.36F (265.75C), the return water temperature, the vapor pressure would instead be 791.6 psia (5454 kPa), converting the tanks into steam-cushioned tanks.

The criteria that define pressure limitations in this example are as follows (See Figs. 8-30 and 8-31):

1. Continuous system operation for process purposes dictates that two tanks, each handling half of system volume expansion, be used. This halves the volume per tank, making it economical to manufacture.

2. System volume expansion is accommodated in the normal temperature range of 510.36-491F (265.75-255C).

3. The initial nitrogen charge must serve as long as possible. Nitrogen collection tanks, compressors and other auxiliaries are not needed, except a small condensate tank.

4. Tank space between "NLL" and "makeup pump on" allows for normal system leakage.

5. When the upper pressure level is reached, the automatic blowdown valve relieves pressure by dropping the water level. When the high pressure level is reached, the operator must decide how to discontinue the process, i.e., stop heat addition to the system. (In district heating systems, the burners ought to be cut out at the high and low pressure levels.)

6. Nitrogen is added to maintain system pressure only when the lower pressure level is reached. On further level drop, indicating a small pipe rupture or serious leakage, the operator at his discretion from the con-

trol board can add both nitrogen and water through the makeup pump or emergency injection system (if there is one) until safe shutdown is achieved.

7. The upper pressure level is used to calculate design vessel pressure and to detect a dangerous condition.

8. System pressure must be maintainable without the benefit of pump head.

An inventory of system volume (not shown) would reveal that total normal system expansion is 156 cu ft (4415 l).

Figure 8-31 indicates various operational conditions.

Expansion per pressurizer = $\dfrac{156}{2}$ = 78 cu ft (2207 l)

Assume pressurizer dia = 6'-6" (1.98 m)
Min. pressurizer height to accommodate expansion

$$= \frac{\text{expansion volume}}{\text{cross sec. area}}$$

$$= \frac{78}{\dfrac{\pi \, 6.5^2}{4}} = 2.35 \text{ ft } (0.72 \text{ m})$$

The next step is to find the following:

(a) Initial nitrogen charge at low pressure level (H_8).
(b) Whether pressure at low pressure level is sufficient to cover minimum system pressure requirements.
(c) Pressure at normal low level (H_5).
(d) Pressure at upper pressure level (H_1).

(a) Assume p_8 = 750 psia (5168 kPa)
Using the ideal gas law $v = RT/p$, we can estimate

Partial N_2 specific volume at H_8, v_8

$$= \frac{55.1 \, \dfrac{\text{lb-ft}}{\text{lb-}^\circ\text{R}} \, (460 + 300)^\circ\text{R}}{750 \, \dfrac{\text{lb}}{\text{in}^2} \times 144 \, \dfrac{\text{in}^2}{\text{ft}^2}}$$

$$= 0.387 \, \frac{\text{ft}^3}{\text{lb}} \, (0.02418 \frac{\text{kg}}{\text{m}^3})$$

Where

R = gas constant, ft-lb/lb-°R
v = specific volume, ft³/lb
V = total volume, ft³
p = gas pressure, psia
°R = degrees Rankine = deg.F + 460 (deg.C + 273.16)

Nitrogen charge = $\dfrac{V_8}{v_8}$ = $\dfrac{597.2 \text{ ft}^3}{0.387 \dfrac{\text{ft}^3}{\text{lb}}}$ = 1543.15 lb (701 kg)

(b) Total pressurizer pressure = partial N_2 press. + vapor press. = 750 psia + 67.01 psia = 817.01, say 817 > 808.0 psia (= 5629 kPa > 5573 kPa), see Fig. 8-30,

where

saturation pressure at 300 F = 67.01 psia (462 kPa)

Even at the low pressure level, minimum pressure is above flashing pressure. However, vapor pressure must be counted upon for its contribution.

(c) $p_5 = \dfrac{RT}{v_5}$

$$v_5 = \frac{V_5}{N_2 \text{ weight}} = \frac{564.06 \text{ ft}^3}{1543.15 \text{ lb}} = 0.365 \, \frac{\text{ft}^3}{\text{lb}}$$

$$p_5 = \frac{55.1 \times 760}{144 \times 0.365} = 796.72 \text{ psia}$$

Total pressurizer pressure = partial N_2 press. + vapor press. = 796.72 + 67.01 = 863.73 psia, say 864 psia > 808.9 psia. Notice that the partial pressure of nitrogen alone is above the minimum required system pressure of 808.9 psia (5573 kPa). Thus, for the normal operating range the entire system is in a very safe pressure range.

(d) Following a calculation procedure identical to that of (c) it is found that

p_1 = 999.31 psia (6885 kPa)

Total pressurizer pressure = 999.31 + 67.01
                           = 1066.32 psia (7374 kPa)

The pump head does not act on the pressurizer if the pressurizer is connected to the system by means of an expansion line, since the point at which the expansion line joins the main pipe is the point of zero pressure change. The pump head should be added to the pressurizer rating, if blockage of a main pipe is expected and the full pump head is forced on the pressurizer. If, on the other hand, the pressurizer is connected in flow-through fashion, the pump head is added to 1066.32 psia for obtaining the rating of equipment connected to the system or the pressurizer.

Method 3A–Nitrogen Pressurization with Water Recovery

The main advantage of this method is that it reduces pressurizer size, but at the expense of adding auxiliary water tanks and pumps. In essence, any expansion water volume that the pressurizer cannot contain is relieved to an atmospheric condensate receiver. In case of volume contraction, the makeup pump injects the stored condensate back into the system.

When the water level in the pressurizer rises, pressure throughout the system builds up rapidly and the power-operated pressure relief valve opens to the condensate receiver. On a drop in pressurizer water level, the level controller activates the makeup pump to return the stored water back into the system. The rapidity of these operations depends on pressurizer size. Minimum length of makeup pump motor operation ought to be at least 5 min so as not to wear out the pump motor. Makeup pump and redundancy of pressure relief valves are required, since system safety depends on these two items. This is perhaps the system's principal weakness. Scheduled testing of makeup pumps and relief valves is recommended. To avoid introducing air into the system, it may be advisable to route cold condensate through a deaerating heater before injection.

A nitrogen-pressurized condensate tank might be considered in lieu of an atmospheric tank, if corrosion of the atmospheric tank is an unacceptable possibility. Nitrogen pressure must be high enough to accommodate high temperature water, in essence acting as a second pressurizer. The advantage is dubious.

An economic analysis can demonstrate whether adequate savings in reduced pressurizer size can offset some of the disadvantages and risks of Method 3A.

Method 3B—Nitrogen Pressurization (Nitrogen Recovery)

Method 3B is identical to Method 3A regarding pressurizer size, since both methods must accommodate full water expansion. However, any overpressure is relieved by nitrogen relief valves.

Relieved nitrogen is conducted to a low pressure receiver. A nitrogen compressor pumps it into the high pressure receiver from where it can be reinjected into the pressurizer.

This system retains the large pressurizer size of Method 3A without its advantages and is saddled with a number of potentially troublesome controls and the cumbersome nitrogen compression system. It is useful only in taking the place of a free-blowing nitrogen relief valve.

Index